Advances in Intelligent Systems and Computing

207

Editor-in-Chief

Prof. Janusz Kacprzyk
Systems Research Institute
Polish Academy of Sciences
ul. Newelska 6
01-447 Warsaw
Poland
E-mail: kacprzyk@ibspan.waw.pl

For further volumes:
http://www.springer.com/series/11156

Smile Markovski and Marjan Gusev (Eds.)

ICT Innovations 2012

Secure and Intelligent Systems

Editors
Prof. Dr. Smile Markovski
Faculty of Information Sciences and Engineering
Ss Cyril and Methodius University
Skopje
Macedonia

Prof. Dr. Marjan Gusev
Faculty of Information Sciences and Engineering
Ss Cyril and Methodius University
Skopje
Macedonia

ISSN 2194-5357 ISSN 2194-5365 (electronic)
ISBN 978-3-642-37168-4 ISBN 978-3-642-37169-1 (eBook)
DOI 10.1007/978-3-642-37169-1
Springer Heidelberg New York Dordrecht London

Library of Congress Control Number: 2013933990

Printed on acid-free paper

Springer is part of Springer Science+Business Media (www.springer.com)

Preface

The Macedonian Society on Information and Communication Technologies (ICT-ACT) organizes the ICT Innovations conferences since 2009, which is one of its primary activities. In such a way ICT-ACT realizes its mission to support and promote scientific research in the field of informatics and information and communication technologies, as well as their application for building information society.

The fourth ICT Innovations conference "Secure and Intelligent Systems" was held in Ohrid, Republic of Macedonia, on September 12–15, 2012. 107 presentations and 21 posters were given at the conference from 203 authors and coauthors, from 24 countries, in the fields of Human Computer Interaction and Artificial Intelligence, Mobile Technologies, Software Engineering, Parallel Processing and High Performance Computing, Computer Networks, Cloud Computing, and Theoretical Computer Science. During the conference 5 special sessions were organized:

- Quasigroups in Cryptology and Coding Theory
- Bioinformatics
- Multimedia and Presentation of Cultural Heritage
- ICT in Education and E-Learning Platforms
- Improving Quality of Life through Biosignal Processing

as well as 2 workshops:

- iKnow - Knowledge Management for e-Services in University Management
- German-Macedonian Initiative on Advanced Audio and Speech Signal Processing (GMI-ASP)

37 papers were selected to be published in this book after an extensive reviewing process, out of 171 submitted. The selection of the papers was realized by a program committee of 83 members. The quality of the conference is based on the hard work done by the reviewers.

The editors are expressing their gratitude to all authors having submitted papers to the conference and to all reviewers for their contributions in enlarging the conference quality. Special thanks to Kiril Kjiroski for his technical support and to Vesna Dimitrova and Hristina Mihajloska for their overall contribution to the conference.

Ohrid,
September 2012

Smile Markovski
Marjan Gusev
Editors

Organization

ICT Innovations 2012 was organized by the the Macedonian Society on Information and Communication Technologies (ICT-ACT).

Conference and Program Chairman

Smile Markovski	UKIM, Macedonia

Program Committee

Nevena Ackovska	UKIM, Macedonia
Azir Aliu, SEEU	Macedonia
Mohammed Ammar	University of Koblenz-Landau, Germany
Ivan Andonovic	University of Strathclyde, UK
Ljupcho Antovski	UKIM, Macedonia
Goce Armenski	UKIM, Macedonia
Hrachya Astsatryan	IIAP, Armenia
Emanouil Atanassov	IPP BAS, Bulgaria
Verica Bakeva	UKIM, Macedonia
Antun Balaz	University of Belgrade, Serbia
Tsonka Baicheva	Bulgarian Academy of Science
Lasko Basnarkov	UKIM, Macedonia
Slobodan Bojanic	UPM, Spain
Dragan Bosnacki	TUE, Netherlands
Stevo Bozhinovski	South Carolina State University, USA
Ivan Chorbev	UKIM, Macedonia
Betim Cico	Polytechnic University, Albania
Danco Davcev	UKIM, Macedonia
Zamir Dika	SEEU, Macedonia
Vesna Dimitrova	UKIM, Macedonia
Nevenka Dimitrova	Philips Research, USA

Ivica Dimitrovski	UKIM, Macedonia
Mihaela Dinsoreanu	Technical University of Cluj, Romania
Martin Drlik	Constantine the Philosopher University, Slovakia
Saso Dzeroski	University of Ljubljana, Slovenia
Victor Felea	UAIC, Romania
Dejan Gjorgjevikj	UKIM, Macedonia
Sonja Gievska	UKIM, Macedonia
Danilo Gligoroski	NTNU, Norway
Norbert Grunwald	Hochschule Wismar, Germany
Jorge Marx Gomez	University of Oldenburg, Germany
Marjan Gusev	UKIM, Macedonia
Liane Haak	University of Oldenburg, Germany
Sonja Filiposka	UKIM, Macedonia
Boro Jakimovski	UKIM, Macedonia
Mirjana Ivanovic	UNS, Serbia
Slobodan Kalajdziski	UKIM, Macedonia
Aneta Karaivanova	CERN, Geneva
Dejan Karastoyanov	Bulgarian Academy of Science
Saso Koceski	UGD, Macedonia
Boicho Kokinov	New Bulgarian University, Bulgaria
Margita Kon-Popovska	UKIM, Macedonia
Ivan Kraljevski	TU Dresden, Germany
Andrea Kulakov	UKIM, Macedonia
Swein Knapskog	NTNU, Norway
Aleksandar Krapez	MISASA, Serbia
Sanja Lazarova-Molnar	UAEU, United Arab Emirates
Vanco Litovski	University of Nis, Serbia
Suzana Loskovska	UKIM, Macedonia
Ana Madevska-Bogdanova	UKIM, Macedonia
Jos Baeten	TUE, Netherlands
Jasen Markovski	TUE, Netherlands
Branko Marovic	University of Belgrade, Serbia
Cveta Martinovska	UGD, Macedonia
Marcin Michalak	Silesian University of Technology, Poland
Marija Mihova	UKIM, Macedonia
Kalinka Mihaylova Kaloyanova	University of Sofia, Bulgaria
Ivan Milentijevic	University of Nis, Serbia
Aleksandra Mileva	UGD, Macedonia
Pece Mitrevski	UKLO, Macedonia
Aleksandar Nanevski	IMDEA Software Institute, Madrid, Spain
Peter Parycheck	Danubo University Krems, Austria
Patel Dilip	LSBU, UK
Shushma Patel	LSBU, UK

Predrag Petkovic	University of Nis, Serbia
Zaneta Popeska	UKIM, Macedonia
Harold Sjursen	NYU poly, USA
Andrej Skraba	University of Maribor, Slovenia
Ana Sokolova	TUE, Netherlands
Dejan Spasov	UKIM, Macedonia
Leonid Stoimenov	University of Nis, Serbia
Georgi Stojanov	American University of Paris, France
Radovan Stojanovic	APEG, Serbia
Igor Stojanovic	UGD, Macedonia
Mile Stojcev	University of Nis, Serbia
Jurij Tasic	University of Ljubljana, Slovenia
Ljiljana Trajkovic	SFU, Canada
Vladimir Trajkovik	UKIM, Macedonia
Igor Trajkovski	UKIM, Macedonia
Francky Trichet	Nantes University, France
Goran Velinov	UKIM, Macedonia
Tolga Yalcin	HGI, Ruhr-University Bochum, Germany
Katerina Zdravkova	UKIM, Macedonia

Organizing Committee

Nevena Ackovska	UKIM, Macedonia
Azir Aliu	SEEU, Macedonia
Ljupcho Antovski	UKIM, Macedonia
Vesna Dimitrova	UKIM, Macedonia
Lidija Gorachinova-Ilieva	FON, Macedonia
Marjan Gusev	UKIM, Macedonia
Andrea Kulakov	UKIM, Macedonia
Hristina Mihajloska	UKIM, Macedonia
Aleksnadra Mileva	UGD, Macedonia
Panche Ribarski	UKIM, Macedoina
Goran Velinov	UKIM, Macedonia
Ustijana Shikovska Reckovska	UIST, Macedonia
Vladimir Trajkovik	UKIM, Macedonia

Contents

Invited Keynote Paper

Proceeding Papers

Controlling Robots Using EEG Signals, Since 1988

Stevo Bozinovski

South Carolina State University
Orangeburg, SC, USA
sbozinovski@scsu.edu

Abstract. The paper considers the emergence of the field of controlling robots using EEG signals. It looks back to the first result in the field, achieved in 1988. From a viewpoint of EEG driven control, it was the first result in controlling a physical object using EEG signals. The paper gives details of the development of the research infrastructure which enabled such a result, including description of the lab setup and algorithms. The paper also gives a description of the scientific context in which the result was achieved by giving a short overview of the first ten papers in the field of EEG driven control.

Keywords: psychokinesis, EEG control of physical objects, EEG control of robots, biosignal processing, contingent negative variation, contingent alpha rhythm variation, probability density distribution, real-time EEG control.

1 Introduction

Telekinesis and psychokinesis are concepts with meaning of moving objects by utilizing energy emanating from a human brain produced by the brain mental processes. One approach to achieve such an effect is using a computer with two interfaces: one toward the brain for EEG signal processing, and the other toward a physical object for example a robot; with today's technology either interface can be wireless.

This paper is looking back to the first result of using EEG signals to control a movement of a physical object, a mobile robot, which was achieved in 1988 [1][2][3][4]. The next section of the paper gives some introductory knowledge on EEG driven control of robots, then this paper describes the details of the realization, and finally it discusses the scientific context in which the result was obtained, by giving brief overview of the first 10 papers in the area of EEG based control.

2 EEG Driven Control of Robots: Basic Concepts

The background knowledge about controlling robots can be presented by the block diagram [1] shown in Fig. 1. The system usually operates in real-time or near-real time.

S. Markovski and M. Gusev (Eds.): *ICT Innovations 2012*, AISC 207, pp. 1–11.
DOI: 10.1007/978-3-642-37169-1_1

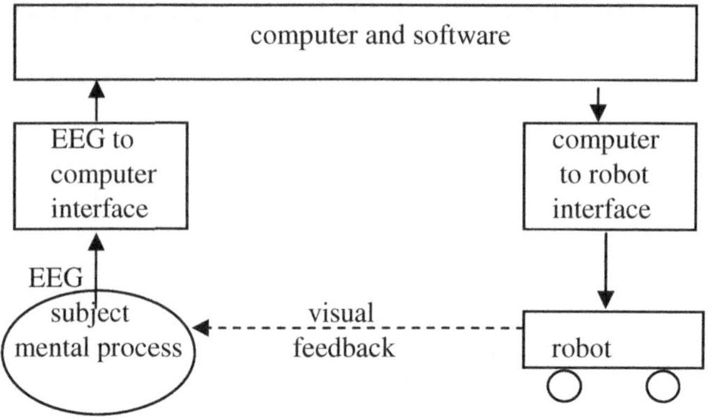

Fig. 1. Conceptual block diagram of EEG driven control of a robot

As Fig.1 shows, a subject generates EEG signals which are amplified and possibly filtered by a EEG amplifier. The generated signal can be result of an intentional mental process, for example, entering willingly a relaxation state can generate higher amplitude of the EEG alpha rhythm. Another way of generating a specific EEG signal is reaction to an event, for example, if a specific pattern is observed on a screen then a brain might react with a P300 potential. Yet another way of generating a EEG signal is before an event, for example, generating CNV signal in expectation of an event. The signal processing software should be able to extract needed features and perform pattern recognition process that would recognize a particular pattern in the EEG signal, for example a CNV shape, or increased amplitude of the alpha rhythm. The pattern classification process often requires a machine learning phase, in which for a given subject the computer learns the template features of a EEG signal pattern, against which, in the exploitation phase, received signal patterns will be compared. Once the pattern recognition process decides that a particular EEG pattern is present, it sends a signal to a robot to perform a predefined behavior. The predefined behavior might be either a simple action such as move or stop, or a rather complex behavior such as wait until some event happens and then turn left. It is often case that a robot executes a default background behavior, such a follow a line on the floor, on which other EEG controlled behaviors are superimposed.

Both single-channel and multi-channel processing might be utilized. Single channel is used if there is a known spot on the scalp where from a particular type of signal can be extracted. Example is alpha rhythm, which can be detected with a single-channel recording from the back of the scalp (occipital, parietal, and temporal area). In some applications at least two-channel recording is needed, for example when brain hemisphere processing difference is utilized, so one channel records from each brain hemisphere. A multi-channel system is often used with ability to give a 2D and 3D spatial distribution of an EEG activity.

3 EEG Potentials

Our interest in EEG patterns recognition started in 1981 and was motivated by the 1964 work of Gray Walter and collaborators [5] describing the CNV (Contingent Negative Variation) potential and the CNV experimental paradigm. We noticed that it is actually a S1-S2-RT paradigm which we knew about since we previously built digital controllers for that paradigm. The CNV potential is related to the processes of expectation and learning. We decided to pursue research on this topic, and the first report was written as a Term paper in 1981 [6]. After some research in the area [7], we introduced [8] a *taxonomy of EEG potentials* in which we distinguished between event related potentials and evoked potentials, and included a new class of potentials, anticipatory potentials containing both the expectation potentials (for example CNV) and preparatory potentials (for example BP Bereitschaftspotenzial [9]

In 1986 we started experimenting with an extension of the classical CNV experimental paradigm by introducing biofeedback in the paradigm [10]. We extended the paradigm beyond CNV appearance, by introducing EEG control of the buzzer that generates the S2 signal. Once CNV is build up, it stops the S2 buzzer, which will cause gradual decay of the CNV signal in the extended paradigm, which in turn will cause the S2 buzzer to be turned on again. In such a way the new experimental paradigm builds an oscillatory expectancy process in the brain. Since the CNV potential in this paradigm changes its shape, an adaptive digital filter was built to extract the variable CNV [8]. In this research we gained experience with adaptive signal processing used later in EEG based robot control. We also built a lab unit devoted to EEG signal processing. Central part was the biopotential amplifier which we obtained from Laboratory of Medical Electronics from Zagreb. It had gain up to 100,000 (10μV/V), adjustable analog filters for band pass filtering, a separate 50Hz filter, long time constant 10s, and input impedance 9MΩ. We requested and obtained a 19” rack system version.

4 Robot Control

The first mobile robot named Adriel-1 we built in 1982 out of a toy car to which we added three tactile sensors and a voltage sensor. The demonstration task was moving in room around a wall and sensing a “door”. The computer used was IBM Series/1 with true multithreading (multitasking) programming language named Event Driven Language (EDL) which utilized explicit commands WAIT and POST for interprocess communication and ENQ (enter queue) and DEQ (depart queue) for resource management. Our first mobile robots were indeed multitasking driven with separate tasks for sensors and motors [11].

In 1984 we purchased from Akihabara market in Tokyo, Japan, a kit for a robot named Elehoby Line Tracer. That was a robot which had own intelligence to follow an arbitrary line drawn on the floor. The external control was a mechanical on/off switch.

5 The Idea of EEG Driven Robot Control

By 1987 we developed a Lab for Intelligent Machines, Bioinformation Systems, and Systems Software in the new annex of the Electrical Engineering Department, University Sts Cyril and Methodius. Inside the lab we built two lab units, one for EEG signal processing (subject chairs, biopotential amplifier rack, and oscilloscope) and one for robotics (polygon rack with several robots and an interface box). The idea came to connect the units and try to control a robot using EEG signals. The subsequent idea that it points toward an engineered solution of the psychokinesis phenomenon looked even more exciting. As result, Fig. 2 shows the lab setup which we built for controlling robots using EEG signals.

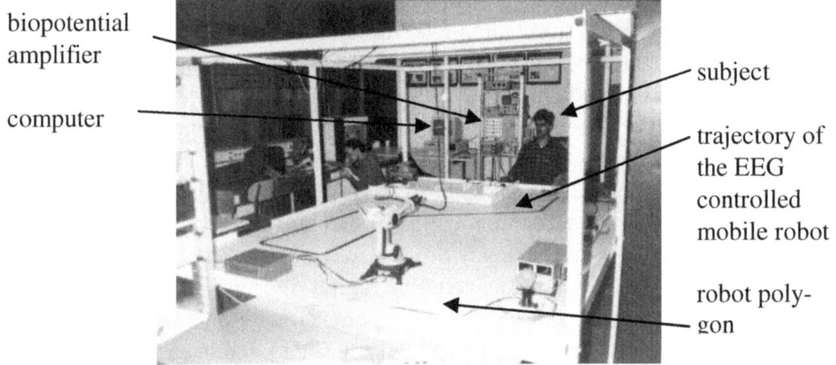

Fig. 2. The lab setup for controlling robots using EEG signals, photo from 1988

Fig. 2 shows the robot polygon we built, containing several mobile and manipulative robots, and a drawn trajectory on the polygon where the Elehoby Line Tracer robot moved. We replaced the mechanical on/off switch of the Elehoby Line Tracer robot with a computer-controlled switch, and connected the same computer with the biopotential amplifier. Fig. 3. shows the drawn trajectory for the robot. The robot was given a "coat" to look as a Flexible Manufacturing System shuttle robot which moves along a closed trajectory.

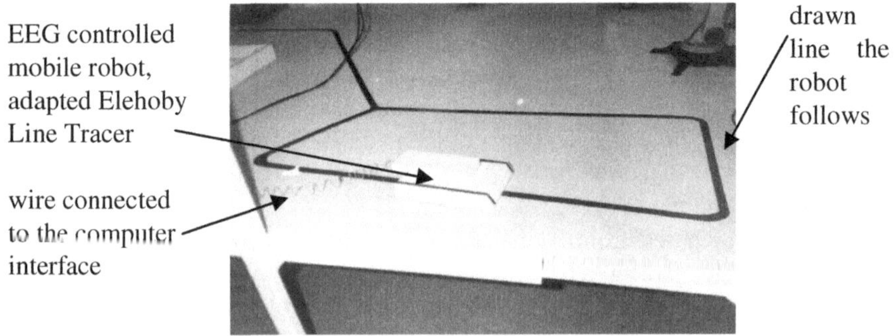

Fig. 3. Robot polygon with trajectory of the EEG controlled mobile robot, 1988

The mental process used for EEG control was relaxation process representing itself by amplitude change of the alpha rhythm, in other words the *contingent alpha rhythm* variation (CαV) of the EEG.

Since we have chosen the alpha rhythm variation, for electrode placement we used spots on the occipital and parietal area, such as O_2 and Pz, where the changes in the frequency band 8-13Hz (alpha rhythm) significantly influence the EEG signal.

The following scenario was introduced: while following a line on the floor, the Elehoby Line Tracer robot will be start to move by alpha rhythm amplitude increase which will be willingly decided by the subject who will close the eyes and relax. Analogously, when the subject decides to open the eyes and consequently reduce alpha rhythm amplitude, the robot will stop at a particular place on the trajectory.

6 The Realization of the Idea

6.1 Robot Control

We already had a robot with own intelligence executing a follow-line behavior. We just needed an EEG switch that will change between robot follow line and robot stop behaviors. Fig. 4. shows the Moore automaton model of the control we used. The events EEGα(+) and EEGα(-) are generated by the subject who willingly increases/decreases the amplitude of the EEG alpha rhythm.

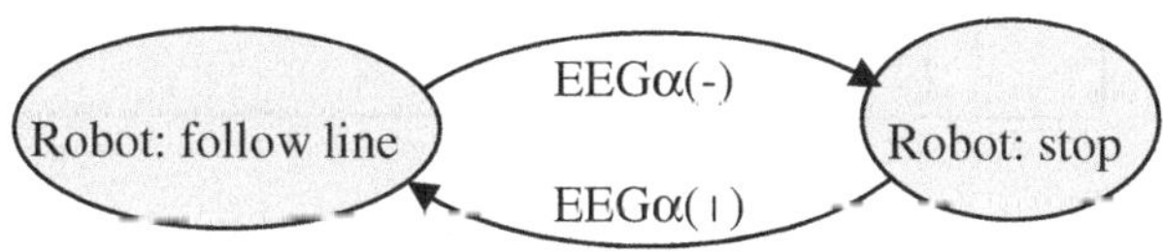

Fig. 4. Line following robot control using EEG switch

6.2 EEG Signal Processing for Robot Control

The signal processing part presented two engineering problems. The first problem was stopping a moving robot at a particular point. If the subject wants to stop the robot at a particular point, the signal processing should be very fast. We decided to find a hard real time algorithm, the one which will execute an action inside the sampling interval of the EEG signal, which in our case was 10ms (100 Hz sampling rate). We needed a procedure that reads an EEG sample, extracts the EEG features, compares them to template feature, and sends command to the robot, all that in less than 10 ms on a 1988 PC/XT computer.

The second problem was variability of the alpha rhythm amplitude across subjects and even for the same subject during a day. In order to adapt to such changes it was obvious that a learning algorithm was needed to be applied before each experiment of alpha rhythm based robot control.

Since we needed an *algorithm that executes inside a sampling interval*, we could use neither a frequency domain processing nor averaging of the EEG signal. So we used analog filtering provided by the biopotential amplifier to extract a frequency band containing the alpha rhythm. Further, we developed an adaptive pattern recognition method consisting of two phases: a learning phase in which the computer learns the EEG features related to increase/decrease of the amplitude of the alpha rhythm, and a pattern recognition phase in which computer compares the just observed features against the template one.

We have chosen 10 seconds of learning procedure in which subject will open and close her/his eyes and generate amplitude change in the alpha rhythm. Since our sampling rate was 100 Hz, we acquired 1000 samples where from the template features will be learned.

Our algorithm used both *changes* of EEG amplitude and *changes* of time intervals between EEG amplitudes (Fig 5).

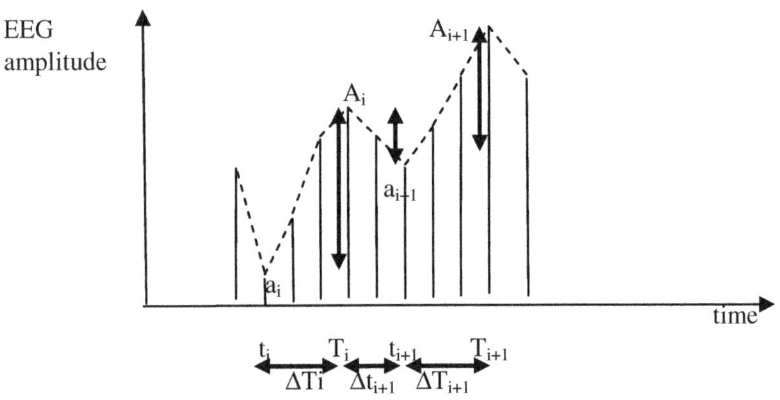

Fig. 5. Obtaining both time difference and amplitude difference for EEG extrema

Those features are fast computable since it needs comparison only with the previous sample to obtain the changes. The learning algorithm scans the EEG samples and looks for local extrema, peaks and valleys of the signal, the points where gradient changes the sign. For each peak, its amplitude is determined relative to the immediate previous valley. Also for each peak the width of the hill is determined as time distance between the previous valley and the peak.

In mathematical terms, whenever change of the sign of gradient of the EEG curve is sensed on a point EEG(t), two differences are computed. One is the time difference between the maxima and minima of the EEG hills. Symbolically $\Delta T_i = T_i - t_i$, is the time difference between the i-th maximum and the i-th minimum, and $\Delta t_{i+1} = t_{i+1} - T_i$ is the time difference between the (i+1)-th minimum and the i-th maximum. For each amplitude extremum, the amplitude difference is computed, $\Delta A_i = A_i - a_i$ and $\Delta a_{i+1} = a_{i+1} - A_i$. Actually we compute the absolute values of the differences.

In the learning process the amplitude differences and time differences are counted and probability density distributions (pdd) are obtained. So for each subject we obtain both the EEG amplitude difference pdd p(A) and EEG time difference pdd p(T). Dute to open and closed eyes each of the pdd's has two instances, so we obtained four pdd's, p(A/open), p(A/closed), p(T/open) and p(T/closed) as shown in Fig. 6. Due to distributions overlap, there are possibilities of false positive and false negative decisions. Decision thresholds should be determined for pair of overlapping distributions, as shown in Fig 6. The decision areas $\Delta T < \theta_{\Delta TO}$ and $\Delta T > \theta_{\Delta TC}$, as well as $\Delta A < \theta_{\Delta AO}$ and $\Delta A > \theta_{\Delta AC}$, are areas where minimum number of false decisions are made.

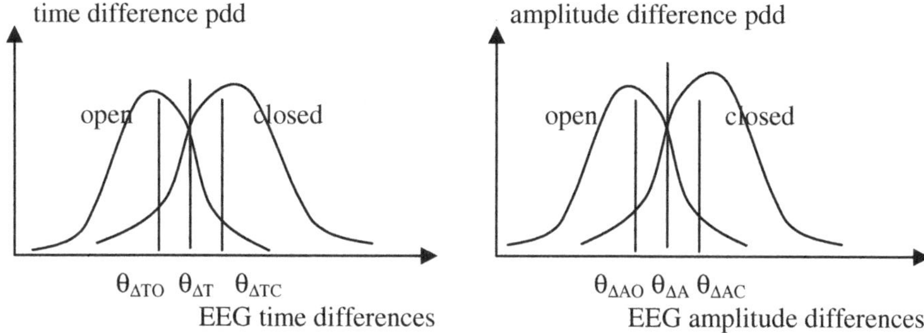

Fig. 6. Probability density distributions for amplitude and time difference

The decision process used *confirmation sequence* of three samples in a row, meaning that in each sample its amplitude difference and time difference should be greater than θΔA and θΔT respectively. So the decision criterion for eyes closed is

if ΔA(t) > θΔAC and ΔT(t) > θΔTC for three consecutive times, then eyes = "closed"

With the obtained probability density distributions, and with determined thresholds, the learning process calibrated the classifier for the pattern classification process that comes in the examination phase and the exploitation phase.

Examination procedure tests the learning process. The subject is given a time, for example 15 seconds, in which s/he will close and open the eyes at least once. The exploitation procedure is the real demonstration of the process of control of a robot using EEG signal. The subject's decision when to close or open the eyes is asynchronous to any external event, and is the subject's choice.

7 Results

Example of a computer screen obtained in our 1988 experiments is shown in Fig. 7. The bottom part of the screen shows an acquired filtered EEG signal in duration of 10-15 seconds. A line below the EEG signal is the segment that will be zoomed. The zoomed segment is shown in the middle of the screen. The pattern recognition algorithm draws a rectangle waveform on the upper part of the screen showing segments where relaxation process is evident (alpha rhythm amplitude has increased) and segments where it is not present.

The software was written in Pascal with some inline sections in assembler. The pseudocode was written in pseudo Cobol [3][4] due to appreciation of Cobol's PERFORM command. [12].

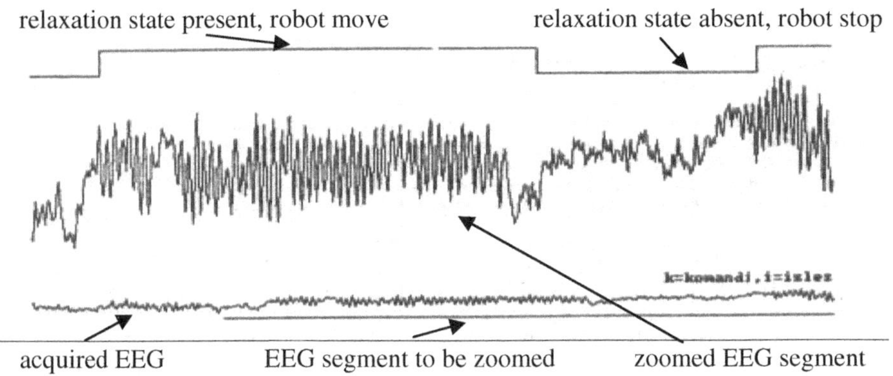

Fig. 7. Example of a asynchronous EEG driven robot control in real time

Two students of Computer Science major successfully carried out the experiments of moving the robot along the closed trajectory and stopping it at a particular place. Four additional students, Computer Science major, were engaged in experiments of moving the robot for a segment of the trajectory. The average learning time was about 30 minutes before successful EEG control was achieved.

8 Research Context

This paper is written almost quarter century after the 1988 result. It is now clear that at that time nobody else attempted an engineering solution of moving a physical object using EEG signals. However, there were efforts of moving objects on a computer screen, of which we were not aware at that time. The first paper that provided a context to us was the paper [13] which came out after we achieved the result. In this section we give a brief overview of the first 10 papers describing control of objects using EEG signals, both physical and objects on computer screen.

The earliest report of using EEG signals for control of external events was given in 1967 [14], about producing Morse code using EEG alpha rhythm. In 1973 the research field of brain-computer communication was established, and the term

Brain-Computer Interface was introduced by Vidal [15]. The challenge was stated of control of objects using EEG signals. Various EEG signals were mentioned as part of the challenge such as EEG rhythms, evoked potentials, P300 potential, and CNV potential among others; EOG signals were also mentioned. In 1977, EEG control of a graphical object on a computer screen was achieved [16]. In 1988, the first control of a physical object was achieved [1][2]. The same year an important computer screen based application was reported [17] in which P300 potential was used to select a letter shown on the screen and write a text from the selected letters. In 1990 a review was given [3] on various types of controlling mobile robots, including EOG control which happened in 1989. In 1990 it was also a report [13] on using EEG difference in brain hemispheres for EEG based control. A control of a one-dimensional cursor movement on a computer screen was given in 1991 [18]. In 1992 a report on controlling a device using CNV potential was given [6]. In addition of solving the CNV-based control of objects as stated in Vidal's challenge [15], it was a first report on interactive EEG-based game between an adaptive brain and an adaptive external device, an adaptive buzzer. In 1992 also a new type of brain potentials were introduced in EEG based control, the steady state visual evoked potentials (SSVEP) [19]. Visual evoked potentials (VEP) were used in 1993 [20].

After the first 10 papers most of the researchers accepted the term Brain-Computer Interface [18][21] proposed in [15]. Let us note that we used the term direct bioelectric control [3].

In addition of listing the first 10 reports in the area of EEG based control, we will also mention the second report on controlling robots using EEG signals [22] which happened in 1999. An invasive EEG recording was applied on open brain, experiments were carried out on monkeys, and a manipulative robot was controlled. The 1988 and 1999 reports were the only ones on EEG controlled robots given in the 20th century, before the year 2000.

9 Conclusion

In 1988 a result was achieved on controlling a physical object, a robot, through EEG signals emanated from a human brain. Although the first report was given in the same year, this paper is more detailed account of the infrastructure developed and signal processing methods used for obtaining the result. This paper also makes an effort to put the 1988 work in the context of the first 10 papers in the area of EEG based control of objects, both physical objects and objects on a computer screen.

Acknowledgement. The work described here was financed in part by Macedonian Association for Scientific Activities which funded the project entitled Adaptive Industrial Robots under grant number 090110384 for the period 1984-1987 and from the Macedonian Ministry of Science which funded the project entitled Adaptive Intelligent Industrial Robots under grant number 08-778 for the period 1989-1992.

References

1. Bozinovski, S., Bozinovska, L., Setakov, M.: Mobile robot control using alpha wave from the human brain. In: Proc. Symp. JUREMA, Zagreb, pp. 247–249 (1988) (in Croatian)
2. Bozinovski, S., Sestakov, M., Bozinovska, L.: Using EEG alpha rhythm to control a mobile robot. In: Harris, G., Walker, C. (eds.) Proc. 10th Annual Conf. of the IEEE Engineering in Medicine and Biology Society, Track 17: Biorobotics, New Orleans, vol. 3, pp. 1515–1516 (1988)
3. Bozinovski, S.: Mobile robot trajectory control: From fixed rails to direct bioelectric control. In: Kaynak, O. (ed.) Proc. IEEE International Workshop on Intelligent Motion Control, Istanbul, vol. 2, pp. 463–467 (1990)
4. Bozinovski, S., Sestakov, M., Stojanov, G.: A learning system for mobile robot control using human head biosignals. Problemji Mashinostroenija i Avtomatizacii 6, 32–35 (1991) (in Russian)
5. Walter, G., Cooper, R., Aldridge, V., McCallum, W.: Contingent Negative Variation: An electric sign of sensory-motor association and expectancy in the human brain. Nature 203, 380–384 (1964)
6. Bozinovska, L.: The CNV paradigm: Electrophysiological evidence of expectation and attention. Unpublished Term Paper. Course PSYCH330 Physiological Psychology, Instructor Beth Powel, Psychology Department, University of Massachusetts/Amherst (1981)
7. Bozinovska, L., Isgum, V., Barac, B.: Electrophysiological and phenomenological evidence of the expectation process in the reaction time measurements. Yugoslavian Physiologica and Pharmacologica Acta, 21–22 (1985)
8. Bozinovska, L., Bozinovski, S., Stojanov, G.: Electroexpectogram: Experimental design and algorithms. In: Proc. IEEE International Biomedical Engineering Days, Istanbul, pp. 58–60 (1992)
9. Kornhuber, H., Deecke, L.: Changes in brain potentials in case of willing movements and passive movements in humans: Readiness potential and reaferent potentials. Pflügers Arch. 284, 1–17 (1965) (in German)
10. Bozinovska, L., Sestakov, M., Stojanovski, G., Bozinovski, S.: Intensity variation of the CNV potential during the biofeedback training guided by a personal computer. Neurologija 37(suppl. 2) (1988) (in Serbian)
11. Bozinovski, S., Sestakov, M.: Multitasking operating systems and their application in robot control. In: Proc. Workshop on Macedonian Informatics, Skopje, pp. 195–199 (1983) (in Macedonian)
12. Walker, T.: Fundamentals of Cobol Programming: A Structured Approach. Allyn and Bacon (1976)
13. Keirn, Z., Aunon, J.: A new mode of communication between man and his surroundings. IEEE Transactions on Biomedical Engineering 37(12), 1209–1214 (1990)
14. Dewan, E.: Occipital alpha rhythm eye position and lens accommodation. Nature 214, 975–977 (1967)
15. Vidal, J.: Toward direct brain-computer communication. Annual Review of Biophysics and Bioengineering, 157–180 (1973)
16. Vidal, J.: Real-time detection of brain events in EEG. Proceedings of the IEEE 65, 633–641 (1977)
17. Farwell, L., Donchin, E.: Talking off the top of your head: a mental prosthesis utilizing event-related brain potentials. Electroencephalography and Clinical Neurophysiology 70, 510–523 (1988)

18. Wolpaw, J., McFarland, D., Neat, G., Forneris, C.: An EEG-based brain-computer interface for cursor control. Electroencephalography and Clinical Neurophysiology 78(3), 252–259 (1991)
19. Sutter, E.: The brain response interface: Communication through visually induced electrical brain responses. Journal of Microcomputer Applications 15, 31–45 (1992)
20. Cilliers, P., Van Der Kouwe, A.: A VEP-based computer interface for C2-quadriplegics. In: Proc. of the 15th Annual International Conf. of the IEEE, p. 1263 (1993)
21. Pfurtscheller, G., Flotzinger, D., Kalcher, J.: Brain Computer Interfaces – A new communication device for handicapped person. Journal of Microcomputer Applications 16, 293–299 (1993)
22. Chapin, J., Moxon, K., Markowitz, R., Nicolelis, M.: Real-time control of a robot arm using simultaneously recorded neurons in the motor cortex. Nature Neuroscience 2, 664–670 (1999)

Hybrid 2D/1D Blocking as Optimal Matrix-Matrix Multiplication

Marjan Gusev, Sasko Ristov, and Goran Velkoski

Ss. Cyril and Methodius University,
Faculty of Information Sciences and Computer Engineering,
Rugjer Boshkovikj 16, 1000 Skopje, Macedonia
{marjan.gushev,sashko.ristov}@finki.ukim.mk, velkoski.goran@gmail.com

Abstract. Multiplication of huge matrices generates more cache misses than smaller matrices. 2D block decomposition of matrices that can be placed in L1 CPU cache decreases the cache misses since the operations will access data only stored in L1 cache. However, it also requires additional reads, writes, and operations compared to 1D partitioning, since the blocks are read multiple times.

In this paper we propose a new hybrid 2D/1D partitioning to exploit the advantages of both approaches. The idea is first to partition the matrices in 2D blocks and then to multiply each block with 1D partitioning to achieve minimum cache misses. We select also a block size to fit in L1 cache as 2D block decomposition, but we use rectangle instead of squared blocks in order to minimize the operations but also cache associativity. The experiments show that our proposed algorithm outperforms the 2D blocking algorithm for huge matrices on AMD Phenom CPU.

Keywords: CPU Cache, Multiprocessor, Matrix Partitioning.

1 Introduction

Matrix multiplication algorithm (MMA) is a basic linear algebra operation used in almost all scientific computations. Different techniques are proposed to speedup the execution. Another very important issue is the selection of appropriate powerful hardware environment for efficient execution. Fused multiply-add (FMA) units within modern CPU architecture execute both addition and multiplication in one clock cycle. Since matrix multiplication is compute intensive algorithm with $O(N^3)$ complexity increasing the processor speed will speedup the execution. It is also a memory demanding algorithm with $O(N^2)$ complexity. However, the most important is the fact that MMA is cache intensive algorithm with $O(N)$ complexity since each matrix element is accessed N times for different computations and the memory access per element vary between several clocks for the lowest L1 cache memory and up to 1000 for main memory [7].

Introducing multi-chip and multi-core CPUs, and many-core GPU processors together with different APIs and libraries for parallelization can significantly speedup the execution since matrix multiplication is excellent algorithm for parallelization and thus can maximize the efficiency. It can achieve almost linear

S. Markovski and M. Gusev (Eds.): *ICT Innovations 2012*, AISC 207, pp. 13–22.
DOI: 10.1007/978-3-642-37169-1_2

speedup. However, there are regions where superlinear speedup can be achieved for multi-chip and multi-core implementations [10]. It is also possible in multi-GPU implementation on Fermi architecture GPU, due to configurable cache [9].

Runtime and platform environments also have huge impact to matrix multiplication performance. Better performance is achieved in shared memory virtual guest environment compared to traditional host operating system, for both sequential and parallel executions of problem sizes that fit in distributed L1 and L2 caches [5]. Adding more processors will not improve the system performance in cloud computing because of the communication overhead [2].

Another focus is optimizing the algorithm if the matrices have some features like symmetric, zero rows or columns, sparse, squared or triangle. The authors in [4] optimized multiplication of small sized matrices. Superlinear speedup was reported for sparse symmetric matrix vector multiplication in [11]. [8] reports also superlinear speedup with parallel execution of MMA using MPI and transposing one source matrix. The authors in [1] improved parallel execution based on Strassen's fast matrix multiplication minimizing communication.

Optimizing the algorithm according to CPU cache, platform and runtime environment is the most appropriate approach. The authors in [12] used padding to the first element of each submatrix to land on equidistant cache sets and to avoid cache associativity performance drawbacks [6]. The authors in [7] propose compiler optimizations with loop interchange and blocking to reduce cache miss rate. The optimal rectangular partitioning can be significantly outperformed by the optimal non-rectangular one on real-life heterogeneous HPC platforms [3].

In this paper we analyze blocking algorithm for matrix multiplication and propose a new hybrid 2D/1D blocking method for matrix multiplication that exploits the maximum of the L1 cache size by reducing the number of compute operations and cache associativity problems. The rest of the paper is organized as follows. Section 2 presents the dense and blocking MMAs. The new proposed Hybrid 2D / 1D blocking MMA is described in Section 3. Section 4 presents the testing methodology and Section 5 presents the results of realized experiments. Finally Section 6 is devoted for conclusion and future work.

2 Existing Dense and 2D Blocking MMAs

This section describes the basic dense MMA and 2D blocking MMA and presents their complexity. We denote that $C_{N \cdot N} = A_{N \cdot N} \cdot B_{N \cdot N}$.

2.1 Dense MMA

This algorithm works with entire matrix A rows and matrix B columns for row major order to calculate one element of matrix C. Figure 1 depicts the algorithm. The relatively small amount of operations clearly makes this algorithm acceptable for small matrices A and B. Problem occurs when the whole matrix B and one row vector from matrix A cannot be placed in the cache and therefore cache misses start to generate. We can conclude that although it performs only

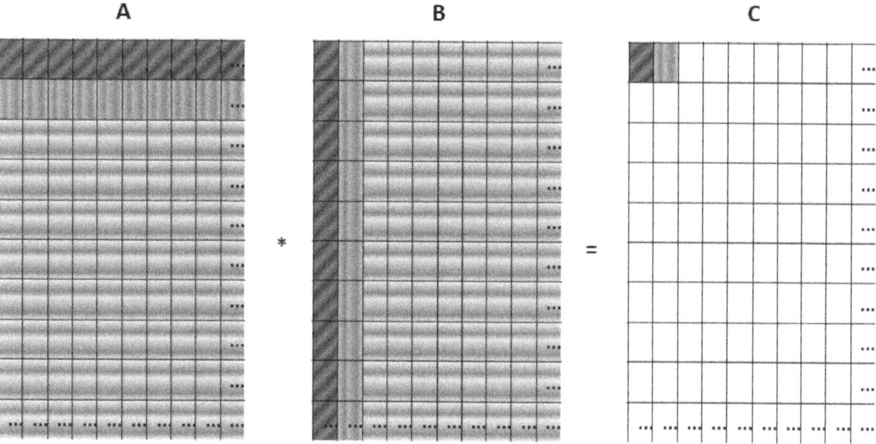

Fig. 1. Dense MMA

minimal $2 \cdot N^3$ operations, it generates a significant number of cache misses for each cache level when using huge matrices with sizes greater than the cache size.

Each element of matrices A and B is accessed N times and thus the overall memory reads are $2 \cdot N^3$. There are N^2 writes in the memory for the elements of the matrix C.

Maximum efficiency for parallel execution of this algorithm is realized when the number of processors P is divisor of the matrix size N.

2.2 2D Blocking MMA

The algorithm described in [7] reduces the number of cache misses. It works on submatrices or blocks of matrices A and B with same size b instead of the entire matrix A rows and matrix B columns. Figure 2 depicts the algorithm. The goal is to maximize the reuse of the data of a block before they are replaced.

The total number of operations in this algorithm is increased. Additional $N/b \cdot N/b \cdot N \cdot b = N^3/b$ are performed for summing the elements of matrix C. Including the necessary $2 \cdot N^3$ operations the total number of floating point operations is $2 \cdot N^3 + N^3/b$. The benefit of these additional operations is the decreased number of memory accesses of $N/b \cdot N/b \cdot N \cdot b \cdot 2 = 2 \cdot N^3/b$ in the worst case.

Maximum efficiency for parallel execution of this algorithm is realized when the product $P \cdot b^2$ is divisor of N^2.

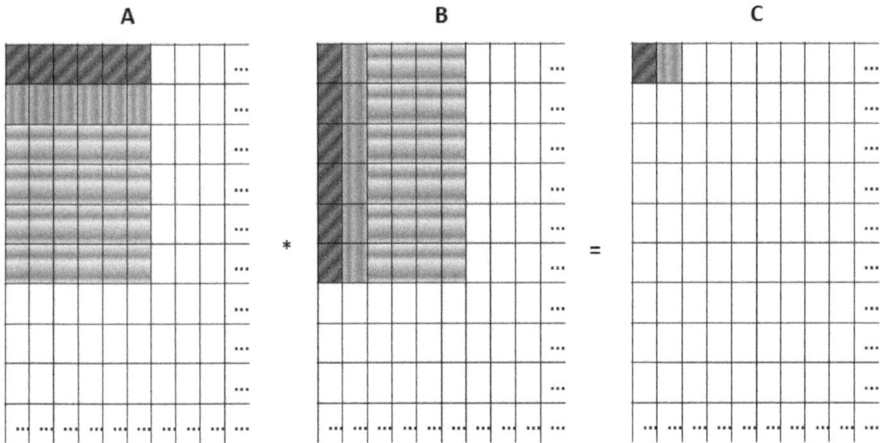

Fig. 2. 2D Blocking MMA

3 Hybrid 2D/1D Blocking MMA

This section describes our new proposed Hybrid 2D/1D Blocking MMA, its complexity, performance, and comparison with the other algorithms. Also we analyze how different cache parameters impact the algorithm performance.

3.1 Decrease the Operations and Memory Accesses

The idea is to exploit the benefits of both algorithms described in Section 2, i.e. to use 2D blocking to minimize the memory access, but in the same time minimizing the additional operations for the elements of matrix C. Figure 3 depicts the algorithm for hybrid blocking matrix blocking matrix multiplication. We propose rectangles with the same area to be used instead of squares for blocks. Next, dense MMA is implemented for multiplication in each block.

Let's denote with b the squared block size of 2D blocking algorithm and with b_X and b_Y the number of rows and columns for block of matrix B correspondingly such that relations (1) and (2) are satisfied

$$b_X > b_Y \tag{1}$$

$$b_X \cdot b_Y = b^2 \tag{2}$$

Then the blocks of matrix A will consists of b_Y rows and b_X columns. It follows that the hybrid blocking algorithm will perform smaller number of memory accesses rather than 2D blocking MMA, i.e. $N/b_Y \cdot N/b_X \cdot N \cdot b_Y \cdot 2 = 2 \cdot N^3/b_X$. It also performs less floating point operations, i.e. $N/b_Y \cdot N/b_X \cdot N \cdot b_Y = N^3/b_X$ which is smaller than 2D blocking since $b_X > b$.

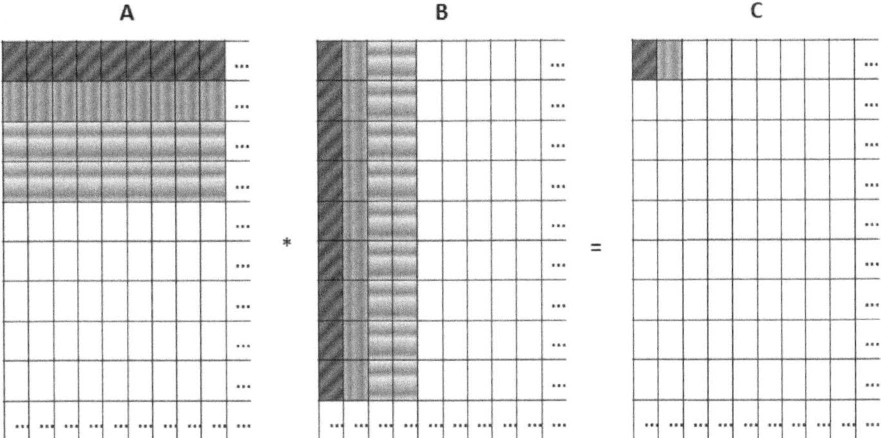

Fig. 3. Hybrid MMA

Maximum efficiency for parallel execution of this algorithm is realized when the product $P \cdot b_X \cdot b_Y$ is divisor of N^2.

3.2 The Algorithm and the Cache Parameters

Since MMA is cache intensive algorithm, lets analyze the cache impact to the new algorithm compared to 2D blocking algorithm. The previous Section 3.1 presents that hybrid algorithm executes smaller number of operations and memory accesses than 2D blocking. But is it enough to be a faster algorithm than the original 2D blocking algorithm?

Cache has several parameters that have a huge impact to the overall performance especially for particular matrix size. Since the block area is the same for both algorithms cache size impact is also the same, i.e. the same cache misses will be generated only when the block is changed. Therefore cache replacement policy will not impact also.

What about cache line? Lets denote with l the number of matrix elements that can be placed in one cache line. Modern CPUs cache line size is 64B, i.e. it can store $l = 8$ matrix elements (double precision) which can be loaded in one operation from the memory. Our new algorithm defines that b_X should be greater as much as possible to reduce the operations and memory accesses, but still (2) should be satisfied. However, to exploit maximum performance the same cache line should be present in the L1 cache, and thus $b_Y < l$.

Further on we analyze *cache associativity problem* that appears in storage of matrix columns and inefficient usage of cache for particular problem size [6]. Matrix B blocks will map onto a smaller group of cache sets than the same 2D blocking MMA and initiate more cache misses. In this case our algorithm will use a smaller group of cache sets in associative memory.

Today's Intel L1 cache is 8-way set associative and AMD L1 cache is only 2-way set associative. For particular matrix sizes N it means that Intel CPU will provide better performance than AMD due to cache set associativity for the same block size $b_X \cdot b_Y$.

3.3 Another Variant of the Algorithm - Modified Hybrid MMA

To avoid cache associativity problem especially for AMD CPU we make another experiment where b_X is smaller and b_Y is maximized. Figure 4 depicts this algorithm. This algorithm has more operations and memory accesses than others, but will be prone to cache size associativity, i.e. L1 cache works as fully associative.

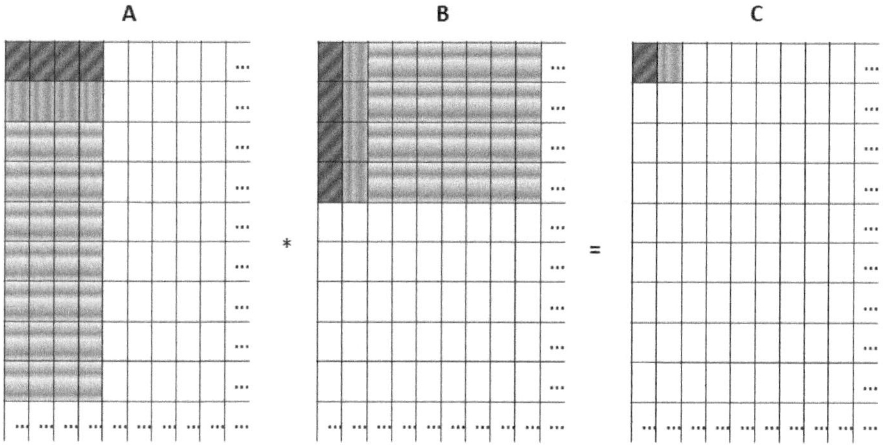

Fig. 4. Modified hybrid MMA

4 The Testing Methodology

The experiments are performed on two multiprocessors with different cache architectures since the hardware can impact to the algorithm performance. The first multiprocessor consists of 2 chips Intel(tm) Xeon(tm) CPU X5680 @ 3.33GHz and 24GB RAM. Each chip has 6 cores, each with 32 KB 8-way set associative L1 data cache dedicated per core and 256 KB 8-way set associative L2 cache dedicated per core. All 6 cores share 12 MB 16-way set associative L3 cache. The second server has one chip AMD Phenom(tm) 9950 Quad-Core Processor @ 2.6 GHz and 8 GB RAM. The multiprocessor has 4 cores, each with 64 KB 2-way set associative L1 data cache dedicated per core, and 512 KB 16-way set associative L2 cache dedicated per core. All 4 cores share 2 MB 32-way set associative L3 cache. The servers are installed with Linux Ubuntu 10.10. C++ with OpenMP for parallelization are used without additional optimizations.

We execute dense, 2D blocking, hybrid and modified hybrid MMAs on each multiprocessor with different matrix sizes to test algorithm behavior in different cache regions. For 2D blocking algorithm we choose $b = 36$ for Intel CPU and $b = 48$ for AMD to satisfy that matrices can be stored in the L1 cache. For our two hybrid algorithm experiments we use $b_x = 162$ and $b_y = 8$ for hybrid MMA and $b_x = 8$ and $b_y = 162$ for modified hybrid MMA for Intel CPU. For AMD CPU $b_x = 288$ and $b_y = 8$, and $b_x = 8$ and $b_y = 288$ are used.

5 The Results of the Experiments

This section presents the results of the experiments realized to measure the performance of the new hybrid 2D/1D MMAs compared to 2D blocking.

Figure 5 depicts the execution time for dense, 2D blocking, hybrid and modified hybrid MMAs on Intel CPU. For $N < 1296$ all MMAs run similar. Dense algorithm increases its execution time for greater matrices more than modified hybrid 8x162. Hybrid 162x8 and blocking 2D are the best MMAs.

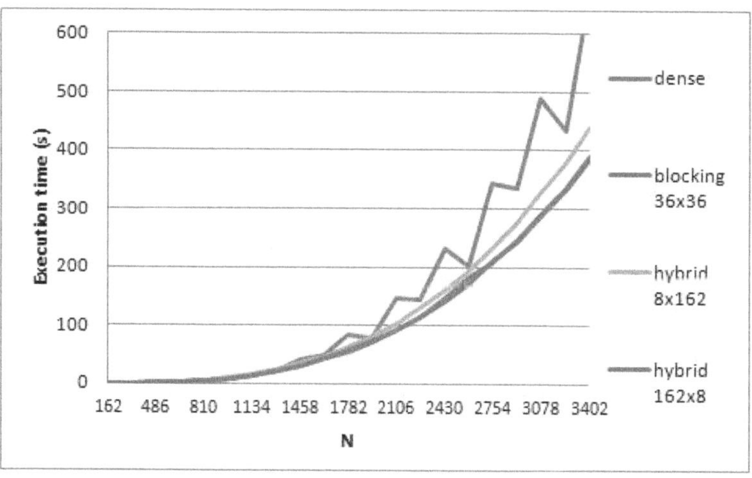

Fig. 5. The execution time for sequential execution on Intel CPU

Figure 6 depicts the speed on Intel CPU. Dense performs the best speed for smaller matrices as expected due to huge L3 size. However, for huge matrices all other MMAs perform better speed compensating the increased number of operations with lower average access time per matrix element. We can conclude that all three other algorithms have constant speed. Our proposed hybrid MMA provides the same speed as 2D blocking, and the modified hybrid MMA is worse.

We can conclude that cache associativity problem does not impact directly on Intel CPU. However, the results do not confirm our hypothesis that our new 2D/1D will be better than 2D blocking. After deep analysis we give the following

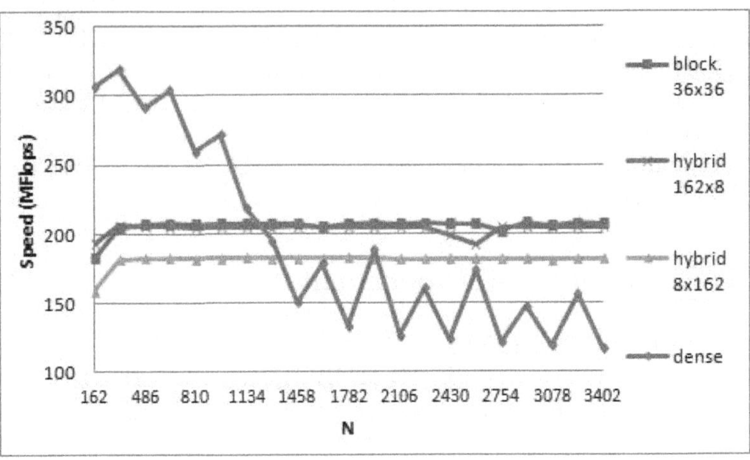

Fig. 6. The speed for sequential execution on Intel CPU

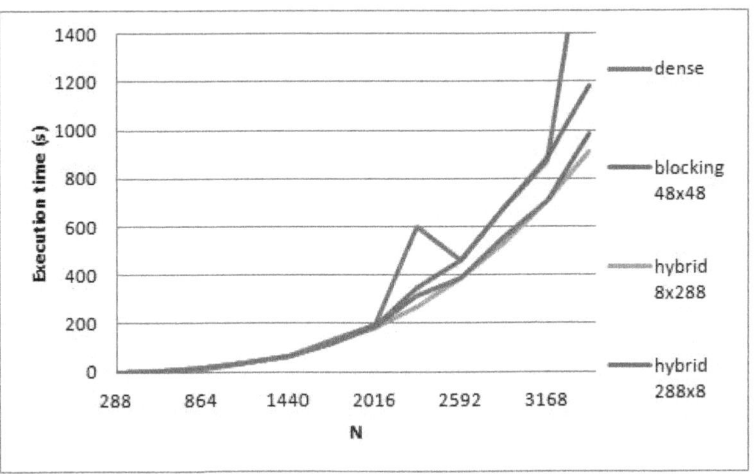

Fig. 7. The execution time for sequential execution on AMD CPU

explanation. The L1 cache is occupied also from operating system. Cache misses appear from the operating system and the whole cache line is replaced.

Figure 7 depicts the execution time for the same four algorithms on AMD CPU. For smaller matrices the execution time is similar for all algorithms, but for greater matrices both our algorithms outperform the 2D blocking and dense.

Figure 8 depicts the speed for better presentation and analysis. As depicted, Dense is the leader in front of 2D blocking, hybrid and modified hybrid for small matrices. However, increasing the matrix size N, the order is opposite: modified hybrid achieves best performance in front of hybrid, 2D blocking and Dense.

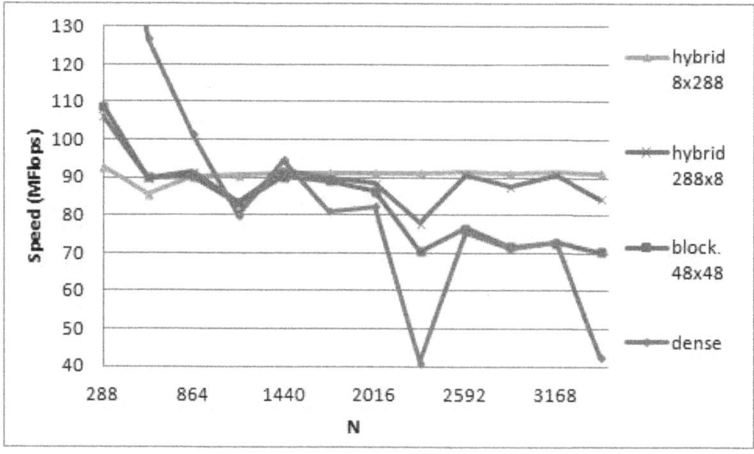

Fig. 8. The speed for sequential execution on AMD CPU

Only our modified hybrid algorithm retains the speed regardless of N. It is prone to the cache associativity problem. All other algorithms achieve drawbacks in particular N due to cache associativity problem. We see that our hybrid MMA has similar performance as modified hybrid in points where there aren't cache associativity performance drawbacks [6].

6 Conclusion and Future Work

This paper presents the new hybrid 2D/1D blocking MMAs with similar performance on Intel CPU as the 2D blocking MMA, and better performance on AMD Phenom CPU. Using theoretical analysis of all cache parameters that can impact the algorithm performance we modify the proposed algorithm and even improved already better performance than 2D blocking MMA. Even more, the modified algorithm is prone to small cache set associativity on AMD CPU caches. The experiments prove the theoretical analysis.

We plan to continue to improve the proposed algorithm in sequential execution for Intel CPU and extend the research for its performance for parallel execution.

References

1. Ballard, G., Demmel, J., Holtz, O., Lipshitz, B., Schwartz, O.: Communication-optimal parallel algorithm for strassen's matrix multiplication. In: Proceedinbgs of the 24th ACM Symposium on Parallelism in Algorithms and Architectures, SPAA 2012, pp. 193–204. ACM, NY (2012)
2. Bataineh, S., Khalil, I.M., Khreishah, A., Shi, J.Y.: Performance evaluation of matrix multiplication on a network of work stations with communication delay. JCIS: Journal of Communication and Information Sciences 1(2), 32–44 (2011)

3. DeFlumere, A., Lastovetsky, A., Becker, B.: Partitioning for parallel matrix-matrix multiplication with heterogeneous processors: The optimal solution. In: 21st International Heterogeneity in Computing Workshop (HCW 2012). IEEE Computer Society Press, Shanghai (2012)
4. Drevet, C.E., Islam, M.N., Schost, E.: Optimization techniques for small matrix multiplication. ACM Comm. Comp. Algebra 44(3/4), 107–108 (2011)
5. Gusev, M., Ristov, S.: Matrix multiplication performance analysis in virtualized shared memory multiprocessor. In: MIPRO, 2012 Proceedings of the 35th International Convention, pp. 264–269. IEEE Conference Publications (2012)
6. Gusev, M., Ristov, S.: Performance gains and drawbacks using set associative cache. Journal of Next Generation Information Technology (JNIT) 3(3), 87–98 (2012)
7. Hennessy, J.L., Patterson, D.A.: Computer Architecture: A Quantitative Approach, 5th edn. (2012)
8. Jenks, S.: Multithreading and thread migration using mpi and myrinet. In: Proc. of the Parallel and Distrib. Computing and Systems, PDCS 2004 (2004)
9. Playne, D.P., Hawick, K.A.: Comparison of gpu architectures for asynchronous communication with finite-di erencing applications. Concurrency and Computation: Practice and Experience 24(1), 73–83 (2012)
10. Ristov, S., Gusev, M.: Superlinear speedup for matrix multiplication. In: Proceedings of the 34th International Conference on Information Technology Interfaces, ITI 2012, pp. 499–504 (2012)
11. So, B., Ghuloum, A.M., Wu, Y.: Optimizing data parallel operations on many-core platforms. In: First Workshop on Software Tools for Multi-Core Systems (STMCS), pp. 66–70 (2006)
12. Williams, S., Oliker, L., Vuduc, R., Shalf, J., Yelick, K., Demmel, J.: Optimization of sparse matrix-vector multiplication on emerging multicore platforms. Parallel Comput. 35(3), 178–194 (2009)

P2P Assisted Streaming for Low Popularity VoD Contents

Sasho Gramatikov[1], Fernando Jaureguizar[1], Igor Mishkovski[2], Julián Cabrera[1], and Narciso García[1]

[1] Grupo de Tratamiento de Imágenes, ETSI Telecomunicación, Universidad Politécnica de Madrid, Spain
{sgr,fjn,julian.cabrera,narciso}@gti.ssr.upm.es
[2] University "Ss Cyril and Methodius", Skopje, Macedonia
igor.mishkovski@finki.ukim.mk

Abstract. The Video on Demand (VoD) service is becoming a dominant service in the telecommunication market due to the great convenience regarding the choice of content items and their independent viewing time. However, due to its high traffic demand nature, the VoD streaming systems are faced with the problem of huge amounts of traffic generated in the core of the network, especially for serving the requests for content items that are not in the top popularity range. Therefore, we propose a peer assisted VoD model that takes advantage of the clients unused uplink and storage capacity to serve requests for less popular items with the objective to keep the traffic on the periphery of the network, reduce the transport cost in the core of the network and make the system more scalable.

Keywords: P2P, streaming, VoD, popularity, cost, scalability.

1 Introduction

The great expansion of the IPTV [1] has made a good ground for the Video on Demand (VoD) to become one of the most popular services. Although VoD is service that is also available on the Internet, it has attracted special attention in the Telecom-managed networks since they are already accustomed for implementation of a variety of TV services. Despite of its numerous advantages from client's point of view, the VoD service is an issue from provider's point of view since it is very bandwidth demanding. Therefore, the design of systems and algorithms that aim at optimal distribution of the content items has become a challenge for many providers. Some of the solutions include hierarchy of cache servers which contain replicas of the content items placed according to a variety of replica placement algorithms [2][3][4]. No matter how good these solutions might be, they all reach a point from where no further improvements can be done because of the resource limitations. One possibility to overcome this problem is the implementation of the classical P2P principles for exchange of files over the Internet for delivering video contents to a large community of users [5][6]. Despite

S. Markovski and M. Gusev (Eds.): *ICT Innovations 2012*, AISC 207, pp. 23–33.
DOI: 10.1007/978-3-642-37169-1_3

the P2P streaming on the Internet has given positive results, its main disadvantage is the reliability of the peers and the Internet. The environment where the implementation of P2P streaming perfectly fits are the telecom-managed IPTV networks. Some of the reasons are that the set-top boxes (STBs) nowadays have considerable storage capacity and the operators have higher control over the devices on the clients premisses, avoiding the reliability issue of the classical P2P systems. Some solutions for the use of P2P in IPTV networks are presented in [7][8][9][10].

Although popular contents generate considerate amount of the overall traffic, there is a large number of contents that are not in the high popularity range, but still take significant part of the streaming traffic. Assuming that in the IPTV networks the content items are distributed according to their popularity, the requests for the less popular contents are a burden for the core of the Telecom-managed network since these contents are stored in servers which are placed further from the clients. Therefore, we propose a solution for a hierarchically structured network with popularity based distribution of contents that aims to reduce the traffic on the servers situated in the core of the network by providing peer assisted streaming of the low popularity contents. This would offload the backbone links of the network and would enable growth of the number of clients subscribed to VoD service without considerate changes and high costs in the core of the network. Despite the common practice to store the most popular contents in the peers for best traffic reduction [10], in our approach, we choose to store the low popularity contents in the peers, thus providing locally close availability of all the popularity range of the videos. We use this approach because the popular contents are already stored in the edge servers which are close to the clients. We tend to reduce the traffic in the core of the network by maintaining replicas on the clients' STBs of those items that are not popular enough to be stored in the servers at the edge of the network. Although the upload capacity of the clients is far below the playback rate of the content items, uninterrupted viewing is achieved by combining a parallel streaming of various parts of the videos by as many peers as it is necessary for obtaining the required quality. With this approach we achieve to concentrate large part of the overall traffic in the periphery of the network and thus, to reduce the transport cost of the traffic from the streaming servers to the clients. Unlike many P2P solutions where the peers self-organize themselves, in our proposed model, the peers have a role of passive contributors to the streaming process having no knowledge of the existence of other peers. All the decisions regarding redirection of the clients are taken by the servers on the edge of the network.

The rest of the paper is organized as follows. In Section 2, we describe the proposed model for peer assisted VoD streaming. In Section 3, we explain the division of the contents for better utilization of the storage capacity of the STBs. In Section 4, we describes the request process for VoD contents and in Section 5, we present the simulation scenarios and the obtained results. Finally we give our conclusions in Section 6.

2 Proposed Model

The model that we propose for optimal distribution of VoD contents is a hybrid solution that unites the advantages of both the IPTV and P2P architectures: the high reliability and scalability of the IPTV architecture and the storage space and unused up-link bandwidth of the P2P architecture. It consists of hierarchically organized streaming servers, management servers and STBs. We consider a managed network owned by a company, which can be managed and configured according to the intensity of the requested traffic. The main streaming functionality is provided by the streaming servers, while the peers have the role to alleviate the overall traffic in the network. Unlike the classical P2P solutions where the clients decide whether to share content or not, in an IPTV managed network, the STBs are owned by the service provider and therefore, part of their storage and streaming capacity can be reserved for the needs of the peer assisted streaming. The hierarchical server architecture is populated with content items in such a way, that the most popular contents are placed in the servers on the edge of the network and the less popular ones are placed in the higher layers of the hierarchy. Since most of the requests are for the popular contents, the traffic will be concentrated on the network periphery, but there will be still a considerable amount of traffic in the backbone of the network for serving the less popular contents. Therefore, we place replicas of these unpopular contents in the peers so that instead of streaming them from the servers in the core of the network, they can be streamed from the peers. In the cases when there are not enough available resources on the peers for streaming the entire content of the low popularity items, the streaming servers are available to serve the missing parts. With this approach, we provide maintaining most of the streaming traffic close to the edge of the network.

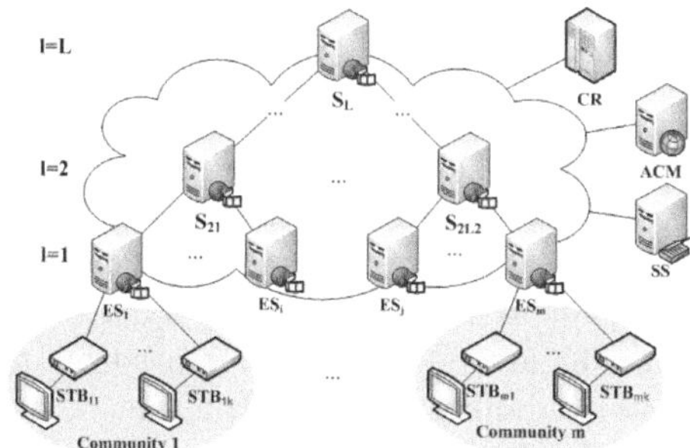

Fig. 1. Model architecture

The streaming servers are organized in a hierarchical structure according to the vicinity to the clients (Figure 1). The servers that are in the edge of the network, called Edge Servers (ES), serve only one group of locally connected clients. All the clients assigned to one ES form a local community. Each peer can serve only clients within the same local community. Each ES keeps track of the popularity of the currently hosted content items. The ES also maintains availability data of the portions of the content items stored in its assigned peers. It uses this data to redirect the clients to other peers whenever there is request for contents that are already stored in the peers.The Central Repository (CR) is highest in the hierarchy and is entry point for new items. It doesn't directly serve the clients, but it supplies the streaming servers with the missing contents whenever it is necessary. The management servers are represented by the ACM and the Service Selection server (SS). The ACM server has the role to monitor the state of the network and to take decisions for a new replica distribution. When necessary, the ACM server runs a redistribution algorithm which decides the number of replicas for each content item and its position in the servers according to the content's popularity and server's utilization data [4]. The SS server is responsible for redirection of the requests to the right servers in a way that the transport cost is minimized and the load between the servers is equally distributed. The SS server is frequently updated by the ACM server with the state of the system and the new position of the replicas.

The clients make requests to their assigned ES. If the ES is not able to serve the client, it addresses it to the SS server, which then redirects it to the most appropriate server. Clients can be served only by servers that are parents of their assigned ES. In the case when there are peers within the same community that contain parts of the requested content item, the ES takes the role of an index server. Additionally, the server redirects the client to the SS server for completing the streaming of the rest of the content. In case of failure of any peer, the missing parts are compensated from other peers or from the streaming servers.

The contents are distributed in the STBs in off-peak hours but we also use the volatile nature of popularity of the content items as an advantage for reduction of the distribution traffic. This property is due to users behaviour regarding repeating a request for a same content. Soon after a video is introduced in the system, it reaches high popularity, but as the time passes, the popularity decays because the clients who already saw the video are unlikely to request it again. Therefore, a content item that is already viewed and stored in the STB of many clients is very likely to be later removed from the ESs as not popular. In such a way, most of the contents with reduced popularity will be already stored in the STBs and available for peer assisted streaming. This saves a lot of additional traffic for distribution of the low popularity contents from the streaming servers to the STBs. The decisions about the content placement on the peers are taken by the ES depending on the distribution determined by the ACM server.

3 Content Division

The limited up-link capacity of the last-mile links that inter-connect the peers is several times smaller than the necessary playback rate of the content items. This capacity is insufficient for immediate and uninterrupted playing of the content items if they are streamed individually by the peers. With such a limitation, the peers cannot act as independent stream suppliers and therefore, the content items are simultaneously streamed by as many peers as it is necessary for reaching its playback rate. When all the streaming portions are delivered to the receiving peer, they are assembled and the content item is played. The necessity of parallel streams requires storing many copies of the same content in many peers. This is quite an inconvenience considering the fact that in our model, we store low popularity content items which represent the majority of the contents present in the system. Storing copies of such a big number of contents on the STBs would require huge storage capacity and also would generate significant amount of traffic for their distribution to the STBs. On the other side, the fact that each peer is streaming only a portion of the entire content makes it reasonable to store in the STB only those portions that the peer is capable to stream. This would contribute to increase the storage efficiency of the peers as well as the contents availability. Therefore, we divide the content items into strips [10], where each strip contains equidistant portions with a predetermined size that will depend on the minimum allowed initial delay. The distance between the portions is k multiple of portions, where k is the number of required peers for uninterrupted streaming. Since the strips are k times smaller in volume than the original content, each peer can store k times more different content items, assuming that all the contents have on average the same size. All the contents that are stored in the STBs are entirely stored in the servers so that they can be delivered whenever the STBs are not able to provide any of the strips.

4 Request Process Description

The requesting process is initiated by the client which sends a request for a content item to its designated ES server. According to the content availability, there are the following cases: the ES already has the content; the server doesn't have the content nor any of the peers, and the ES doesn't have the content but it knows which peers partially contains it. In the first case, the ES sends acknowledgement to the client which is followed by a direct streaming session. In the second case, the ES redirects the client to the SS server which then chooses the best server to serve it and sends it the address of the chosen server. Once the client has the address, the process is the same as the first case. In the case when some strips are stored in the peers, the ES looks up in its availability table and sends a strip-peer list of the available strips and their location. If there is not sufficient number of strips available on the peers, the ES redirects the client to the SS server. Just like in the previous case, the SS redirects the client to the best streaming server for the delivery of the missing strips. When the client receives the availability data of all the strips, it initiates streaming sessions with each peer of the list obtained by the ES and at the same time initiates streaming session for the missing strips with the server assigned by the SS. The ES keeps track of available streaming capacity of each peer.

5 Simulations and Results

We developed a simulation environment for testing the behaviour of the proposed model. In our experiments we consider a network of $S = 13$ streaming servers organized in a tree structure with $L = 3$ levels (Figure 1), such that the lowest level consists of 10 ES, the next level has 2 servers and the highest level has one server. Each of the ES forms a local community of $N = 400$ clients. The streaming capacities of the servers at each level are such that all the requests for the contents they contain can be immediately served. The links that interconnect the servers have enough capacity to support the maximum streaming load of all the servers. The streaming servers host $C = 1500$ Standard Definition (SD) quality contents with playback rate $r_s = 2$ Mbps and average duration of 90 min. For P2P streaming purposes, the contents are divided in $k = 10$ strips. The clients posses STB with capacity to store the entire length of 5 content items. The STBs are connected to the network with links that have download capacity much higher than the playback rate of the SD video quality and uplink capacity of three simultaneous strip streams which is $u = 600$ kbps.

We consider that the popularity of the content items obeys the Zipf-Mandelbrot distribution and that they are previously ranked according to past request data and estimation of the recently inserted items. According to this distribution, the relative frequency (popularity) of the content item with i-th rank in the system is defined as:

$$f(i) = \frac{(i+q)^{-\alpha}}{\sum_{c=1}^{C}(c+q)^{-\alpha}} \tag{1}$$

where q is shifting constant and α is real number that typically takes values between 0.8 and 1.2. In our simulation scenarios, the shifting coefficient $q = 10$ and $\alpha = 0.8$.

We divide the contents according to their popularity in two groups: popular and unpopular contents. We consider that the first 20% of the videos are popular. This division is based on the pareto distribution where 80% of all the requests are a aimed for the first 20% most popular contents and is a common practice for classifying the contents in many related works, although the requests process does not obey the pareto distribution. The process of generating requests for VoD contents is modelled as a Poisson process with average waiting time between two request of 20 min.

The contents are previously distributed on the servers and on the STBs in the hours of the day when there is very low activity of the clients and there is plenty of unused available link capacity that can be dedicated for distribution of the contents. The popular content items are stored in the ES, and the unpopular contents are stored in the higher levels of the hierarchy. The classification of the contents in popularity groups and the decisions on how to be distributed each of the contents in the servers are done by the ACM server.

In the simulations, we considered several different scenarios. The first scenario is the reference for comparison and consists of the simple case when the streaming process is completely done by the servers. In the second scenario, the clients are capable of serving part of the requests for the low popularity content items. In order to compare the results of the proposed model and a related work [10], we consider the case when only the most popular contents are distributed on the STBs. We also include the scenario when both the popular and unpopular contents are equally distributed on the STBs. We assume that the contents are distributed uniformly, i.e. the number of replicas of any content is the same, no matter its popularity. Since each STB contains different strips of the content items, in order to obtain equilibrium of the uplink utilization of each STB, we distribute the strips in such a way that the average popularity of all the strips stored in one STB is the same.

The overall server streaming traffic and the backbone traffic (the streaming traffic that is originated from the servers at the higher levels of the hierarchy) are shown in Figure 2. As expected, the introduction of any type of P2P streaming reduces the overall streaming traffic. However, there is a remarkable distinction in the reduction of the backbone traffic in each of the scenarios. We can see that the distribution of only the popular contents does not change the amount of generated backbone traffic compared to the case when there is no P2P at all, while the distribution of the unpopular contents considerably reduces this traffic. When the peers are serving the requests for unpopular contents, a significant part of the backbone traffic is redirected to the peers, keeping the ES equally busy as in the case of pure server streaming. Thus, a larger amount of the overall traffic is concentrated in the periphery of the network. When peers are serving the requests for the popular contents, a significant part of the ES traffic is reduced, which is not of a great importance, since the ES servers are not optimally used and there is no reduction of the backbone traffic. The only advantage of the later case is that a slightly better reduction of the overall server traffic is achieved. The reason for this is the fact that, although the popular contents are represented by only 20% of the total contents, they generate approximately 56% of the total traffic in the system (1) and therefore, serving the requests for these contents by the peers will reduce more the overall traffic compared to serving the unpopular contents. The values for the overall server and backbone traffic in the last case when all the contents are equally distributed on the STBs are in-between the values of the previously considered P2P distributions.

Since one of our goals is to concentrate most of the traffic within the local communities, we introduce a value called traffic locality that will note the level of localization of the traffic in the system. We define this value as percent of the overall streaming traffic in the system streamed by any member of a local community i.e. a peer or ES. This value is mostly dependant on the amount of the traffic that are able to serve the peers. Therefore, we show the traffic locality for the previously considered cases of distribution of the contents in the STBs for various streaming capacities of the peers. The values of the streaming capacity of each peer vary from bandwidth enough to stream only one strip to bandwidth for streaming $k = 10$ simultaneous strips, which is same as the play rate r_s of the video. The results are shown in Figure 3.

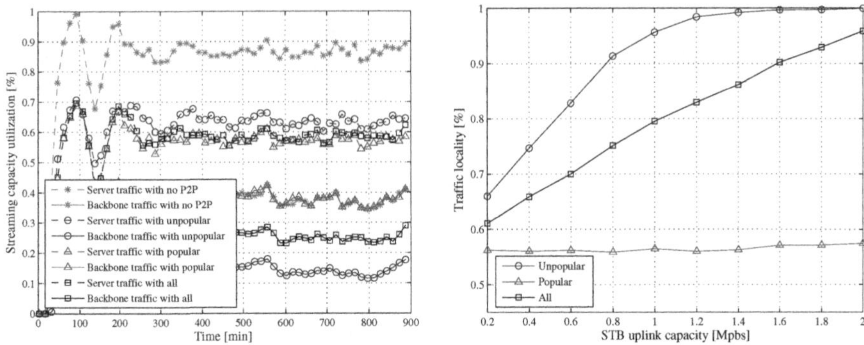

Fig. 2. Throughput Utilization

Fig. 3. Traffic locality

In the figure we can observe that when only the popular contents are distributed in the STBs, there is no change of the traffic locality, and moreover, it keeps the same value as the case when no P2P is implemented at all. This result is quite expected because although the peers serve a significant part of the traffic, they only relieve the ES servers, and the traffic generated for serving the popular contents by the ES is now passed to the peers. In that case, the traffic of the unpopular contents remains the same and there is no improvements in the locality. On the contrary, the distribution of only the unpopular contents considerably improves the traffic locality even for small values of the STBś streaming capacity. As the streaming capacity of the STBs grows, the traffic locality rapidly grows and finally reaches 100% local coverage of service. In the case of distribution of all the contents, the traffic locality increases linearly with the STBs uplink streaming capacity, but it has lower values than the case of distribution of unpopular contents and never reaches full local coverage of service.

In order to estimate the contribution of each of the considered distributions, we define a transport cost for delivering the content items to the clients. This measure is mainly based on the distance of the servers from the clients they are serving and their current load.

$$Cost = \sum_{s=1}^{S} d(s)u(s) \tag{2}$$

where $d(s)$ is the distance of server s from the local communities it is serving and $u(s)$ is its current streaming rate. Since the P2P streaming is done over the unused uplink rate of the clients, we do not include it in the overall cost function.

Figure 4 shows the average transport cost for P2P assisted streaming for various streaming rates of the STBs, relative to the case when all the streaming is done by the servers. We can see that both the curves of only popular and only unpopular contents have a similar behaviour. In the lower range of streaming rates, the traffic cost decreases as the rate increases, and than asymptotically converges towards a certain limit value. The difference between these two distributions is

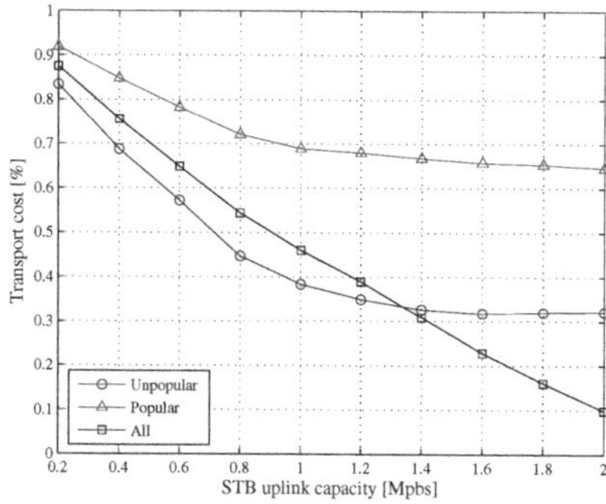

Fig. 4. Transport cost

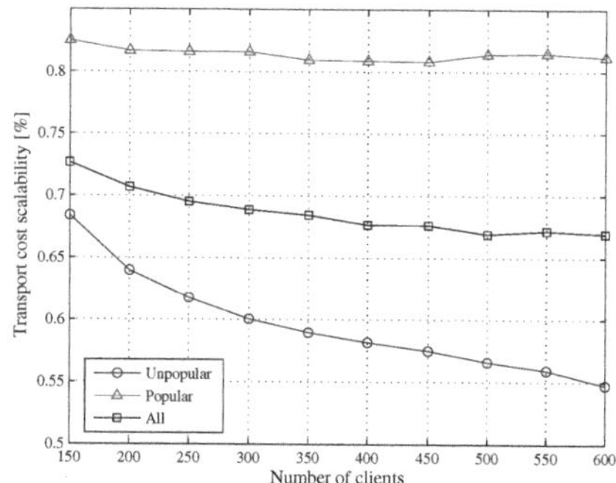

Fig. 5. Transport cost scalability

that distributing the unpopular contents contributes to a lower transport cost. Moreover, it gives lowest values of all the distributions for a smaller values of the streaming rates of the STBs. For higher values it is over-performed by the uniform distribution of all the contents because in this case, the traffic on each level of the hierarchy reduces as all the contents are both placed in the servers and in the STBs. In the cases of distribution of only certain range of popularity, there is only reduction of the traffic on the edge of the network or in the higher levels. We should also keep in mind that with the current ADSL technology most of the operators offer limited uplink capacity of its users which reaches values of up to 1 Mbps. This is actually the range in which the distribution of unpopular contents gives the best results.

One of the advantages of the P2P systems, in general, is their scalability. The more clients are present in the system, the more streaming and storage capacity is available. However, the increased number of clients also implies more requests, which in our model for P2P assisted streaming will also imply more streaming traffic for the servers. Our objective is to see how the different distributions will affect the scalability of the system since the higher request rate will also require increasing the number of streaming servers and the capacities of the links in the core of the network. In Figure 5 we present the dependence of the transport cost of the system relative to the case when no P2P is implemented on the number of clients in the system. In this simulation scenario, the STBs have streaming rate of $u = 600$ kbps. The distribution of the popular contents has no improvement in the scalability of the transport cost with the growth of the system size. This means that the cost relative to the system with no P2P distribution will not reduce although there are more available resources. The distribution of the unpopular contents appears to give far better results, since the higher number of clients will reduce the transport cost relative to the case when no P2P is implemented. As in the previous cases, the distribution of all the contents gives a moderate contribution that is between the contributions of a single range popularity distribution.

Apart from the considered transport cost, we should not forget that there is a certain cost for installation of new servers in any level of the network when the size of the system increases. According to the results discussed before, installation of new servers at any of the levels in any of the distributions would be inevitable. What makes the distribution of the unpopular contents favourable, is the fact that installation of new ES is the only price that has to be paid for increasing the number of the clients. This contributes to serve more clients with the same streaming capacity of the servers in the core of the network. In the case of distribution of popular contents, however, the growth of the number of clients will not require only new servers, but it will also require installation of new links in the core of the networks, which is quite an expensive process.

6 Conclusions

In this work we proposed a P2P assisted VoD streaming model for privately managed network that takes the advantage of the unused storage and uplink capacities of the STBs to serve the less popular content items. We made analysis for different distributions of the contents on the STBs and showed that by storing the unpopular contents it can be achieved to concentrate most of the streaming traffic within the service area of the edge servers. The localization of the streaming traffic close to the clients contributes to reduce the backbone traffic of the network, which consequently reduces the traffic transport cost. Other advantage that comes with the reduction of the backbone traffic is the increased scalability of the system since the same links of the core of the network are able to serve higher number of new clients with no additional upgrades. The only cost that has to be paid is the installation of new server on the edge of the network, which is an inevitable cost of any P2P distribution.

References

1. Simpson, W., Greenfield, H.: IPTV and Internet Video: Expanding the Reach of Television Broadcasting. Elsevier Science & Technology (2009)
2. De Vleeschauwer, D., Laevens, K.: Performance of caching algorithms for iptv on-demand services. IEEE Transactions on Broadcasting 55(2), 491–501 (2009)
3. Verhoeyen, M., De Vleeschauwer, D., Robinson, D.: Content storage architectures for boosted IPTV service. Bell Lab. Tech. J. 13(3), 29–43 (2008)
4. Gramatikov, S., Jaureguizar, F., Cabrera, J., Garcia, N.: Content delivery system for optimal vod streaming. In: Proceedings of the 2011 11th International Conference on Telecommunications (ConTEL), pp. 487–494 (June 2011)
5. Carlsson, N., Eager, D.L., Mahanti, A.: Peer-assisted on-demand video streaming with selfish peers. In: Fratta, L., Schulzrinne, H., Takahashi, Y., Spaniol, O. (eds.) NETWORKING 2009. LNCS, vol. 5550, pp. 586–599. Springer, Heidelberg (2009)
6. Huang, C., Li, J., Ross, K.W.: Peer-Assisted VoD: Making Internet Video Distribution Cheap. In: IPTPS 2007 (2007)
7. Cha, M., Rodriguez, P., Moon, S., Crowcroft, J.: On next-generation telco-managed p2p tv architectures. In: Proceedings of the 7th International Conference on Peer-to-peer Systems, IPTPS 2008, p. 5. USENIX Association, Berkeley (2008)

8. Suh, K., Diot, C., Kurose, J., Massoulie, L., Neumann, C., Towsley, D.F., Varvello, M.: Push-to-peer video-on-demand system: Design and evaluation. IEEE Journal on Selected Areas in Communications 25(9), 1706–1716 (2007)
9. Chen, Y.F., Huang, Y., Jana, R., Jiang, H., Rabinovich, M., Rahe, J., Wei, B., Xiao, Z.: Towards capacity and profit optimization of video-on-demand services in a peer-assisted iptv platform. Multimedia Syst. 15(1), 19–32 (2009)
10. Chen, Y.F., Jana, R., Stern, D., Wei, B., Yang, M., Sun, H., Dyaberi, J.: Zebroid: using IPTV data to support STB-assisted VoD content delivery. Multimedia Systems (May 2010)

Cross-Language Acoustic Modeling for Macedonian Speech Technology Applications

Ivan Kraljevski[1], Guntram Strecha[1], Matthias Wolff[2], Oliver Jokisch[1], Slavcho Chungurski[3], and Rüdiger Hoffmann[1]

[1] TU Dresden, Chair for System Theory and Speech Technology, Dresden, Germany
{ivan.kraljevski,guntram.strecha,oliver.jokisch, ruediger.hoffmann}@tu-dresden.de

[2] BTU Cottbus, Electronics and Information Technology Institute, Cottbus, Germany
matthias.wolff@tu-cottbus.de

[3] FON University, Faculty of Informatics, Skopje, Macedonia
slavcho.chungurski@fon.edu.mk

Abstract. This paper presents a cross-language development method for speech recognition and synthesis applications for Macedonian language. Unified system for speech recognition and synthesis trained on German language data was used for acoustic model bootstrapping and adaptation. Both knowledge-based and data-driven approaches for source and target language phoneme mapping were used for initial transcription and labeling of small amount of recorded speech. The recognition experiments on the source language acoustic model with target language dataset showed significant recognition performance degradation. Acceptable performance was achieved after Maximum a posteriori (MAP) model adaptation with limited amount of target language data, allowing suitable use for small to medium vocabulary speech recognition applications. The same unified system was used again to train new separate acoustic model for HMM based synthesis. Qualitative analysis showed, despite the low quality of the available recordings and sub-optimal phoneme mapping, that HMM synthesis produces perceptually good and intelligible synthetic speech.

Keywords: speech recognition, speech synthesis, cross-language bootstrapping.

1 Introduction

With recent advances in the speech recognition technology and the usage of faster and more available hardware, the state-of-the-art Automatic Speech Recognition (ASR) and synthesis technology becomes more accessible for many speech enabled human-computer interface (HCI) applications. However, in order to develop speech enabled applications with highest possible level of performance and usability, acquisition of large quantity of carefully prepared and transcribed acoustic data is required.

This requirement is especially emphasized in the case of under-resourced languages, where little or no electronic speech and language resources exist. This is the case with the languages used by small, but also by large speaker population languages where sparse language resources exist. The acquisition and transcription of speech

S. Markovski and M. Gusev (Eds.): *ICT Innovations 2012*, AISC 207, pp. 35–45.
DOI: 10.1007/978-3-642-37169-1_4

data is costly and time consuming process and presents one of the major limiting factors for speech enabled applications development. Transcription process of newly recorded speech material requires manual labeling by human language experts, usually 20-40 times the real-time where the labeling process is inconsistent and error prone. Therefore, the availability of sophisticated ASR and speech synthesis technology is limited on the most wide spread worlds languages. Many approaches support rapid creation and development of speech enabled systems for under-resourced languages. The most common is the exploitation of acoustical similarities between the sounds in different languages. Some methods as cross-language transfer uses generalized set of acoustic models capable to perform speech recognition on multiple languages without the need of any target language training data.

The authors in [1] used ASR systems on six different source languages to build a system on Vietnamese with cross-language transfer, unsupervised training and bootstrapping. In [2] the authors examined the performance in the scenario of cross-language transfer of Large Vocabulary Continuous Speech Recognition (LVCSR) systems to other languages when no training data is available at all. The bootstrapping method is used where the target language acoustic models are initialized with models developed for another language, and these initial acoustic models could be recreated or re-adapted using target language speech data, [3] [4]. Other methods, known as cross-language adaptation are based on adapting of existing acoustic models of one language in order to improve discriminatory ability on a other language using very small amount of available transcribed data [5] [6].

In this paper, cross-language modeling and adaptation framework is presented where the source acoustic model used as a bootstrapping platform was trained on a large German speech database. This model, with the knowledge-based phoneme mapping, was used as a first step to transcribe and align small amount of available speaker dependent speech recordings on Macedonian. Afterward, the recognition performance was evaluated and compared between the source and adapted acoustic model.

Using the derived knowledge, data-driven approach was applied to locate phonemes with highest confusion score and refine the phoneme mapping. The newly transcribed and aligned speech data was used as a development set for source acoustic model adaptation which enables possible implementation of speaker and language dependent speech recognition system with small to medium vocabulary size. The same labeled dataset was used again to create and train completely new specialized acoustic model for HMM based speech synthesis. The experimental results showed that cross-language acoustic modeling with non-related languages like German and Macedonian could be performed successfully even on small amount of speaker and language dependent data. This provides a foundation toward creation of more sophisticated speech resources and rapid development of practical speech technology applications on Macedonian language. The remainder of the paper is organized as follows, first an description of the cross-language acoustic modeling concepts is given, including phoneme differences and similarities, knowledge based phoneme mapping, data-driven phoneme mapping and automatic transcription. Next section presents the experimental setup with and acoustic model adaptation, as well as the results of the experiments carried for speech recognition and speech synthesis.

The last section summarizes the overall performance and concludes the paper along with directions for further developments and improvements.

2 Cross-Language Acoustic Modeling

Based on the assumption that the articulatory production of phonemes is very similar among different languages, most of them could be considered independent of specific language. For cross-language acoustic modeling, phoneme units' similarities across languages have to be investigated in order to create appropriate phone mapping tables.

In this paper the German phoneme set with 43 acoustic models including "silence" and "garbage" models, was used for phoneme matching with the Macedonian phoneme inventory with 33 units. If labeled and transcribed training data from the target language does not exist for successful acoustic model training, then the phoneme mapping is determined by human expert according to the IPA conventions, choosing the most similar IPA symbol of the source language as a match for the target language phoneme (*Knowledge based phoneme mapping*).

It is possible to achieve reasonable mappings for each language, but there could be significant variations in the level of detail used in the source language phonetic inventory. In this case, German vowels are represented with larger number of distinct symbols like, checked vowels (SAMPA): I, E, a, O, U, Y, 9, free vowels: E: a:, o:, u:, y:, 2:, and three free diphthongs: aI, aU and OY. While for Macedonian vowels there are only 6 phonemes: i, e, a, @, o and u. That would lead to unacceptable phoneme error rate as observed in the speech recognition experiments. Table 1 presents the phoneme mapping (only where difference exists) between target phonemes (Macedonian) noted in IPA and SAMPA conventions and the resulting knowledge-based (KB) phoneme and data-driven (DD) mapping in the source language (German).

Table 1. Knowledge-based and data-driven mapping table (only where difference exists)

Macedonian Cyrilic	IPA Target	SAMPA Target	SAMPA German KB	SAMPA German DD	Macedonian Cyrilic	IPA Target	SAMPA Target	SAMPA German KB	SAMPA German DD
а	/a/	a	a	a:	њ	/ɲ/	J	nj	nj
ѓ	/ɟ/	J\	gj	gj	о	/ɔ/	O	O	o:
е	/ɛ/	E	E	e:	ќ	/c/	c	kj	kj
ж	/ʒ/	Z	z	z	у	/u/	u	u:	u:
ѕ	/dz/	dz	dz	dz	ч	/tʃ/	tS	tS	tS
и	/i/	i	i:	i:	џ	/dʒ/	dZ	dz	dz
љ	/ʎ/	L	lj	lj					

Twenty of thirty three units found in Macedonian phonetic inventory, which were used in the speech synthesis system TTS-MK [7], could be directly mapped with exact SAMPA German counterpart. The remaining 13 phonemes were replaced with the nearest possible match or replaced by phoneme combinations. The former is applied for the Macedonian phonemes “ѓ” (IPA: /ɟ/, SAMPA: J\), “ѕ” (IPA: /dz/, SAMPA: dz), “љ” (IPA: /ʎ/, SAMPA: L), “ќ” (IPA: /c/, SAMPA: c).

Because there are no similar phonemes in German language, these phonemes could be successfully modeled by combining two other existing phonemes from the source language. Post-alveolar phoneme “ж” (IPA: /ʒ/, SAMPA: Z) is also present in the German phonetic inventory, but it is rarely found in available training data and the existing acoustic model does not contain this phoneme, therefore, it is matched with the alveolar “z” (IPA: /z/, SAMPA: z). The phonemes “ѕ” (IPA: /dz/, SAMPA: dz) and “џ” (IPA: /dʒ/, SAMPA: dZ) were modeled as a combination of the existing model for /d/ and /z/. As a result, acoustic models were tied and the final phoneme inventory for the target language is reduced from 33 to 24. For the recognition problem, this suboptimal acoustic modeling could be successfully compensated using proper language modeling either by constrained formal grammar or statistical language models. Opposite, for the speech synthesis problem these compromises will introduces false pronunciation for the mentioned phoneme units. However, these compromises were made regarding that the knowledge-based phoneme mapping is always the first step toward more detailed data-driven analysis.

Using this phoneme mapping, the available speech recordings on Macedonian were accordingly transcribed and labeled using the German ASR system. The labeling process was performed by Viterbi forced alignment with known speech transcriptions and pronunciations. Then, the labeled speech data was used for speech recognition evaluation and the accuracy performances were observed.

When phonetically labeled and aligned target language speech data becomes available, using a supervised method becomes possible for automatic phoneme mapping (*Data-driven phoneme mapping*). The confusion matrix between referenced (Macedonian) and the hypothesized (German) phoneme models represents the likelihood of the confusion between two phonemes and therefore could be used for phoneme similarities evaluation. The confusion matrix is computed by comparing frame by frame (or phoneme labels) and counts the matches between the source and target language units. The phoneme entries are normalized by the summed frequency of the hypothesized phoneme [3]. The data-driven mappings were derived by selecting each target phoneme with the appropriate source phoneme which has the highest confusion score.

The results are given in the column "SAMPA German DD" of Table 1. In the speech recognition experiments, the Frame Sequence Correctness (FSC), as well as Label Sequence Correctness (LSC) and Accuracy (LSA) which essentially corresponds to the count of correct hypothesized phonemes, were used to evaluate the quality of the acoustic models. From phoneme mapping presented the Table 1, the consistency in consonants mapping was confirmed also from the confusion matrix, it could be noticed that there is a larger confusions scores for the vowels /a/, /e/ and /o/. The reason is that the vowels are more sensitive to co-articulation effects and the consonants are less confusable across different languages [2].

3 Speech Recognition and Synthesis Experiments

In our experiments, the UASR (Unified Automatic Speech Recognition and Synthesis) [8] system was used for speech recognition. UASR is a speech dialogue system where the speech recognition process and the speech synthesis components use common databases at each processing level. During the recognition process, prosodic information is separated from the data flow and later it can be used for synthesis, improving the naturalness of the synthesized speech. The UASR system uses arc-emission HMMs with one single Gaussian density per arc and an arbitrary topology. The structure is built iteratively during the training process by state splitting from an initial HMM model [9]. The resulting acoustic model structure consists of 42 monophonic HMMs plus one pause and one garbage model. During the training phase on clean speech, the phoneme states were split in 3 iterations thus giving 24 Gaussian distributions per phoneme. For training and evaluation of the source model, the real data subset VM2 of Verbmobil German Database was used [10]. For the target model creation and assessment, small speaker dependent dataset on Macedonian language designed for diphone concatenation based TTS system was used [7]. It consists of 758 single word sequences recorded in less controlled (office) conditions with present background noise and the following characteristics: constant volume, without intonation, average pronunciation speed, clear articulation and single recording session.

The text corpus consists of 500 existing and 258 meaningless words for every theoretically possible diphone found in Macedonian language. This gives a dataset which provides good phoneme units coverage and therefore could be used for cross-language recognition performance assessment, speaker dependent acoustic model adaptation or new model training. Because the dataset consists of only speech recordings and non-aligned phonetic transcriptions, segmentation and labeling have to be performed automatically by taking advantage of the existing German speech recognizer. Since the majority of phonemes could be mapped heuristically with their corresponding pair in the source language, Viterbi force alignment was successfully performed and phoneme aligned labeled dataset was produced. Manual segmentation and labeling would introduce segmentation errors as well and by using automatic labeling the errors will be at least consistent throughout the whole dataset.

In the recognition experiments all 758 (1115 second duration) available sentences on Macedonian language were used. For model adaptation, small development set was created by randomly choosing ~10% (76 sentences, 111 seconds duration) of the dataset, the remained sequences (682 sentences, 1004 seconds duration) forms the training set for the new speaker dependent HMM synthesis acoustic model training.

3.1 Model Adaptation

While both the knowledge-based and data-driven phone mappings can be used without modification of the original source language acoustic models, the recognition accuracy could be improved by using conventional adaptation methods like Maximum a posteriori (MAP) [11]. MAP algorithm updates the trained acoustic model parameters by joining the old with the new statistics parameters derived from the adaptation data. The adaptation process consists of two stages. In the first step, the statistics required for computing the distribution weight, mean and covariance are gathered. In the second step the statistics from adaptation data are combined with the old statistics from the HMM model using a data-dependent weighting coefficient. The data dependency is designed to weight the statistics with higher population toward new parameters and with lower population toward the original parameters. The new mean and covariance for the distribution j presents weighted sum of the old and the new statistics:

$$\hat{\mu}_j = \frac{n_j}{n_j + \rho_\mu}\tilde{\mu}_j + \frac{\rho_\mu}{n_j + \rho_\mu}\mu_j,\ \hat{\Sigma}_j = \frac{n_j}{n_j + \rho_\Sigma}\tilde{\Sigma}_j + \frac{\rho_\Sigma}{n_j + \rho_\Sigma}\Sigma_j \tag{1}$$

The data-dependency of the weighting coefficients is realized by the relevance factor ρ_μ and ρ_Σ. Theirs values mark the point where the data count of the adaptation data has the same weight as the old parameter. Higher values of the relevance factors give more weight to the prior information i.e. the old parameters. It requires a relatively large amount of adaptation data in order to be effective for sparsely occupied Gaussian distributions. In this case, since a diphone balanced dataset is available, using MAP with lower data-dependent weighting coefficient will provide good coverage and adaptation.

3.2 Speech Recognition Experiments

Baseline performance evaluation of the target language dataset on UASR system was performed initially to assess the effects of knowledge-based phoneme mapping. The evaluation parameters accuracy *(A)* and correctness *(C)* of the recognized frame and label sequences were calculated using the number of phonemes in the reference sequence *(N)*, the number of removed phonemes *(D)*, the number of substituted phonemes *(S)* and the number of inserted phonemes *(I)*. These numbers are calculated over a sequence alignment using Levenstein distance:

$$C = (N - D - S)/N,\ A = (N - D - S - I)/N \tag{2}$$

Table 2 presents the speech recognition results, at first the test set of VM2 database consisted of 1223 sentences with total duration of 9133 seconds was used to evaluate the baseline performance the source language dataset.

Table 2. Frame and Label Sequences Correctness and Label Sequences Accuracy

	FSC (%)	± (%)	LSC (%)	± (%)	LSA (%)	± (%)
Baseline German acoustic model	54.90	0.70	60.20	0.70	47.70	1.00
Baseline (knowledge based) AM	28.30	1.50	41.00	1.10	17.30	1.80
Adap. (KB) ρ_μ=dim, ρ_Σ=dim^2	78.30	0.80	68.10	1.10	59.10	1.30
Adap. (KB) ρ_μ=10, ρ_Σ=100	81.40	0.70	72.80	1.00	65.90	1.20
Baseline (data driven) AM	39.80	1.50	53.30	1.10	29.40	1.70
Adap. (DD) ρ_μ=dim, ρ_Σ=dim^2	80.80	0.70	70.20	1.00	61.90	1.30
Adap. (DD) ρ_μ=10, ρ_Σ=100	82.70	0.60	73.10	1.00	66.50	1.20

Afterwards, the knowledge-based phoneme mapping was performed and the target language dataset was accordingly labeled and aligned. The baseline recognition performance was observed with significant recognition accuracy degradation. From the resulting confusion matrix it was clear which phonemes introduces significant recognition errors (Table 3).

Table 3. Phoneme confusions after knowledge-based (KB) and data-driven (DD) mapping

Target IPA	a	E	O	b	d	g	p	z
Source IPA (KB)	a:	e:	o:	v	t	v	v	l
Source IPA (DD)				v	v, t	v	v	

Model adaptation was performed on the same dataset without any phoneme mapping and labeling changes. MAP algorithm was used to adapt mean and covariance of the Gaussian distributions with adaptation set for real data consisted of 76 sentences with total duration of 111 seconds. After adaptation 444 Gaussian distributions were updated with the new target language speech data. Because of the very small amount of adaptation data, given the fact that $\rho_\mu = dim(features) = 24$ and $\rho_\Sigma = dim(features)^2 = 576$ and according to the equations (1) those thresholds marks the point where the new and the old parameters equally influences the resulting Gaussian distribution. The knowledge-based observations from Table 3 served as a basis for new data-driven phoneme mapping and consequently new labeling and alignment. Same set of experiments were conducted and after adaptation (439 distributions were updated) it could be noticed from the confusion matrix that there are still some phoneme confusion occurrences. They all belong to the group of stop consonants - labiodental plosives: /b/, /p/, alveolar plosive: /d/ and velar plosive /g/, mostly confused with the labiodental fricative: /v/. The reasons are the erroneous automatic labeling process due to shorter stop consonants durations and the strong presence of background noise. From the results presented in Table 2 it is obvious that speech recognition on source acoustic model for both phoneme mapping approaches performs in the expected range given the fact that available target language dataset is characterized with background non-stationary noise energy level > 25 dB. After data-driven phoneme remapping it could be seen that the phoneme recognition performance or label sequence accuracy (LSA) improves from 17.30 ± 1.80 % to 29.40 ± 1.70 %, same apply for the other parameters FCS and LSC.

Model adaptation resulted in further improvements, inversely proportional with the data-dependent weighting coefficients. Despite the small target language dataset, the adapted model achieved phoneme recognition accuracy of 65.90 % for knowledge-based and 66.50 % for data-driven phoneme mapping with ρ_μ=10 and ρ_μ=100. This gives an indication that the adapted acoustic model with appropriate phoneme mapping could be successfully used in speaker dependent recognition applications with small to medium vocabulary and constrained language model.

3.3 HMM Synthesis Experiments

The second sets of experiments are related to HMM speech synthesis using the same unified ASR and synthesis system. This parametric acoustic synthesis known as HMM synthesis has some interesting features like: good quality, low footprint, well suited for embedded applications and finally provides better framework for personalization compared to diphone concatenation based systems [13]. The available speech dataset was originally created for the purpose of development for a diphone concatenation based TTS system [7]. As a result from the speech recognition experiments and the corresponding phoneme mapping, phoneme labels alignment becomes available, therefore giving the possibility to create new acoustic model specialized only for HMM synthesis tasks.

New HMM models were constructed for 25 phoneme with 3 states per model. The used units represent phonetic subset with total phoneme count smaller as found in Macedonian language. This is a potential problem for the synthesis of the phonemes /z/, "ж" (IPA: /ʒ/, SAMPA: Z), "ѕ" (IPA: /dz/, SAMPA: dz) and "џ" (IPA: /dʒ/, SAMPA: dZ) because of the acoustic model tying. This could be compensated during the acoustic model training by using larger number of Gaussian distributions per phoneme model. To synthesize speech from the arc-emission HMMs with one single Gaussian density per arc and an arbitrary topology, an algorithm was used for extraction of the appropriate feature vector sequence depending on a given target phone duration, i.e. the number of feature vectors to be produced. The selection algorithm finds the best path through the model by considering the self-transition probabilities and additional statistics for modeling the state duration within the model [9] and the resulting parameter sequence is synthesized by a MLSA filter [12]. Combined with proper linguistic processing, the control parameters like duration and fundamental frequency were derived from the multilingual TTS system DRESS [14]. Preliminary tests showed that the quality of synthesized speech using arc-emission HMMs with single Gaussian density is slightly better than speech synthesized using classic HMMs synthesis approach. The resulting speech is intelligible, but still lacks naturalness of the original signal. That could be improved with better intonation and duration modeling using statistics derived from Macedonian language analysis.

The acoustic model was trained on all available speech data and evaluated on 10% of the used training set. The objective is to "over-train" the acoustic model in order to cover the all possible feature vectors occurrences and transition paths that could be found in the training data. Contrary, increasing the number of Gaussian distributions introduces the problem of data sparseness and significant number of them will be rendered as invalid. The training process was performed with 5 splits producing total number of 2400 Gaussian distributions, 96 different per phoneme model, where 1508 of them were considered as valid according the occurrence count in the training data. The objective was to achieve accuracy values as closer as possible to 100%, ensuring the highest possible signal synthesis quality. The condition that must be fulfilled is the existence of good quality speech recordings with precisely determined phoneme alignment. Because, the currently available data does not comply with the given requirements the achieved performance is close but not equal to the desired objective (FSC=99.10 ± 0.10 %, LSC=99.70 ± 0.10 %, LSA=98.80 ± 0.40 %).

In order to create the diphone inventory, as a first step, pitch mark analysis was performed and the intonation statistics for each speech sequences were estimated. Using these statistics and the speaker dependent acoustic model with the highest quality (split 5, iteration 1) the diphone inventory was produced. The acoustic model was used to perform recognition for each unique diphone found in the dataset and the recognition path throughout the feature space described by the phoneme models was stored. If more than one example of the same diphone unit occurred, the path of the recognized diphone with the highest accuracy will be preserved [9].

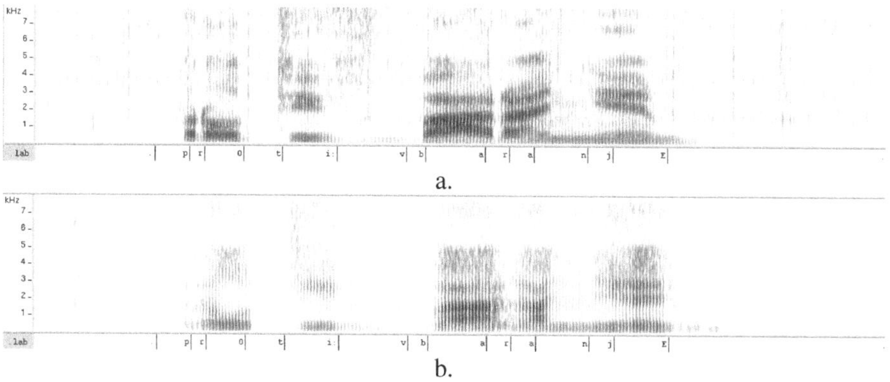

Fig. 1. Spectrogram of the original (a) and the synthetic (b) speech (IPA: prɔtivbaraɲɛ)

Figure 1 presents the spectrogram of the original (a) and the synthetic (b) speech sequence obtained using intonation and phoneme duration statistics acquired during the pitch mark analysis. It could be noticed, that the spectral features were successfully reconstructed where the intonation and duration models providing relatively good quality compared against the original sequence. The overall perceptual quality could be improved with more carefully prepared and higher quality studio recordings, larger amount of speech sequences and by increasing the number of phoneme models for the target language.

4 Conclusions

This paper presents a cross-language development of speech recognition and synthesis applications for a new target language (Macedonian) using bootstrapping approach from a German acoustic model. Both knowledge-based and data-driven approaches for phoneme sets mapping between the source and target languages were used for automatic transcription and labeling. The recognition experiments on the source acoustic model showed significant performance degradation due to language, speaker and environmental conditions. Using very limited amount speech data in the target language (tens of minutes) very acceptable speech recognition performance could be achieved for speaker dependent applications. The same unified ASR and synthesis system, with the already transcribed data, was used again to create specialized acoustic model for HMM based synthesis.

Qualitative analysis showed, despite the low quality of the available recordings and the sub-optimal phoneme mapping, that HMM synthesis produces perceptually good and intelligible synthetic speech. Further improvements in the recognition performance could be achieved using larger amount (several hours) of speaker-independent studio quality speech recordings. Beside this, for speech synthesis better and optimal phoneme acoustic modeling is needed, as well as including appropriate intonation and phoneme duration modeling based on the statistics derived from the Macedonian language.

References

1. Vu, N.T., Kraus, F., Schultz, T.: Rapid building of an ASR system for Under-Resourced Languages based on Multilingual Unsupervised training. In: Interspeech 2011, Florence, Italy, August 28 (2011)
2. Schultz, T., Waibel, A.: Experiments on Cross-language Acoustic Modeling. In: Proceedings of the 7th European Conference on Speech Communication and Technology, Eurospeech 2001, Aalborg, Denmark, p. 2721 (2001)
3. Le, V.B., Besacier, L.: First steps in fast acoustic modeling for a new target language: application to Vietnamese. In: ICASSP 2005, Philadelphia, USA, March 19-23, vol. 1, pp. 821–824 (2005)
4. Martin, T., Sridharan, S.: Cross-language acoustic model refinement for the Indonesian language. In: International Conference on Acoustics, Speech, and Signal Processing, vol. 1, pp. 865–868 (March 2005)
5. Lööf, J., Gollan, C., Ney, H.: Cross-language Bootstrapping for Unsupervised Acoustic Model Training: Rapid Development of a Polish Speech Recognition System. In: Interspeech, pp. 88–91 (September 2009)
6. Le, V.B., Besacier, L., Schultz, T.: Acoustic-Phonetic Unit Similarities for Context Dependent Acoustic Model Portability. In: IEEE International Conference on Acoustics, Speech and Signal Processing, ICASSP 2006 (2006)
7. Chungurski, S., Kraljevski, I., Mihajlov, D., Arsenovski, S.: Concatenative speech synthesizers and speech corpus for Macedonian language. In: 30th International Conference on Information Technology Interfaces, Dubrovnik, Croatia, June 23-26, pp. 669–674 (2008)
8. Hoffmann, R., Eichner, M., Wolff, M.: Analysis of verbal and nonverbal acoustic signals with the Dresden UASR system. In: Esposito, A., Faundez-Zanuy, M., Keller, E., Marinaro, M. (eds.) COST Action 2102. LNCS (LNAI), vol. 4775, pp. 200–218. Springer, Heidelberg (2007)
9. Strecha, G., Wolff, M.: Speech synthesis using HMM based diphone inventory encoding for low-resource devices. In: 2011 IEEE International Conference on Acoustics, Speech and Signal Processing (ICASSP), May 22-27, pp. 5380–5383 (2011)
10. Bub, T., Schwinn, J.: VERBMOBIL: The Evolution of a Complex Large Speech-to-Speech Translation System. In: Int. Conf. on Spoken Language Processing, Philadelphia, PA, USA, vol. 4, pp. 2371–2374 (October 1996)

11. Gauvain, J.-L., Lee, C.-H.: Maximum A Posteriori Estimation for Multivariate Gaussian Mixture Observations of Markov Chains. IEEE Transactions on Speech and Audio Processing 2(2), 291–298 (1994)
12. Imai, S., Sumita, K., Furuichi, C.: Mel log spectrum approximation (MLSA) filter for speech synthesis. Trans. IECE J66-A, 122–129 (1983)
13. Tokuda, K., et al.: Speech parameter generation algorithms for HMM-based speech synthesis. In: ICASSP. Proc. IEEE Int. Conf. on Acoustics, Speech, and Signal Processing, Istanbul, June 5-9, vol. III, pp. 1315–1318. IEEE Computer Society Press, Los Alamitos (2000)
14. Hoffmann, R., Hirschfeld, D., Jokisch, O., Kordon, U., Mixdorff, H., Mehnert, D.: Evaluation of a multilingual TTS system with respect to the prosodic quality. In: Proc. 14th Intern. Congress of Phonetic Sciences (ICPhS), San Francisco, USA, August 1-7, pp. 2307–2310 (1999)

Efficient Classification of Long Time-Series

Josif Grabocka[1], Erind Bedalli[2], and Lars Schmidt-Thieme[1]

[1] Information Systems and Machine Learning Lab
Samelsonplatz 22, 31141 Hildesheim, Germany
[2] Ph.D. Candidate in University of Tirana,
Lecturer in University of Elbasan, Albania
{josif,schmidt-thieme}@ismll.de, erindbedalli@gmail.com

Abstract. Time-series classification has gained wide attention within the Machine Learning community, due to its large range of applicability varying from medical diagnosis, financial markets, up to shape and trajectory classification. The current state-of-art methods applied in time-series classification rely on detecting similar instances through neighboring algorithms. Dynamic Time Warping (DTW) is a similarity measure that can identify the similarity of two time-series, through the computation of the optimal warping alignment of time point pairs, therefore DTW is immune towards patterns shifted in time or distorted in size/shape. Unfortunately the classification time complexity of computing the DTW distance of two series is quadratic, subsequently DTW based nearest neighbor classification deteriorates to quartic order of time complexity per test set. The high time complexity order causes the classification of long time series to be practically infeasible. In this study we propose a fast linear classification complexity method. Our method projects the original data to a reduced latent dimensionality using matrix factorization, while the factorization is learned efficiently via stochastic gradient descent with fast convergence rates and early stopping. The latent data dimensionality is set to be as low as the cardinality of the label variable. Finally, Support Vector Machines with polynomial kernels are applied to classify the reduced dimensionality data. Experimentations over long time series datasets from the UCR collection demonstrate the superiority of our method, which is orders of magnitude faster than baselines while being superior even in terms of classification accuracy.

Keywords: Machine Learning, Data Mining, Time Series Classification, Dimensionality Reduction.

1 Introduction

Time-series classification is one of the most appealing domains of machine learning due to the abundance of application domains ranging from medicine to finance. The nearest neighbor classifier empowered with a distance metric known as Dynamic Time Warping holds the primate among accurate classifiers. However, the nearest neighbor suffers from a major drawback in terms of classification time complexity which deteriorates to quartic time for the whole dataset. In order

S. Markovski and M. Gusev (Eds.): *ICT Innovations 2012*, AISC 207, pp. 47–57.
DOI: 10.1007/978-3-642-37169-1_5

to overcome the scalability problems associated with nearest neighbor we propose a fast matrix factorization in order to reduce the data dimensionality and then Support Vector Machines for classification. We show through experimental results that the both the accuracy and the classification time is significantly improved, while the training time is competitive to a standard Support Vector Machines training duration.

2 Related Work

2.1 Time-Series Classification

The recent decades have witnessed a plethora of methodologies addressed at the classification of time series. The methodologies vary from Neural Networks [1], Bayesian Networks [2] to SVMs [3].

Dynamic Time Warping. Among the most successful methods proposed within the scope of time-series classification are distance based similarity metrics. Various distance metrics have been proposed, however the most widely recognized and accurate metric is the so-called Dynamic Time Warping (DTW) [4]. Dynamic Time Warping is able to detect distortions between series of the same class via computing the minimum warping alignment path of the respective time points. The minimum time points alignment is computed through a dynamic algorithm, which constructs a cost matrix where each cell represent the cumulative distance for the partial alignment up to the indexes of its coordinates [5,6]. DTW is used in combination with the nearest neighbor classifier, denoted DTW-NN. Recent studies have pointed out that DTW-NN is a hard-to-beat baseline in terms of classification accuracy [7]. In order to speed up the classification time of DTW, a warping window concept has been introduced in order to reduce the number of computations of the cost matrix cells, by omitting candidate warping paths having deviations from the matrix diagonal, i.e from the euclidean alignment of time points [6]. In this study we are comparing our method against the DTW-NN.

2.2 Matrix Factorization

Matrix factorization is a variant of dimensionality reduction that projects data into a reduced/latent/hidden data space which usually consists of lower dimensions than the original space [8,9]. Different approaches have been unified under a generalized Bregman divergence theory [10] Matrix factorization has been applied in domains involving time-series data as in music transcription [11], up to EEG processing [12] In comparison we are going to use very fast variations of matrix factorization with very low dimensions, fast learning rate and early stopping.

2.3 Support Vector Machines (SVM) Classification

Support Vector Machines (SVM) are considered to be one of the best off-the-shelf classifier for a wide application domains of machine learning. The success of support vector machines is based on their principle to find the maximum margin decision boundary hyperplane that accurately splits class regions [13]. The training process of SVMs is done by optimizing the maximum margin objective, usually in the dual representation of the objective function. In order to overcome problems with non-linear separability of certain datasets, introduction of slack variables has been applied to allow regularized disobedience from the decision boundary. In addition kernel theory has been combined with the dual learning of SVMs in order to offer various types of non-linear expressiveness to the decision boundary [14]. We are applying SVMs for classifying the projected data in the latent space.

3 Motivation

The time complexity of classifying long time series is obviously determined by the runtime complexity of the method performing such classification, while the complexities of those methods are functions of series lengths. In this section we provide more insight into the diagnosis and the cure for time complexity issues of lengthy series. However, before starting the analysis we need to provide a brief description of DTW-NN.

3.1 Dynamic Time Warping and Nearest Neighbor

A warping path between two series $A = (A_1, ..., A_n)$ and $B = (B_1, ..., B_m)$, denoted as $\pi^{A,B}$ is defined as an alignment $\tau^{A,B} = (\tau_1^{A,B}, \tau_2^{A,B})$ between the elements of A and B. The alignment starts and ends with extreme points,

$$\begin{aligned} P &= |\tau^{A,B}| \\ 1 = \tau^{A,B}(1)_1 &\leq ... \leq \tau^{A,B}(P)_1 = n \\ 1 = \tau^{A,B}(1)_2 &\leq ... \leq \tau^{A,B}(P)_2 = m \end{aligned}$$

while involving incremental alignment of adjacent pairs as:

$$\begin{pmatrix} \tau^{A,B}(i+1)_1 - \tau^{A,B}(i)_1 \\ \tau^{A,B}(i+1)_2 - \tau^{A,B}(i)_2 \end{pmatrix} \in \left\{ \begin{pmatrix} 0 \\ 1 \end{pmatrix}, \begin{pmatrix} 1 \\ 0 \end{pmatrix}, \begin{pmatrix} 1 \\ 1 \end{pmatrix} \right\}$$

The overall distance of the points aligned by a warping path is computed as the sum of distances of each aligned pair. Such distance is called the Dynamic Time Warping distance [6].

$$\mathrm{DTW}(A, B) = \underset{\tau^{A,B}}{\operatorname{argmin}} \sum_{p=1}^{|\tau^{A,B}|} \left(A_{\tau^{A,B}(p)_1} - B_{\tau^{A,B}(p)_2} \right)^2$$

Dynamic Time Warping distance is practically computed by a dynamic algorithm, which is computed via a cost matrix, denoted W. Each cell of the cost matrix is computed as follows:

$$\begin{aligned} \text{DTW}(A, B) &= \mathbf{W}_{\text{length}(\mathbf{A}),\text{length}(\mathbf{B})} \\ \mathbf{W}_{1,1} &= (A_1 - B_1)^2 \\ \mathbf{W}_{i,j} &= (A_i - B_j)^2 + \min(\mathbf{W}_{i-1,j}, \mathbf{W}_{i,j-1}, \mathbf{W}_{i-1,j-1}) \end{aligned} \quad (1)$$

The nearest neighbor classifier based on DTW distance metric is described in Algorithm 1 and basically predicts the label of a test instance as the label of the closest train instance.

Algorithm 1. DTW-NN

Require: Training set D, Test instance I
Ensure: Predicted label of I
  nearestNeighbor $\leftarrow \text{argmin}_{I' \in D} \text{DTW}(I, I')$
  return nearestNeighbor.label

3.2 The *Curses* of Dimensionality and Complexity

The computation of the DTW distance from Equation 1 between two series A and B requires the computation in total of (length(A) $\times$ length(B)) cells which make the distance metric an $O(n^2)$ operation. In order to classify a series instance using the nearest neighbor classifier from Algorithm 1, we will have to compare its distance against all training set instances which requires $O(n)$ calls to DTW computation. Therefore the overall classification time complexity of classifying an instance is $O(n^3)$, while the classification of a whole test set becomes $O(n^4)$. Cubic and/or quadric time complexities creates prohibitive possibilities for low computational devices, or systems where the response time is critical. Please note that especially in long time series the DTW computation tend to become expensive, therefore we can metaphorically call this behavior as *the curse of complexity*.

SVMs are much faster in terms of classification speed, however a learning phase step is required first, because a maximum margin decision boundary has to be created. In contrast, the nearest neighbor classifier requires no training step at all. Even worse the classification and learning phases of SVMs can be time consuming in case the number of features (here series length) is large. Such phenomenon is known as *curse of dimensionality*. The purpose of this study is to make the classification time of SVMs faster through first projection the data into a latent space having much less features with even better accuracy. This speedup comes with the additive learning time cost of the projection method, therefore we will propose a very fast projection technique. The overall learning and classification time of our method is shown to be much faster than DTW-NN. In the end of Section 4.3 we explain that our method has an $O(n)$ classification time complexity.

4 Proposed Method

Throughout this study we propose e method that aims at solving the problem of time-series classification as formalized in section 4.1. Our method applies a Fast Matrix Factorization, described in section 4.2, in order to project the data to a lower dimensionality. Finally SVMs classifier, section 4.3, is applied to the projected data for classification.

4.1 Problem Description

Given a training time series dataset $X_{train} \in \mathbb{R}^{N \times M}$ consisting of N time series each of M points length and observed target variables $Y_{train} \in \mathbb{N}$, then the task is to predict the labels Y_{test} of a given test set of series $X_{test} \in \mathbb{R}^{N' \times N}$.

4.2 Fast Matrix Factorization

Matrix Factorization is a technique to decompose a matrix as the dot product of other matrices having typically lower dimensionality. In our study, the time-series dataset $X \in \mathbb{R}^{(N+N') \times M}$ will be approximated by the dot product of the projected latent data $\Phi \in \mathbb{R}^{(N+N') \times K}$ and the matrix $\Psi \in \mathbb{R}^{K \times M}$ as in Equation 2. The value of K determine the dimensionality of the projected space.

$$X \approx \Phi \cdot \Psi \tag{2}$$

Such an approximation is converted to an objective function that can be written in terms of a minimization objective function, denoted L, as in Equation 3. The loss terms are euclidean distances and can be expanded as in Equation 4. The loss terms on the right preceded by λ coefficients are regularization terms which prevent the over-fitting of Φ and Ψ.

$$\underset{\Phi,\Psi}{\operatorname{argmin}}\ L(X,\Phi,\Psi) = ||X - \Phi \cdot \Psi||^2 + \lambda_\Psi ||\Psi||^2 + \lambda_\Phi ||\Phi||^2 \tag{3}$$

$$\begin{aligned}\underset{\Phi,\Psi}{\operatorname{argmin}}\ L(X,\Phi,\Psi) = &\sum_{i=1}^{N+N'} \sum_{j=1}^{M} \left(X_{i,j} - \sum_{k=1}^{K} \Phi_{i,k} \Psi_{k,j} \right)^2 \\ &+ \lambda_\Phi \sum_{i=1}^{N+N'} \sum_{k=1}^{K} \Phi_{i,k}^2 + \lambda_\Psi \sum_{k=1}^{K} \sum_{j=1}^{M} \Psi_{k,j}^2\end{aligned} \tag{4}$$

The solution of the objective function is carried through stochastic gradient descent where we randomly correct the loss created by each cell $X_{i,j}$. The corresponding partial derivatives can be derived as follows:

$$\text{Let: } e_{i,j} = X_{i,j} - \sum_{k=1}^{K} \Phi_{i,k} \Psi_{k,j} \tag{5}$$

$$\frac{\partial L_{X_{i,j}}}{\partial \Phi_{i,k}} = -2 e_{i,j} \Phi_{i,k} + 2 \lambda_\Phi \Phi_{i,k} \tag{6}$$

$$\frac{\partial L_{X_{i,j}}}{\partial \Psi_{k,j}} = -2 e_{i,j} \Psi_{k,j} + 2 \lambda_\Psi \Psi_{k,j} \tag{7}$$

In order to obtain final versions of the latent matrices Φ, Ψ we learn them through a stochastic gradient descent learning as shown in Algorithm 2. The algorithm iterates through every cell $X_{i,j}$ and updates the cells of the latent matrices until the loss is minimized. In the end of the learning the latent matrix Φ is fed to the classifier.

Algorithm 2. Stochastic Gradient Descent Learning

Input: Time-series dataset X, Learning rate η, Maximum Iterations $MaxIterations$
Output: Φ, Ψ

Initialize Φ, Ψ randomly
$previousLoss \leftarrow MaxValue$
$currentLoss \leftarrow L(X, \Phi, \Psi)$
$numIterations \leftarrow 0$
while $currentLoss < previousLoss \wedge numIterations \leq MaxIterations$ **do**
  for $i = 1$ to $N + N'$ **do**
    for $j = 1$ to M **do**
      for $k = 1$ to K **do**
        $\Phi_{i,k} \leftarrow \Phi_{i,k} - \eta \cdot \frac{\partial L_{X_{i,j}}}{\partial \Phi_{k,j}}$
        $\Psi_{k,j} \leftarrow \Psi_{k,j} - \eta \cdot \frac{\partial L_{X_{i,j}}}{\partial \Psi_{k,j}}$
      end for
    end for
  end for
  $previousLoss \leftarrow currentLoss$
  $currentLoss \leftarrow L(X, \Phi, \Psi)$
  $numIterations \leftarrow numIterations + 1$
end while
return Φ, Ψ

In order to speed up the factorization there are three main steps that can be taken. The latent dimensionality, parameter K specifying latent dimensionality of matrices Φ, Ψ, is selected to be as low as $K = c \times cardinality(Y)$, meaning a small multiple c of the number of labels. In addition the learning rate η is set to be large, which forces the optimization to quickly converge towards the global minimum of the quadratic objective function in Equation 4. The final element that speeds the convergence relies on stopping the iterations via a limited maximum iterations count.

4.3 Support Vector Machines

The Support Vector Machines, hereafter denoted as SVMs, are a classifier that aims at finding the maximum margin separating decision boundary among class regions [15]. The decision boundary lies in the form of a hyperplane, denoted $\mathbf{w}$. For binary classification the target variable y is binary $y \in \{-1, +1\}$. The classification of a test instance is computed the sign of the dot product between the hyperplane and the instance vector, as in Equation 8.

$$\hat{y}_{test} = \text{sign}(\langle w, x_{test} \rangle + w_0) \tag{8}$$

The maximum margin hyperplane is computed by solving the optimization function in Equation 9. Such formulation is known as the soft-margin primal form [13].

$$\begin{aligned} \text{minimize} \quad & \frac{1}{2}||w||^2 + C\sum_{i=1}^{n}\xi_i \\ & \text{subject to:} \\ & y_i(\langle w, x_i \rangle + w_0) \geq 1 - \xi_i, \text{and } \xi_i \geq 0, \quad i = 1, ..., n \end{aligned}$$

The so called slack variables, defined in Equation 9 represent the violation of each series instance from the boundary with the objective aim of minimizing the violations.

$$\xi_i = \max(0, 1 - y_i(\langle w, x_i \rangle + w_0)) \tag{9}$$

The primal form objective function is manipulated by expressing the inequality conditions via Lagrange multipliers denoted α_i, one per instance. Then the objective function is solved for w and w_0 and the dual form is yield as shown in Equation 10.

$$\begin{aligned} \max \quad & \sum_{i=1}^{n}\alpha_i - \frac{1}{2}\sum_{i=1}^{n}\sum_{j=1}^{n}\alpha_i\alpha_j y_i y_j K(x_i, x_j) \\ \text{subject to:} \quad & 0 \leq \alpha_i \leq C, \; i = 1, ..., n \\ & \sum_{i=1}^{n}\alpha_i y_i = 0 \end{aligned} \tag{10}$$

In order to classify datasets exhibiting non-linear separation, the dot product present in the dual objective function, $\langle x_i, x_j \rangle$, is substituted by the so called kernel trick which is shown as $K(x_i, x_j)$ [14]. A typical kernel, the inhomogeneous polynomial one, is presented in Equation 11, where d is the degree of the polynomial and c a constant.

$$\langle x_i, x_j \rangle \rightarrow K(x_i, x_j) = (\langle x_i, x_j \rangle + c)^d \tag{11}$$

The purpose of the kernel trick approach is to express the feature space in terms of a higher dimensionality space, such that a linear decision boundary in the high dimensional space would represent a non-linear boundary in the original one. The solution of the dual problem is carried through dedicated optimization algorithms. The ultimate decision boundary can be formed as a function of the solved Lagrange multipliers as depicted by Equation 12. Please note that the non-zero α_j coefficients correspond to the so-called support vector instances (x_j, y_j).

$$y_i = \text{sign}(\sum_{j=1}^{n}\alpha_j y_j K(x_j, x_i)) \tag{12}$$

Once the decision boundary is learn then a new instance can be classified as a summation iteration over the support vectors, which make the operation of magnitude $O(a \times b)$, where a is the number of support vectors and b the complexity of the dot product of two instance's features. Please note that a is a fraction of the number of instances, so still $O(n)$ at worst-case, while b is linearly proportional to the number of features. So SVM classification time deteriorates to $O(n^2)$ in worst case scenario. In our method we will project the data into a very low dimensionality, so b will be quasi $O(1)$ and aggregatively our method will have a worst-case classification time complexity of $O(n)$.

5 Experimental Setup

The analysis and the experiments served to analyze the proposed methods are tested on the longest five datasets of the UCR time-series data collection [1]. The statistics of the selected datasets are found in Table 1.

Table 1. Statistics of Datasets

Dataset	Number of Instances	Series Length	Number of Labels
CinC_ECG_Torso	1420	1639	4
Haptics	462	1092	5
InlineSkate	650	1882	7
Mallat	2400	1024	8
StarLightCurves	9236	1024	3

In order to test our method, which we denote as Fast Matrix Factorization and SVM, hereafter denoted as FMF-SVM, we run experiments against the following implemented baselines:

- **DTW-NN:** The nearest neighbor with Dynamic Time Warping is a hard to beat classifier in the time-series domain.
- **E-NN:** The nearest neighbor with Euclidean Distance is a fast version classifier compared to DTW-NN.
- **SVM:** Support Vector Machines are a strong off-the-shelf classifier.

The hyper-parameters of our method are searched via grid search using only the validation set in a 5-cross validation fashion. The values of the learning rate η are searched from a range of $\{10^{-1}, 10^{-2}, 10^{-3}\}$, the latent dimensionality among $\{1, 2, 3, 4\} \times \mathit{cardinality}(Y)$, parameter C of SVMs is selected from $\{10^{-1}, 1, 10^{1}\}$, the maximum iterations from $\{100, 200, 300\}$, and finally the degree of the polynomial kernel is chosen one of $\{2, 3, 4\}$. The combination yielding the minimum error on the validation split, is tested over the test split.

[1] `www.cs.ucr.edu/~eamonn/time_series_data`

6 Results

Our proposed method FMF-SVM exhibits excellent classification accuracy by producing the smallest misclassification rates compared to the baselines in all the datasets, as demonstrated in Table 2.

Table 2. Misclassification Rate Results Table *(5-fold cross validation)*

Dataset	FMF-SVM		DTW-NN		E-NN		SVM	
	mean	*st.dev.*	*mean*	*st.dev.*	*mean*	*st.dev.*	*mean*	*st.dev.*
CinC_ECG_Torso	**0.0007**	0.0016	**0.0007**	0.0001	0.0014	0.0020	**0.0007**	0.0035
Haptics	**0.4903**	0.0118	0.5484	0.0025	0.5745	0.0750	0.5162	0.0405
InlineSkate	**0.5098**	0.0151	0.5131	0.0018	0.5603	0.0297	0.5197	0.0638
Mallat	**0.0150**	0.0027	0.0162	0.0001	0.0163	0.0063	0.0196	0.0084
StarLightCurves	**0.0633**	0.0074	0.0652	0.0001	0.1139	0.0044	0.0933	0.0449

The proposed method requires an additional matrix factorization step which elongates the overall training time of building the model. Nevertheless the learning time is not prohibitive, and it is practically feasible as Table 3 shows.

Table 3. Learning Times (seconds)

Dataset	FMF-SVM		SVM	
	mean	*st.dev.*	*mean*	*st.dev.*
CinC_ECG_Torso	2.44	0.17	10.74	1.14
Haptics	223.55	3.86	2.80	0.11
InlineSkate	214.74	5.17	10.85	0.85
Mallat	840.81	30.42	10.69	0.32
StarLightCurves	2354.30	568.88	1033.51	656.81

Finally, in addition to being superior in classification, our method is also extremely faster in terms of classification time. As Table 4 proves, FMF-SVM is by far superior in terms of classification time regarding new test instances. The presented results support our theoretic complexity analysis of Section 3.2 and Section 4.3. (Note that $M = 10^6$).

Table 4. Classification Times Results (milliseconds)

Dataset	FMF-SVM		DTW-NN		E-NN		SVM	
	mean	*st.dev.*	*mean*	*st.dev.*	*mean*	*st.dev.*	*mean*	*st.dev.*
CinC_ECG_Torso	**0.21**	0.01	182900.92	3403.49	72.22	1.36	11.49	2.45
Haptics	**0.68**	0.03	24985.02	116.01	16.64	0.48	3.48	0.08
InlineSkate	**0.73**	0.04	98908.95	354.88	34.92	0.41	11.09	0.27
Mallat	**0.64**	0.02	115535.66	572.01	75.61	1.52	9.05	0.31
StarLightCurves	**0.57**	0.05	256.87M	4.54M	286.76	9.09	15.99	4.76

7 Conclusion

Throughout this study we presented a new approach addressing the problem of classifying long time series. In comparison to state-of-art similarity based nearest neighbor classifiers, which deteriorate up to cubic orders of classification time complexity, we propose a fast data projection to low dimensions and then a SVMs classification on the latent space. Overall, our method's classification time complexity is only $O(n)$ at worst-case scenario. Experiments over long time-series datasets demonstrate that our method is clearly, both the fastest, and the most accurate compared to selected state-of-art baselines.

Acknowledgement. Funded by the Seventh Framework Programme of the European Commission, through project REDUCTION (# 288254). `www.reduction-project.eu`

References

1. Kehagias, A., Petridis, V.: Predictive modular neural networks for time series classification. Neural Networks 10(1), 31–49 (1997)
2. Pavlovic, V., Frey, B.J., Huang, T.S.: Time-series classification using mixed-state dynamic bayesian networks. In: CVPR, p. 2609. IEEE Computer Society (1999)
3. Eads, D., Hill, D., Davis, S., Perkins, S., Ma, J., Porter, R., Theiler, J.: Genetic Algorithms and Support Vector Machines for Time Series Classification. In: Proc. SPIE 4787; Fifth Conference on the Applications and Science of Neural Networks, Fuzzy Systems, and Evolutionary Computation; Signal Processing Section; Annual Meeting of SPIE (2002)
4. Ding, H., Trajcevski, G., Scheuermann, P., Wang, X., Keogh, E.J.: Querying and mining of time series data: experimental comparison of representations and distance measures. PVLDB 1(2), 1542–1552 (2008)
5. Berndt, D.J., Clifford, J.: Finding patterns in time series: A dynamic programming approach. In: Advances in Knowledge Discovery and Data Mining, pp. 229–248 (1996)
6. Keogh, E.J., Pazzani, M.J.: Scaling up dynamic time warping for datamining applications. In: KDD, pp. 285–289 (2000)
7. Keogh, E.J., Ratanamahatana, C.A.: Exact indexing of dynamic time warping. Knowl. Inf. Syst. 7(3), 358–386 (2005)
8. Koren, Y., Bell, R.M., Volinsky, C.: Matrix factorization techniques for recommender systems. IEEE Computer 42(8), 30–37 (2009)
9. Srebro, N., Rennie, J.D.M., Jaakola, T.S.: Maximum-margin matrix factorization. In: Advances in Neural Information Processing Systems 17, pp. 1329–1336. MIT Press (2005)
10. Singh, A.P., Gordon, G.J.: A unified view of matrix factorization models. In: Daelemans, W., Goethals, B., Morik, K. (eds.) ECML PKDD 2008, Part II. LNCS (LNAI), vol. 5212, pp. 358–373. Springer, Heidelberg (2008)
11. Smaragdis, P., Brown, J.C.: Non-negative matrix factorization for polyphonic music transcription. In: IEEE Workshop on Applications of Signal Processing to Audio and Acoustics, pp. 177–180 (2003)

12. Rutkowski, T.M., Zdunek, R., Cichocki, A.: Multichannel EEG brain activity pattern analysis in time-frequency domain with nonnegative matrix factorization support. International Congress Series, vol. 1301, pp. 266–269 (2007)
13. Cristianini, N., Shawe-Taylor, J.: An Introduction to Support Vector Machines and Other Kernel-based Learning Methods. Cambridge University Press (2010)
14. Scholkopf, B., Smola, A.J.: Learning with Kernels: Support Vector Machines, Regularization, Optimization, and Beyond. MIT Press, Cambridge (2001)
15. Cortes, C., Vapnik, V.: Support-vector networks. Machine Learning 20(3), 273–297 (1995)

Towards a Secure Multivariate Identity-Based Encryption

Simona Samardjiska and Danilo Gligoroski

Department of Telematics, NTNU, Trondheim, Norway
{simonas,danilog}@item.ntnu.no

Abstract. We investigate the possibilities of building a Multivariate Identity-Based Encryption (IBE) Scheme, such that for each identity the obtained Public Key Encryption Scheme is Multivariate Quadratic (MQ). The biggest problem in creating an IBE with classical MQ properties is the possibility of collusion of polynomial number of users against the master key or the keys of other users. We present a solution that makes the collusion of polynomial number of users computationally infeasible, although still possible. The proposed solution is a general model for a Multivariate IBE Scheme with exponentially many public-private keys that are instances of an MQ public key encryption scheme.

Keywords: Identity-Based Encryption (IBE), Public Key Encryption Schemes, Multivariate Quadratic Schemes (MQ schemes).

1 Introduction

Identity-Based Encryption (IBE) is a kind of Public Key Encryption (PKE) in which the public key of a user is some unique information about the identity of the user that is publicly known. (e.g. a user's email address). The corresponding private key is generated by a Trusted Third Party, called Private-Key Generator (PKG), that possesses a master secret key. The idea was first proposed by Shamir in 1984 [13], but it wasn't until 2001 when the first two solutions emerged: Boneh and Franklin's [3] scheme based on the Bilinear Diffie-Hellman problem, and Cocks' [7] scheme based on the quadratic residue assumption. The schemes following the first direction are now known as pairing-based IBE schemes [4,10,18]. Although rather inefficient, the second direction was also further developed in [8,5]. Recently, a new type of IBE emerged, based on hard problems on lattices [11,1]. This construction, while quite elegant, is also less efficient than the pairing-based IBE systems.

In the open literature there have been other attempts of building IBE solutions. In the area of Multivariate Public Key Cryptography, the development is in its very beginnings. To our knowledge, there has been only one proposal for IBE scheme [6] that combines multivariate principles and bilinear maps. It was soon proven to be highly vulnerable to collusion in [17,2]. Other ID-Based solutions in multivariate cryptography are rather scarce too. In [19] an ID-Based Zero-Knowledge identification protocol was proposed that is also susceptible to

S. Markovski and M. Gusev (Eds.): *ICT Innovations 2012*, AISC 207, pp. 59–69.
DOI: 10.1007/978-3-642-37169-1_6

collusion. Later, in [15] the solution was generalized and improved to avoid the collusion. Recently, a more efficient ZKP was proposed [14], and although not ID-Based, it can be extended to one, and to ID-Based signatures as well. Again, a very recent paper [12], where a new multivariate assumption was introduced, opens the possibility for a IBE construction in the future, due to similarities with assumptions for lattices. However, a concrete solution is yet to be developed.

1.1 Contributions and Organization of the Paper

We investigate the possibilities of building a Multivariate IBE Scheme, such that for each identity the obtained PKE Scheme is Multivariate Quadratic (MQ). The biggest problem arising in this attempt is the possibility of collusion of polynomial number of users against the master key or the keys of other users.

We show that a Multivariate IBE in which the users obtain a classical MQ public-private key pair is highly vulnerable to collusion, and can not be made secure for any set of reasonable parameters in an colluding environment. We locate the reasons for this, and as a result propose a solution that is not completely immune to collusion, but for realistic parameters the attack becomes infeasible, even if highly distributed among the colluders. The proposed solution is a general model for a Multivariate IBE Scheme with exponentially many users.

The paper is organized as follows. In Section 2 we provide the basic definitions for IBE and MQ cryptosystems. In Section 3 we describe a natural construction of a multivariate IBE and locate the reasons why this construction is insecure for any practical parameters. We propose our new model in Section 4, and analyze its security in Section 4.1. The open questions regarding a practical implementation of the model are discussed in Section 4.2. We conclude the paper in Section 5.

2 Preliminaries

2.1 Identity Based Encryption

Definition 1. ([3]) *An Identity-Based Encryption scheme $\mathcal{E}_{ID}$ is specified by four randomized algorithms* Setup, Extract, Encrypt, Decrypt*:*

Setup : *Is run by the PKG one time for creating the whole IBE environment. It accepts a security parameter k and outputs a set of public system parameters* params *including the message space $\mathcal{M}$ and ciphertext space $\mathcal{C}$, and a master key* msk *that is kept secret by the PKG.*

Extract : *Is run by the PKG on a user's private key request. It takes as input* params, msk, *and $ID \in \{0,1\}^*$ and returns the private key s_{ID} for user ID.*

Encrypt : *Takes as input* params, *a message $m \in \mathcal{M}$ and $ID \in \{0,1\}^*$ and outputs the ciphertext $c \in \mathcal{C}$.*

Decrypt : *Accepts* params, *s_{ID}, and $c \in \mathcal{C}$ and returns $m' \in \mathcal{M}$.*

The algorithms have to satisfy the standard ***Correctness constraint:***

$$\forall m \in \mathcal{M}, ID \in \{0,1\}^* : \mathsf{Decrypt}(\mathsf{params}, s_{ID}, \mathsf{Encrypt}(\mathsf{params}, ID, m)) = m.$$

We consider an adversary from a standard security model for IBE (see for ex. [3]). Let Sec be a PKE security model. Sec−ID is a security model for IBE in which the adversary can query all the oracles of Sec for any identity $ID \in \{0,1\}^*$, and additionally has access to a key extraction oracle, that on a query ID returns the private key s_{ID} for the identity ID. Such an adversary can issue polynomially many different queries, under the constrain that she does not query the key extraction oracle on the identity ID^* under attack. This situation is equivalent to a collusion of polynomial number of users, and is a realistic threat in practice.

2.2 Multivariate Quadratic (MQ) Cryptosystems

Let $\mathbb{F}_q$ be a finite field of order q, where q is a prime power. We will use the notations $\mathbf{u} = (u_1, u_2, \ldots, u_n) \in \mathbb{F}_q^n$, and $\mathbf{x} = (x_1, x_2, \ldots, x_n) \in \mathbb{F}_q[x_1, \ldots, x_n]$ where $n \in \mathbb{N}$. Let $P(\mathbf{x}) = (p_1(\mathbf{x}), p_2(\mathbf{x}), ..., p_m(\mathbf{x})) \in \mathbb{F}_q^m[x_1, x_2, \ldots, x_n]$ be a system of m polynomials of degree $d, d \geq 2$. Let $\mathbf{v} \in \mathbb{F}_q^m$.

The problem of Simultaneous Multivariate Equations over the field $\mathbb{F}_q$ consists of finding a solution $\mathbf{u} \in \mathbb{F}_q^n$ to the system of equations $\mathbf{v} = (p_1(\mathbf{x}), ..., p_m(\mathbf{x}))$ over $\mathbb{F}_q$. It has been shown that for every $d \geq 2$ this problem is NP complete [9]. The case of $d = 2$ has been most exploited and it is called the MQ-problem.

A typical MQ public key scheme relies on the knowledge of a trapdoor for a particular system of polynomials $P \in \mathbb{F}_q^m[x_1, x_2, \ldots, x_n]$. Usually, the system P is created as a composition of three polynomial transformations: two affine mappings S and T and one quadratic P' as $P(\mathbf{x}) = T \circ P' \circ S(\mathbf{x})$. Without loss of generality, it can be assumed that the private key is $s = (S, P', T)$.

3 A Natural Construction of Multivariate IBE

As our goal is to create a Multivariate IBE that relies only on the principles in multivariate cryptography, a natural first step would be to require that the related PKC of each of the exponentially many users in the IBE scheme is a secure classical MQ PKC. We next formally define such multivariate IBE.

Let Π be a MQ PKE scheme over $\mathbb{F}_q$ with message space $\mathcal{M}$ and ciphertext space $\mathcal{C}$. For simplicity, we take the public key to be a system P of n polynomials in n variables over $\mathbb{F}_q$, and the private key to be $s = (S, P', T)$. (The case of m polynomials in n variables is analogous.) We also neglect any additional parts of the scheme, as well as the encryption $\mathcal{E}$ and decryption $\mathcal{D}$ specifics, since they are not important in the context. Let Π be secure in a given security model Sec.

We define an IBE scheme $\mathcal{E}_{IDclassic}$ by the following four algorithms:

Setup :

Given a security parameter $k \in \mathbb{N}$ perform the steps:

1. Let the message space $\mathcal{M}$ and ciphertext space $\mathcal{C}$ be the same as for Π. Let $l \in \mathbb{N}$ be a system parameter and $H : \{0,1\}^* \to \mathbb{F}_q^l$ a cryptographic hash function.
2. Generate at random two multivariate mappings $M_S : \mathbb{F}_q^l \times \mathbb{F}_q^n \to \mathbb{F}_q^n$ and $M_T : \mathbb{F}_q^l \times \mathbb{F}_q^n \to \mathbb{F}_q^n$ of degrees $d_S \geq 2$ and $d_T \geq 2$ respectively, such that $M_S(\mathbf{a}, \mathbf{x}) = S_{\mathbf{a}}(\mathbf{x})$ and $M_T(\mathbf{a}, \mathbf{x}) = T_{\mathbf{a}}(\mathbf{x})$ are affine mappings for every $\mathbf{a} \in \mathbb{F}_q^l$.

3. Generate at random a multivariate mapping $M_{P'} : \mathbb{F}_q^l \times \mathbb{F}_q^n \to \mathbb{F}_q^n$ of degree $d_{P'} \geq 2$ such that $M_{P'}(\mathbf{a}, \mathbf{x}) = P'_{\mathbf{a}}(\mathbf{x})$ is a central mapping for some MQ PKC $\Pi_{\mathbf{a}}$ that is an instance of the scheme Π, for every $\mathbf{a} \in \mathbb{F}_q^l$.
4. Construct the system of pol. $P_{pub}(\mathbf{x}_{ID}, \mathbf{x}) = M_T(\mathbf{x}_{ID}, M_{P'}(\mathbf{x}_{ID}, M_S(\mathbf{x}_{ID}, \mathbf{x})))$ from $\mathbb{F}_q^n\left[x_1^{(ID)}, x_2^{(ID)}, \ldots, x_l^{(ID)}, x_1, x_2, \ldots, x_n\right]$.

Output $\mathsf{params} = (\mathcal{M}, \mathcal{C}, H, P_{pub})$ and the master key $\mathsf{msk} = (M_S, M_{P'}, M_T)$.

Extract :

Given the input $\mathsf{params} = (\mathcal{M}, \mathcal{C}, H, P_{pub})$, $\mathsf{msk} = (M_S, M_{P'}, M_T)$, and an identifier $ID \in \{0,1\}^*$, create the mappings $S_{H(ID)}(\mathbf{x}) = M_S(H(ID), \mathbf{x})$, $T_{H(ID)}(\mathbf{x}) = M_T(H(ID), \mathbf{x})$ and $P'_{H(ID)}(\mathbf{x}) = M_{P'}(H(ID), \mathbf{x})$.
Output the private key $s_{ID} = (S_{H(ID)}, P'_{H(ID)}, T_{H(ID)})$ for user ID.

Encrypt :

Take as input $\mathsf{params} = (\mathcal{M}, \mathcal{C}, H, P_{pub})$, a message $m \in \mathcal{M}$ and $ID \in \{0,1\}^*$ and:

1. Evaluate $P_{H(ID)}(\mathbf{x}) = P_{pub}(H(ID), \mathbf{x}) = T_{H(ID)} \circ P'_{H(ID)} \circ S_{H(ID)}(\mathbf{x})$.
2. Use the algorithm $\mathcal{E}$ of Π with public key $P_{H(ID)}$ to encrypt the message m.

Output the ciphertext $c = \mathcal{E}(m, P_{H(ID)})$.

Decrypt :

Input $\mathsf{params} = (\mathcal{M}, \mathcal{C}, H, P_{pub})$, $s_{ID} = (S_{H(ID)}, P'_{H(ID)}, T_{H(ID)})$, and $c \in \mathcal{C}$ and use the decryption algorithm $\mathcal{D}$ of Π with private key s_{ID} to output $m' = \mathcal{D}(c, s_{ID})$.

Clearly, every identity ID has a public-private key pair that can be created using the Key generation algorithm of Π, and the encryption and decryption are performed using the $\mathcal{E}$ and $\mathcal{D}$ algorithms of Π. So each user has a PKC that is an instance of the scheme Π. Hence, this scheme satisfies our initial requirements.

We must note that since the system of polynomials P_{pub} is a public parameter, and should be evaluated during encryption, in order for the scheme to be efficient, the polynomials must be of some fixed low degree. This implies that $deg(M_S) = d_S$, $deg(M_T) = d_T$ and $deg(M_{P'}) = d_{P'}$ have to be of low degree as well.

Also, since for a given ID, its PKC is an instance of Π, it is secure in the sense of Sec. Note that we have not defined what "Sec-secure" means. This is because no matter the requirements of Sec that each of the MQ PKCs satisfy, the IBE system is vulnerable to collusion of a small number of users.

Proposition 1. *A standard adversary from* Sec$-$ID *can find the master key of $\mathcal{E}_{IDclassic}$ making $\mathcal{O}(l^d)$ key extraction queries where $d = max(d_S, d_T, d_{P'})$. The computational load for finding the* msk *is $\mathcal{O}(l^{\omega \cdot d})$ sequential field operations, where ω is the linear algebra constant.* □

Note that an MQ IBE scheme does not have to be defined by $\mathcal{E}_{IDclassic}$, as we will see in the next section (certainly, other solutions are possible). But if we want the scheme to have our initial requirements, then this is necessary the case.

Proposition 2. *Let $\mathcal{E}_{ID}$ be an IBE scheme with a system of polynomials over a finite field as a public key, and let for each identity, $\mathcal{E}_{ID}$ turn into a classical MQ cryptosystem. Then $\mathcal{E}_{ID}$ can be defined by the four algorithms of $\mathcal{E}_{IDclassic}$.* □

Propositions 1 and 2 imply that the natural construction of a Multivariate IBE is vulnerable to collusion and thus insecure if enough colluders are present, despite the fact that each of the users in the system have a secure PKC system[1]. In practice, collusion is a serious threat for many different systems (not just IBE), and as solutions without collusion are sometimes difficult to find it is important to investigate whether a collusion scenario is feasible for practical parameters[2].

For $\mathcal{E}_{IDclassic}$, the most desirable parameters for practical use are $d_S = 2$, $d_T = 2$, $d_{P'} = 2$ for which $deg(P_{pub}) = 5$. If we assume $l = n$, it can be calculated that the number of needed colluders is $\frac{n(n+1)}{2} + n + 2$. Now, even if all the computational load of the attack is centralized (which is rarely the case in practice), we need $n \approx 3200$ to obtain a security level of $\mathcal{O}(2^{80})$. This clearly implies that such a scheme can not provide a secure implementation in a colluding environment.

4 Our New Model

In Section 3 we constructed and analyzed a rather natural but also insecure MQ IBE scheme $\mathcal{E}_{IDclassic}$. Not only a collusion of a small number of users is possible and leads to discovery of the secret msk, but also the attack can be mounted using rather low computational resources. The reason for this can be located in the fixed small degree of the secret key $\mathsf{msk} = (M_S, M_{P'}, M_T)$, and the fact that the users ID_i are given $s_{ID_i} = (S_{H(ID_i)}, P'_{H(ID_i)}, T_{H(ID_i)})$ as their private key.

The small degree of $M_S, M_{P'}$, and M_T can not be avoided, since otherwise, the public parameter P_{pub} would be of high degree, and the scheme would not be efficient. What can be done is to give the private keys to the users in a form that will hide the structure of the building parts. In MQ public key cryptosystems, such technique is used in the construction of the public key, which is formed as a composition of the private trapdoor mappings. When the composition is given in an explicit form as a system of polynomials, the security of the system lies in the inability to decompose it in the secret building parts.

So, instead of the private key $s_{ID} = (S_{H(ID)}, P'_{H(ID)}, T_{H(ID)})$ for user ID, the user can be given the key $s_{ID} = S^{-1}_{H(ID)} \circ P'^{-1}_{H(ID)} \circ T^{-1}_{H(ID)}$. This raises a new problem, since the mapping s_{ID} can in general be of any degree, thus making the scheme inefficient. (Since $S_{H(ID)}$ and $T_{H(ID)}$ are affine, so are $S^{-1}_{H(ID)}$ and $T^{-1}_{H(ID)}$, but the inverse of the quadratic $P'_{H(ID)}$ can be of high degree, and therefore have an exponential number of terms.) In order for this technique to work, $P'^{-1}_{H(ID)}$ has to be of some small degree $d_{P'^{-1}}$, as well. We next define a new model of a Multivariate IBE using the mentioned approach.

Let $\mathcal{E}_{ID-MQ}$ be defined by the following four probabilistic algorithms:

[1] The proofs will be given in an extended version of the paper.

[2] In some applications collusion is dealt with by constraining the number of users that can collude or by using tamper resistant hardware for storing the secret key.

Setup :

Given a security parameter $k \in \mathbb{N}$ perform the steps:

1. Let $n, \rho \in \mathbb{N}$ where ρ is a small integer. Let $\mathcal{M} = \mathbb{F}_q^{n/2}$ and $\mathcal{C} = \mathbb{F}_q^{n-\rho}$ be the message and ciphertext space respectively. Let $l \in \mathbb{N}$ be a system parameter and $H : \{0,1\}^* \to \mathbb{F}_q^l$, $G : \mathbb{F}_q^{3n/4} \to \mathbb{F}_q^{n/4}$ two cryptographic hash functions.
2. Generate at random two multivariate mappings $M_S, M_T : \mathbb{F}_q^l \times \mathbb{F}_q^n \to \mathbb{F}_q^n$ of degrees $d_S \geq 2$ and $d_T \geq 2$ respectively, such that:
   - $M_S(\mathbf{a}, \mathbf{x}) = S_{\mathbf{a}}(\mathbf{x})$ and $M_T(\mathbf{a}, \mathbf{x}) = T_{\mathbf{a}}(\mathbf{x})$ are affine mappings for every $\mathbf{a} \in \mathbb{F}_q^l$.
   - M_S^{inv} and M_T^{inv} defined by $M_S^{inv}(\mathbf{u}, \mathbf{v}) = \mathbf{w} \Leftrightarrow M_S(\mathbf{u}, \mathbf{w}) = \mathbf{v}$, $M_T^{inv}(\mathbf{u}, \mathbf{v}) = \mathbf{w} \Leftrightarrow M_T(\mathbf{u}, \mathbf{w}) = \mathbf{v}$, for every $\mathbf{u} \in \mathbb{F}_q^l$ and $\mathbf{v}, \mathbf{w} \in \mathbb{F}_q^n$, are of degree $\mathcal{O}(n)$.
3. Generate at random a multivar. mapping $M_{P'} : \mathbb{F}_q^l \times \mathbb{F}_q^n \to \mathbb{F}_q^n$, $d_{P'} \geq 2$ such that:
   - $M_{P'}(\mathbf{a}, \mathbf{x}) = P'_{\mathbf{a}}(\mathbf{x})$ is a bijective quadratic transformation for every $\mathbf{a} \in \mathbb{F}_q^l$.
   - The transformation $M_{P'}^{inv}$ defined by $M_{P'}^{inv}(\mathbf{u}, \mathbf{v}) = \mathbf{w} \Leftrightarrow M_{P'}(\mathbf{u}, \mathbf{w}) = \mathbf{v}$, for every $\mathbf{u} \in \mathbb{F}_q^l$ and $\mathbf{v}, \mathbf{w} \in \mathbb{F}_q^n$, is of degree $\mathcal{O}(n)$.
   - $M_{P'}^{inv}(\mathbf{a}, \mathbf{x}) = P'^{-1}_{\mathbf{a}}(\mathbf{x})$ is of small degree $d_{P'^{-1}} \geq 2$ for every $\mathbf{a} \in \mathbb{F}_q^l$, and can be efficiently found given $P'_{\mathbf{a}}(\mathbf{x})$.
4. Construct the system of pol. $P_{pub\_full}(\mathbf{x}_{ID}, \mathbf{x}) = M_T(\mathbf{x}_{ID}, M_{P'}(\mathbf{x}_{ID}, M_S(\mathbf{x}_{ID}, \mathbf{x})))$ from $\mathbb{F}_q^n\left[x_1^{(ID)}, x_2^{(ID)}, \ldots, x_l^{(ID)}, x_1, x_2, \ldots, x_n\right]$.
5. Let P_{pub} consist of the first $n - \rho$ polynomials of P_{pub_full}.

Output params $= (\mathcal{M}, \mathcal{C}, H, G, P_{pub})$, and the master key msk $= (M_S, M_{P'}, M_T)$.

Extract :

Input params $= (\mathcal{M}, \mathcal{C}, H, G, P_{pub})$, msk $= (M_S, M_{P'}, M_T)$, and an $ID \in \{0,1\}^*$,

1. Create the mappings $S_{H(ID)}(\mathbf{x}) = M_S(H(ID), \mathbf{x})$, $T_{H(ID)}(\mathbf{x}) = M_T(H(ID), \mathbf{x})$ and $P'_{H(ID)}(\mathbf{x}) = M_{P'}(H(ID), \mathbf{x})$. Find the inverses $S^{-1}_{H(ID)}$, $P'^{-1}_{H(ID)}$, and $T^{-1}_{H(ID)}$.
2. Construct the system of polynomials $s_{ID}(\mathbf{x}) = S^{-1}_{H(ID)} \circ P'^{-1}_{H(ID)} \circ T^{-1}_{H(ID)}(\mathbf{x})$.

Return s_{ID} as the private key for user ID.

Encrypt :

Input params $= (\mathcal{M}, \mathcal{C}, H, G, P_{pub})$, $m \in \mathcal{M}$, random $r \in \mathbb{F}_q^{n/4}$, and $ID \in \{0,1\}^*$ and:

1. Evaluate $G(m, r)$. Let $\mathbf{m} = (m, r, G(m, r))$. Set $P_{H(ID)}(\mathbf{x}) = P_{pub}(H(ID), \mathbf{x})$.
2. Evaluate $c = P_{H(ID)}(\mathbf{m})$.

Output the ciphertext c.

Decrypt :

Input params $= (\mathcal{M}, \mathcal{C}, H, G, P_{pub})$, s_{ID}, and $c \in \mathcal{C}$ and
For all $c_1 \in \mathbb{F}_q^{\rho}$ perform the steps:

1. Set $\mathbf{c} = (c, c_1)$.
2. Evaluate $\mathbf{m}' = s_{ID}(\mathbf{c})$. Let $\mathbf{m}' = (m', r', g')$, where $m' \in \mathbb{F}_q^{n/2}$, $r' \in \mathbb{F}_q^{3n/4}$, $g' \in \mathbb{F}_q^{n/4}$.
3. **If** $G(m', r') = g'$ **then break**;

Output the plaintext m'.

4.1 Security Analysis of $\mathcal{E}_{ID-MQ}$

First of all, we should emphasize that the four algorithms of $\mathcal{E}_{ID-MQ}$ presented in the previous section describe a model of an IBE. Details regarding concrete implementation of a scheme, i.e. the construction of the mappings $M_S, M_{P'}$, and M_T are not given. A complete security analysis strongly depends on their characteristics. Therefore, we will not consider the internal structure and possible weaknesses that could arise from a particular implementation of $\mathcal{E}_{ID-MQ}$.

Let Π_{ID-MQ} be the related PKE scheme arising from $\mathcal{E}_{ID-MQ}$ for a fixed identity ID. Let $P_{full}(\mathbf{x}) = P_{pub\_full}(H(ID), \mathbf{x})$, and (P, s) be a public-private key pair from Π_{ID-MQ}, where $P = P_{H(ID)}$ and $s = s_{ID}$. Then $deg(P) = 2$ and $deg(s) = d_{P'^{-1}}$ is a small integer.

Proposition 3. *Given the public key P of Π_{ID-MQ}, the degree $deg(s)$ of s and no other implementation specifics of Π_{ID-MQ}, the private key s can not be recovered in polynomial time.*

Proof. First, we show that the problem is polynomial time reducible to the problem of finding the missing ρ polynomials from P_{full}.

Suppose $P_{full} : \mathbb{F}_q^n \to \mathbb{F}_q^n$ is known. We evaluate $P_{full}(m_1^{(i)}, \dots, m_n^{(i)}) = (c_1^{(i)}, \dots, c_n^{(i)})$ for arbitrary $(m_1^{(i)}, \dots, m_n^{(i)}) \in \mathbb{F}_q^n$, $i \in \{1, \dots, N+1\}$, where $N = \binom{n+deg(s)-1}{deg(s)} + \dots + \binom{n-1}{0}$ for $q \neq 2$ and $N = \binom{n}{deg(s)} + \dots + \binom{n}{0}$ for $q = 2$. Now, we can obtain, for every $k \in \{1, \dots, n\}$, the system of equations

$$\sum_{j_1=1}^{n}\sum_{j_2=j_1}^{n}\cdots\sum_{j_{deg(s)}=j_{deg(s)-1}}^{n} a^{(k)}_{j_1\dots j_{deg(s)}} c^{(i)}_{j_1}\dots c^{(i)}_{j_{deg(s)-1}} + \dots + \sum_{j=1}^{n} a_j^{(k)} c_j^{(i)} + a^{(k)} = m_k^{(i)}, \; i \in \{1, \dots, N+1\}$$

in the unknown coefficients $a^{(k)}_{j_1\dots j_{deg(s)}}, a_j^{(k)}, a^{(k)}$ of s, that is a full rank system with overwhelming probability of $1 - \frac{1}{q^{N+1}}$. The k obtained systems can be solved in time $\mathcal{O}(N^{\omega}) = \mathcal{O}(n^{deg(s)\cdot\omega})$, thus recovering completely the private key s.

Similarly, if the private key s is known, using the same technique, the missing ρ polynomials from P_{full} can be recovered in time $\mathcal{O}(n^{2\omega})$. Thus, if P_{full} can be found, than s can be found (in polynomial time) as well, and vice-versa.

Now, it is enough to see that P_{full} can not be recovered without s. Indeed, since no information about the internal structure of the system is known, the only way to recover the missing ρ polynomials from P_{full} is by polynomial interpolation which is not possible without s, since $c_{n-\rho+1}^{(i)}, \dots, c_n^{(i)}$ can not be discovered for any $(m_1^{(i)}, \dots, m_n^{(i)}) \in \mathbb{F}_q^n$. □

Proposition 3 basically says that a powerful adversary that has access to any standard PKC oracle can not find s, provided Π_{ID-MQ} is seen in a black-box manner. The situation is completely different in an IBE environment, where the adversary has access to a key extraction oracle (or, in other words, in a colluding environment). In fact, we will show that the new model $\mathcal{E}_{ID-MQ}$ is not completely immune to collusion. However, we will also show that polynomial number of colluding users can only find out other users' private key, and not the master key. Also, although polynomial, the computational load of the attack is rather big, making it infeasible for realistic implementation parameters.

Proposition 4. *A standard polynomial time IBE adversary without knowledge of the internal structure of $\mathcal{E}_{ID-MQ}$ needs at least $\mathcal{O}(l^{2d_S+d_{P'}+d_T-3})$ calls to the key extraction oracle and at least $\mathcal{O}(l^{(2d_S+d_{P'}+d_T-3)\omega})$ sequential field operations in order to discover the secret key of a legitimate user in $\mathcal{E}_{ID-MQ}$.*

Proof. Let the components of M_S, $M_{P'}$ and M_T be given by

$$M_I^{(k)}(\mathbf{x}_{ID},\mathbf{x}) = f_I^{(k)}(\mathbf{x}_{ID}) + \sum_{j=1}^{n} g_I^{(k,j)}(\mathbf{x}_{ID})x_j, \quad I \in \{S,T\}, \tag{1}$$

$$M_{P'}^{(k)}(\mathbf{x}_{ID},\mathbf{x}) = f_{P'}^{(k)}(\mathbf{x}_{ID}) + \sum_{j=1}^{n} g_{P'}^{(k,j)}(\mathbf{x}_{ID})x_j + \sum_{i=1}^{n}\sum_{j=i}^{n} h_{P'}^{(k,i,j)}(\mathbf{x}_{ID})x_i x_j, \tag{2}$$

where $deg(f_I^{(k)}) = d_I$, $deg(g_I^{(k,j)}) = d_I - 1$, for $I \in \{S, P', T\}$, and $deg(h_{P'}^{(k,i,j)}) = d_{P'} - 2$. From (1) and (2), the components of P_{pub_full} can be represented as:

$$P_{pub\_full}^{(k)}(\mathbf{x}_{ID},\mathbf{x}) = F_1^{(k)}(\mathbf{x}_{ID}) + \sum_{j=1}^{n} F_2^{(k,j)}(\mathbf{x}_{ID})x_j + \sum_{j=1}^{n}\sum_{i=j}^{n} F_3^{(k,i,j)}(\mathbf{x}_{ID})x_i x_j, \tag{3}$$

where $F_1^{(k)}$, $F_2^{(k,j)}$ and $F_3^{(k,i,j)}$ are polynomials in l variables for all $k,i,j \in \{1,\dots,n\}$ with degrees $deg(F_1^{(k)}) = d_{F_1} = 2d_S + d_{P'} + d_T - 3$, $deg(F_2^{(k,j)}) = 2d_S + d_{P'} + d_T - 4 = d_{F_1} - 1$ and $deg(F_1^{(k,i,j)}) = 2d_S + d_{P'} + d_T - 5 = d_{F_1} - 2$.

From the proof of Proposition 3, finding the secret key of a legitimate user is polynomial time equivalent to finding the full system P_{full} of the user. Since the adversary is not aware of any internal weakness of $\mathcal{E}_{ID-MQ}$, in order to find P_{full} of a user, she first needs to find the missing polynomials of P_{pub_full}.

From (3), the number of coefficients of $F_1^{(k)}$ is $N_{F_1} = \binom{l+d_{F_1}-1}{d_{F_1}} + \binom{l+d_{F_1}-2}{d_{F_1}-1} + \dots + \binom{l-1}{0}$ when $q \neq 2$ and $N_{F_1} = \binom{l}{d_{F_1}} + \binom{l}{d_{F_1}-1} + \dots + \binom{l}{0}$ when $q=2$, for all $k \in \{1,\dots,n\}$. Similar reasoning holds for the number N_{F_2} (i.e. N_{F_3}) of coefficients of $F_2^{(k,j)}$ (i.e. $F_3^{(k,i,j)}$). Now, in order to find the coefficients of $F_1^{(k)}$ (i.e. $F_2^{(k,j)}$ and $F_3^{(k,i,j)}$) via interpolation (which is the only possible way) the adversary needs at least N_{F_1} (N_{F_2}, N_{F_3}) different evaluations of the mappings. Note that less can not produce a full rank system necessary for interpolation. Hence, the minimal number of calls to the key extraction oracle is N_{F_1}.

So, first, the adversary queries the key extraction oracle $N_{F_1} + 1$ times to obtain the private keys $s_{ID_1}, \dots, s_{ID_{N_{F_1}+1}}$. Using the technique from the proof of Proposition 3 she can recover $P_{pub\_full}(H(ID_i),\mathbf{x})$ of each of the users ID_i, $i \in \{1,\dots,N_{F_1}+1\}$, performing $\mathcal{O}(n^{2\omega})$ field operations for each of the users.

Now, for a fixed $k \in \{n-\rho+1,\dots,n\}$, for user ID_t the k th component of $P_{pub\_full}(H(ID_t),\mathbf{x})$ can be represented as

$$P_{pub\_full}^{(k)}(H(ID_t),\mathbf{x}) = a_{1,ID_t}^{(k)} + \sum_{j=1}^{n} a_{2,ID_t}^{(k,j)} x_j + \sum_{j=1}^{n}\sum_{i=j}^{n} a_{3,ID_t}^{(k,i,j)} x_i x_j,$$

and thus from (3), the following $1 + n + \binom{n+1}{2}$ systems (of full rank with overwhelming probability) can be formed:

$$\begin{aligned} & F_1^{(k)}(H(ID_t)) = a_{1,ID_t}^{(k)}, \quad t \in \{1, ..., N_{F_1} + 1\}, \\ \forall j \in \{1, ..., n\} : \quad & F_2^{(k,j)}(H(ID_t)) = a_{2,ID_t}^{(k,j)}, \quad t \in \{1, ..., N_{F_2} + 1\}, \\ \forall j \in \{1, ..., n\}, i \in \{j, ..., n\} : \quad & F_3^{(k,i,j)}(H(ID_t)) = a_{3,ID_t}^{(k,i,j)}, \quad t \in \{1, ..., N_{F_3} + 1\}, \end{aligned}$$

which can be solved in $\mathcal{O}(N_{F_1}^{\omega})$, $\mathcal{O}(N_{F_2}^{\omega})$, $\mathcal{O}(N_{F_3}^{\omega})$ field operations, respectively.

Repeating the same procedure for every $k \in \{n - \rho + 1, \ldots, n\}$, P_{pub_full} can be fully recovered. Let ID^* be any identity not queried previously. Then using $P_{pub\_full}(H(ID^*), \mathbf{x})$, and the proof of Proposition 3, s_{ID^*} can be fully recovered in $\mathcal{O}(n^{d_{P'-1} \cdot \omega})$ field operations. The entire attack requires $\mathcal{O}(l^{(2d_S + d_{P'} + d_T - 3)\omega})$ sequential field operations. □

In the attack described in Proposition 4 the key point is the recovery of P_{pub_full}. Once it is found, attacking users can be done quite fast. However, since the mappings M_S^{inv}, $M_{P'}^{inv}$ and M_T^{inv} are of degree $\mathcal{O}(n)$, the master key msk can not be recovered in polynomial time. Indeed, in this case, the number of equations needed for interpolation is $\mathcal{O}(2^n)$. We can state this as:

Proposition 5. *A standard polynomial time IBE adversary without the knowledge of the internal structure of $\mathcal{E}_{ID-MQ}$, can not discover the master secret key* msk *of the PKG.* □

The recovery of P_{pub_full} is not only the most important, but also brings the biggest computational load for the adversary. Thus, it is crucial to see whether this step can be made infeasible for realistic, practical parameters. Again, as for $\mathcal{E}_{IDclassic}$, the most desirable parameters for practical use are $d_S = 2$, $d_T = 2$, $d_{P'} = 2$ for which $deg(P_{pub}) = 5$. Then, the minimal number of needed colluders is $\binom{l+4}{5} + \binom{l+3}{4} + \cdots + \binom{l-1}{0}$ when $q \neq 2$ and $\binom{l}{5} + \binom{l}{4} + \cdots + \binom{l}{0}$ when $q = 2$. Now, solving a system of linear equations of the same size is the most consuming part. It can be calculated that, using the fastest algorithm for Gaussian elimination of dense systems with $\omega = log_2 7$, for $l \geq 144$, the complexity of this operation is $> 2^{80}$ for any q. Hence, for $l \geq 144$, the described attack is infeasible, even if we assume that the complete computational load is distributed among the colluders. This suggests, that the IBE model $\mathcal{E}_{ID-MQ}$ can be used in practice and provide a satisfactory level of computational security.

4.2 Open Questions

As it was already emphasized in the text, the described $\mathcal{E}_{ID-MQ}$ is an IBE model, and the security analysis made in Section 4.1 makes sense only if the internal structure of an implementation does not introduce additional weaknesses that can compromise the security of Π_{ID-MQ}, and thus also of $\mathcal{E}_{ID-MQ}$. Hence, the possibility of building a secure implementation can be investigated in two stages:

Efficient construction of the mappings M_S, $M_{P'}$ and M_T. Note that the three mappings can be seen as generalized left quasigroups (bivariate mappings that are bijections in the second variable), with the property that both the quasigroup and the parastrophe (generalized inverse in the second variable) are of low degree in the second variable. The theory from [16] provides an efficient construction of a special class of such mappings. Finding other classes that can be easily constructed, as well as investigating their and the already known one's suitability for use in our setting is an open research problem.

Security analysis of Π_{ID-MQ}. At the moment, it is not clear whether a secure multivariate encryption scheme with small degree of P'^{-1} can be constructed. This is even a more challenging research direction, since building a secure MQ encryption scheme has proven to be a hard task, and currently, in the open literature, there are no MQ encryption schemes considered secure.

5 Conclusions

In this paper we proposed a new model for Multivariate Identity-Based encryption. Each of the exponentially many possible users has a public-private key pair that is an instance of a PKC system similar to a classical MQ system. The main focus was put on the security against collusion attack, which seems to be an inherent problem for any scheme that has classical MQ properties. Our new model, also, is not immune to collusion. However, we prove that the attack becomes infeasible for practical values of the parameters, and for length of the identity of $l \geq 144$, the complexity of the attack is $> 2^{80}$ sequential field operations. A concrete implementation of the model is an open research problem.

References

1. Agrawal, S., Boneh, D., Boyen, X.: Efficient lattice (H)IBE in the standard model. In: Gilbert, H. (ed.) EUROCRYPT 2010. LNCS, vol. 6110, pp. 553–572. Springer, Heidelberg (2010)
2. Albrecht, M.R., Paterson, K.G.: Breaking an Identity-Based Encryption Scheme Based on DHIES. In: Chen, L. (ed.) Cryptography and Coding 2011. LNCS, vol. 7089, pp. 344–355. Springer, Heidelberg (2011)
3. Boneh, D., Franklin, M.: Identity-based encryption from the Weil pairing. In: Kilian, J. (ed.) CRYPTO 2001. LNCS, vol. 2139, pp. 213–229. Springer, Heidelberg (2001)
4. Boneh, D., Boyen, X.: Secure identity based encryption without random oracles. In: Franklin, M. (ed.) CRYPTO 2004. LNCS, vol. 3152, pp. 443–459. Springer, Heidelberg (2004)
5. Boneh, D., Gentry, C., Hamburg, M.: Space-efficient identity based encryption without pairings. In: FOCS 2007, pp. 647–657 (2007)
6. Chen, Y., Charlemagne, M., Guan, Z., Hu, J., Chen, Z.: Identity-based encryption based on DHIES. In: ASIACCS 2010, pp. 82–88. ACM (2010)
7. Cocks, C.: An identity based encryption scheme based on quadratic residues. In: Honary, B. (ed.) Cryptography and Coding 2001. LNCS, vol. 2260, pp. 360–363. Springer, Heidelberg (2001)
8. Di Crescenzo, G., Saraswat, V.: Public key encryption with searchable keywords based on Jacobi symbols. In: Srinathan, K., Rangan, C.P., Yung, M. (eds.) INDOCRYPT 2007. LNCS, vol. 4859, pp. 282–296. Springer, Heidelberg (2007)

9. Garey, M.R., Johnson, D.S.: Computers and Intractability - A Guide to the Theory of NP-Completeness. W.H. Freeman and Company (1979)
10. Gentry, C.: Practical identity-based encryption without random oracles. In: Vaudenay, S. (ed.) EUROCRYPT 2006. LNCS, vol. 4004, pp. 445–464. Springer, Heidelberg (2006)
11. Gentry, C., Peikert, C., Vaikuntanathan, V.: Trapdoors for hard lattices and new cryptographic constructions. In: STOC 2008, pp. 197–206. ACM (2008)
12. Huang, Y.-J., Liu, F.-H., Yang, B.-Y.: Public-Key Cryptography from New Multivariate Quadratic Assumptions. In: Fischlin, M., Buchmann, J., Manulis, M. (eds.) PKC 2012. LNCS, vol. 7293, pp. 190–205. Springer, Heidelberg (2012)
13. Shamir, A.: Identity-Based Cryptosystems and Signature Schemes. In: Blakely, G.R., Chaum, D. (eds.) Advances in Cryptology - CRYPT0 1984. LNCS, vol. 196, pp. 47–53. Springer, Heidelberg (1985)
14. Sakumoto, K., Shirai, T., Hiwatari, H.: Public-Key Identification Schemes Based on Multivariate Quadratic Polynomials. In: Rogaway, P. (ed.) CRYPTO 2011. LNCS, vol. 6841, pp. 706–723. Springer, Heidelberg (2011)
15. Samardjiska, S., Gligoroski, D.: Identity-Based Identification Schemes Using Left Multivariate Quasigroups. In: NISK 2011, Tapir, pp. 19–30 (2011)
16. Samardjiska, S., Gligoroski, D.: Left MQQs whose left parastrophe is also quadratic. Commentat. Mathematicae Un. Carolinae. 53(3), 397–421 (2012)
17. Susilo, W., Baek, J.: On the Security of the Identity-based Encryption based on DHIES from ASIACCS 2010. In: ASIACCS 2011, pp. 376–380. ACM (2011)
18. Waters, B.: Efficient identity-based encryption without random oracles. In: Cramer, R. (ed.) EUROCRYPT 2005. LNCS, vol. 3494, pp. 114–127. Springer, Heidelberg (2005)
19. Wolf, C., Preneel, B.: MQ*-IP: An Identity-based Identification Scheme without Number-theoretic Assumptions. Cryptology ePrint Archive, 2010/087 (2010)

Optimal Cache Replacement Policy for Matrix Multiplication

Nenad Anchev, Marjan Gusev, Sasko Ristov, and Blagoj Atanasovski

Ss. Cyril and Methodious University, Faculty of Computer Science and Engineering, Rugjer Boshkovikj 16, 1000 Skoipje, Macedonia
nenad_ancev@hotmail.com, {marjan.gushev,sashko.ristov}@finki.ukim.mk, blagoj.atanasovski@gmail.com

Abstract. Matrix multiplication is compute intensive, memory demand and cache intensive algorithm. It performs $O(N^3)$ operations, demands storing $O(N^2)$ elements and accesses $O(N)$ times each element, where N is the matrix size. Implementation of cache intensive algorithms can achieve speedups due to cache memory behavior if the algorithms frequently reuse the data. A block replacement of already stored elements is initiated when the requirements exceed the limitations of cache size. Cache misses are produced when data of replaced block is to be used again. Several cache replacement policies are proposed to speedup different program executions.

In this paper we analyze and compare two most implemented cache replacement policies First-In-First-Out (FIFO) and Least-Recently-Used (LRU). The results of the experiments show the optimal solutions for sequential and parallel dense matrix multiplication algorithm. As the number of operations does not depend on cache replacement policy, we define and determine the average memory cycles per instruction that the algorithm performs, since it mostly affects the performance.

Keywords: FIFO, HPC, LRU, Performance, Speedup.

1 Introduction

CPU runs a particular program by accessing data from the memory, executing basic operations addition or multiplication and storing the results in the memory. The main bottleneck in the process is the data access in memory which is approximately up to 1000 times slower than floating point operation execution [7]. Introducing memory hierarchy based on caches in CPU speeds up the execution of programs that reuse the same data, i.e. cache intensive algorithms. This paper focuses on dense matrix multiplication algorithm.

Most modern multiprocessors use three layer n-way associative cache memory to speedup main memory access. The cache size grows but the access time and miss penalty rise going from the lowest L1 to L3 cache. The effect of exploiting last level shared cache affinity is considerable, due its sharing among multiple threads and high reloading cost [14]. Intel introduces Intel Smart Cache into their newest CPUs to improve their performance [8].

S. Markovski and M. Gusev (Eds.): *ICT Innovations 2012*, AISC 207, pp. 71–80.
DOI: 10.1007/978-3-642-37169-1_7

However, cache memory speeds up the execution only when matrix data fits in the cache. When the problem size exceeds a particular cache size then cache misses are being generated and the performance decreases. Two drawbacks appear in this case [5]. If a cache block is replaced after a particular matrix element is accessed then the next access to this element will generate a cache miss. The second drawback refers to a situation when there is an access to another matrix element from the same cache block (line) but in meantime the block was replaced. Inefficient usage of cache is possible if matrix elements map onto a small group of same cache sets and initiate a significant number of cache misses due to cache associativity [17].

Cache replacing policy also impacts the algorithm performance, i.e. which cache line will be replaced from some cache set to place the requested data from some of next level caches or main memory into the cache that generated miss. Three basic cache replacement policies are suggested: Random, Least-Recently-Used (LRU) and First-In-First-Out (FIFO) [7]. Several improvements are proposed for LRU. LRU Insertion Policy (LIP) places the incoming line in the LRU position instead of the MRU [15]. The authors in [2] propose even better replacement policy, i.e. a Score-Based Memory Cache Replacement Policy. Adaptive Subset Based Replacement Policy for High Performance Caching is proposed in [6], i.e. to divide one cache set into multiple subsets and victims should be always taken from one active subset when cache miss occurs. Map-based adaptive insertion policy estimates the data reuse possibility on the basis of data reuse history [9]. The authors in [11] propose Dueling CLOCK cache replacement policy that has low overhead, captures recency information in memory accesses and exploits the frequency pattern of memory accesses compared to LRU. A new replacement algorithm PBR_L1 is proposed for merge sort which is better than FIFO and LRU [3]. The authors in [12] propose LRU-PEA replacement policy that enables more intelligent replacement decisions due to the fact that some types of data are less commonly accessed depending on which bank they reside in. The authors in [10] propose cache replacement using re-reference interval prediction to outperform LRU in many real world game, server, and multimedia applications. However, improving replacing policies requires either additional hardware or modification of existing. PAC-PLRU replacing policy utilizes the prediction results generated by the existing stride prefetcher and prevents these predicted cache blocks from being replaced in the near future [18].

In this paper we focus on two most common cache replacement policies, LRU and FIFO and their impact on dense matrix multiplication performance. A tool for automatic computation of relative competitive ratios for a large class of replacement policies, including LRU, FIFO, and PLRU can be found in [16]. LRU has a gap of 50% optimal replacement policies [1]. We realize series of experiments for sequential and parallel execution of dense matrix multiplication on different hardware infrastructure with LRU and FIFO cache replacement policies. The rest of the paper is organized as follows. Section 2 describes the dense matrix multiplication algorithm and its parallel implementation and Section 3 the hardware infrastructure and runtime environment for the experiments. In

Section 4 we present the results of the experiments and analyze which replacement policy is better for both for sequential and parallel dense matrix multiplication algorithm. Section 5 is devoted to conclusion and future work.

2 The Algorithm

We choose squared matrices with dimension N. For all $i, j = 0, 1, \ldots, N-1$ the result product matrix $C_{N \cdot N} = [c_{ij}]$ is defined in (1) by multiplying the multiplier matrix $A_{N \cdot N} = [a_{ij}]$ and the multiplicand matrix $B_{N \cdot N} = [b_{ij}]$. More details about algorithm complexity are given in [4]. To exploit maximum performance for parallel execution on P processing elements we use dynamic schedule directive of OpenMP with $chunk = 1$ [13].

$$c_{ij} = \sum_{k=0}^{N-1} a_{ik} \cdot b_{kj} \tag{1}$$

2.1 Algorithm Definitions and Analysis

In this section we analyze the algorithm execution. For better presentation and analysis we use CPU clock cycles instead of execution time. Relation (2) derives the total execution clock cycles (TC) as a sum of clock cycles needed for operation execution (CC) and clock cycles needed for accessing the matrix elements (MC) [7].

$$TC = CC + MC \tag{2}$$

CC does not depend neither of CPU architecture nor cache size, associativity and replacement policy, but directly depends of matrix size N. CPU executes N^3 sums and N^3 multiplications or total $2 \cdot N^3$ floating points operations. MC is more interesting for analysis. It depends on matrix size N, but also on cache size, associativity and replacement policy.

More important parameters for analysis are the average values of TC, MC and CC defined in the next three definitions.

Definition 1 (Average Total Cycles Per Instruction). *$CPI_T(N)$ for particular matrix size N is defined as a ratio of total number of clock cycles and total number of instructions given in (3).*

$$CPI_T(N) = \frac{TC}{2 \cdot N^3} \tag{3}$$

Definition 2 (Average Memory Cycles Per Instruction). *$CPI_M(N)$ for particular matrix size N is defined as a ratio of total number of memory cycles and total number of instructions given in (4).*

$$CPI_M(N) = \frac{MC}{2 \cdot N^3} \tag{4}$$

Definition 3 (Average Calculation Cycles Per Instruction). $CPI_C(N)$ *for particular matrix size N is defined as a ratio of total number of calculation cycles CC and total number of instructions given in (5).*

$$CPI_C(N) = \frac{CC}{2 \cdot N^3} \tag{5}$$

We measure speed, TC, CC, MC for each matrix size, number of cores in defined testing environments. We calculate $CPI_T(N)$, $CPI_M(N)$ and $CPI_C(N)$ and analyze the distribution of $CPI_M(N)$ in $CPI_T(N)$. All the experiments are realized both for sequential and parallel execution.

2.2 Measurement Methodology

This section describes how we measure TC, CC, MC to calculate $CPI_T(N)$, $CPI_M(N)$ and $CPI_C(N)$. We measure total execution time TT for each experiment with algorithm described in (1) and then calculate TC as defined in [7] and calculate $CPI_T(N)$ using (3).

To measure MC we developed another algorithm defined in (6). This algorithm performs the same floating point operations on constant operands and writes the results in matrix C elements. The difference is that it does not read from memory or some cache the elements of matrices A and B.

$$c_{ij} = \sum_{k=0}^{N-1} a \cdot b \tag{6}$$

Executing this algorithm we measure its execution time CT for each experiment and then calculate the difference from TC and CT. Then we calculate MC as defined in [7] using CPU speed for particular processor and calculate $CPI_M(N)$ using (4).

CC and $CPI_C(N)$ are calculated as defined in (7) and (5).

$$MC = TC - CC \tag{7}$$

3 The Testing Environment

Two servers with different CPUs with different cache replacement policies are used: FIFO and LRU. Both servers are installed with Linux Ubuntu 10.10. C++ with OpenMP support is used for parallel execution.

FIFO testing hardware infrastructure consists of one Intel(R) Xeon(R) CPU X5680 @ 3.33GHz and 24GB RAM. It has 6 cores, each with 32 KB 8-way set associative L1 and 256 KB 8-way set associative L2 cache. All 6 cores share 12 MB 16-way set associative L3 cache. Each experiment is executed using different matrix size N for different number of cores from 1 to 6. Tests are performed by unit incremental steps for matrix size and number of cores.

LRU testing hardware infrastructure consists of one CPU Quad-Core AMD Phenom(tm) 9550. It has 4 cores, each with 64 KB 2-way set associative L1 and 512 KB 16-way set associative L2 cache. All 4 cores share 2 MB 32-way set associative L3 cache. Each experiment is executed using different matrix size N on different number of cores from 1 to 4. Tests are performed by unit incremental steps for matrix size and number of cores.

4 Results of the Experiments

This section presents and compares the results of the experiments on two CPUs with different replacement policies.

4.1 Results for CPU with FIFO Cache Replacement Policy

Figure 1 depicts the results of measured speed. $SpeedT(i)$ denotes the speed in gigaFLOPS for algorithm execution on i cores where $i = 1, 2, ..., 6$.

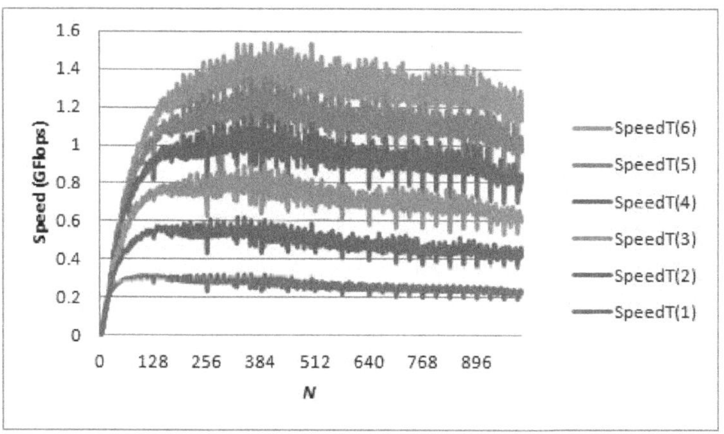

Fig. 1. Speed for execution on FIFO CPU

$CPI_T(N)$ presents another perspective of the experiment. Figure 2 depicts the results for algorithm execution on $1, 2, ..., 6$ cores for each matrix size $128 < N < 1000$. We can conclude that executing the dense matrix multiplication algorithm on more cores needs more average cycles per core for each matrix size N. Also, the speed decreases by increasing the matrix size N.

The next experiment analyzes the decomposition of the average total cycles per instruction on average calculation cycles per instructions and average memory cycles per instruction. Figure 3 depicts the decomposition of $CPI_T(N)$ on $CPI_M(N)$ and $CPI_C(N)$ for sequential execution. The left graph depicts the absolute decomposition of $CPI_T(N)$. The conclusion is that $CPI_C(N)$ is almost

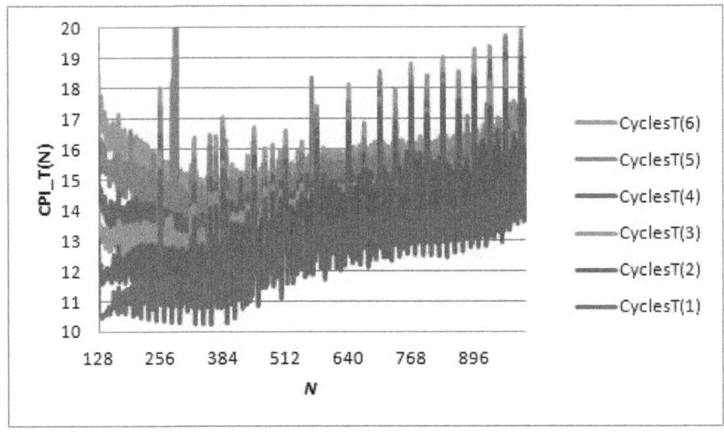

Fig. 2. $CPI_T(N)$ for execution on FIFO CPU

constant with average value of 4.93 cycles per instruction. More important is that $CPI_M(N)$ follows $CPI_T(N)$, i.e. $CPI_T(N)$ depends directly of average memory cycles per instruction. The right graph depicts the relative value of $CPI_M(N)$ to $CPI_T(N)$. $CPI_M(N)$ has a trend to equalize with $CPI_T(N)$ as N grows.

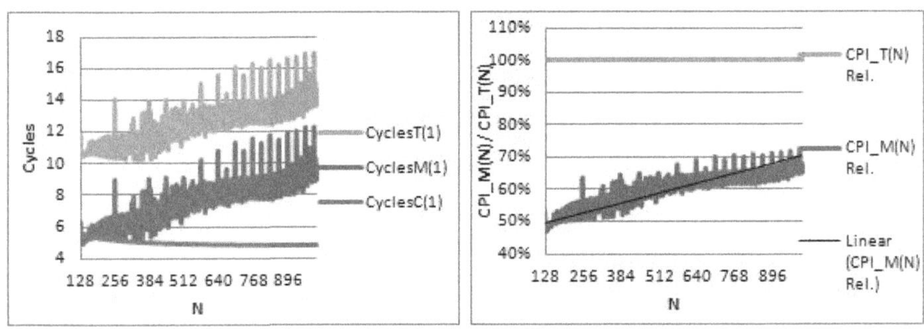

Fig. 3. Decomposed $CPI_T(N)$ for sequential execution on FIFO CPU, absolute (left) and relative (right)

4.2 Results for CPU with LRU Cache Replacement Policy

Figure 4 depicts the results of measured speed. $SpeedT(i)$ denotes the speed in gigaFLOPS for algorithm execution on i cores where $i = 1, 2, ..., 6$. We can conclude that there is a huge performance drawback after $N > 362$ which is entrance in the L_4 region, i.e. the region where elements of matrices A and B cannot be placed in L3 cache and thus producing L3 cache miss.

$CPI_T(N)$ presents better the information. Figure 5 depicts results for executions on 1, 2, 3 and 4 cores for each matrix size $128 < N < 1000$. We can see 2

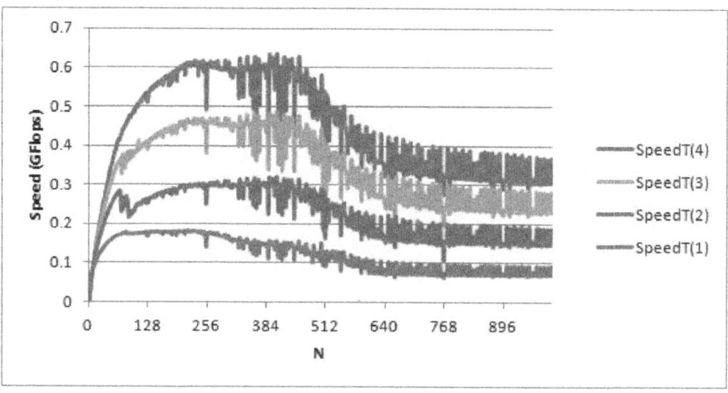

Fig. 4. Speed for execution on LRU CPU

regions, *Region 1* for $N < 362$ and *Region 2* for $N > 362$. The former presents the L_1 , L_2 and L_3 cache regions, i.e. low-cycle accesable regions where matrices can be stored completely in some of the caches. In this region sequential execution provides the worst $CPI_T(N)$ compared to parallel execution. The latter presents L_4 region, i.e. main memory region where matrices cannot be stored completely in L3 cache. In this region sequential execution provides the best $CPI_T(N)$ compared to parallel execution.

Figure 6 depicts the decomposition of $CPI_T(N)$ on $CPI_M(N)$ and $CPI_C(N)$ for sequential execution. The left graph depicts the absolute decomposition of $CPI_T(N)$. $CPI_C(N)$ is almost constant to the average value of 7.17 cycles per instruction. More important is that $CPI_M(N)$ follows $CPI_T(N)$, i.e. $CPI_T(N)$ depends directly of average memory cycles per instruction. The right graph depicts the relative value of $CPI_M(N)$ to $CPI_T(N)$. As depicted, $CPI_M(N)$ has a trend to equalize with $CPI_T(N)$ as N grows for $N \cdot (N+1) < 2MB$. This is the case when matrix $B_{N \cdot N}$ and one row od matrix $A_{1 \cdot N}$ can be placed in the L3 cache. $CPI_M(N)$ relative remains constant for greater N.

4.3 LRU and FIFO Cache Replacement Policy Comparison

In this section we compare the results between performance of FIFO and LRU cache replacement policies.

Speed Comparison. Comapring figures 1 and 4 we can conclude that both infrastructures have a region around entrance to L_4 region when the speed begins to fall down to a local maximum. The graphs show that the speed decrease is more emphasized in LRU rather than FIFO. However, it is because L_4 region in LRU begins for $N > 362$ and for FIFO CPU for $N > 886$. Therefore the real comparison should be the regions $N > 362$ on LRU CPU with $N > 886$ on FIFO CPU, which are the beginning of L_4 region.

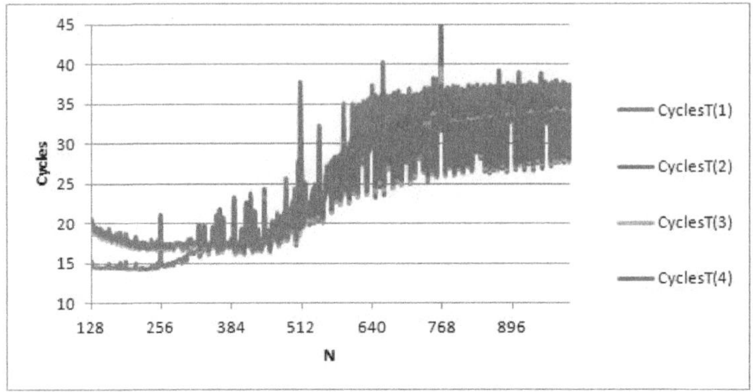

Fig. 5. $CPI_T(N)$ for execution on LRU CPU

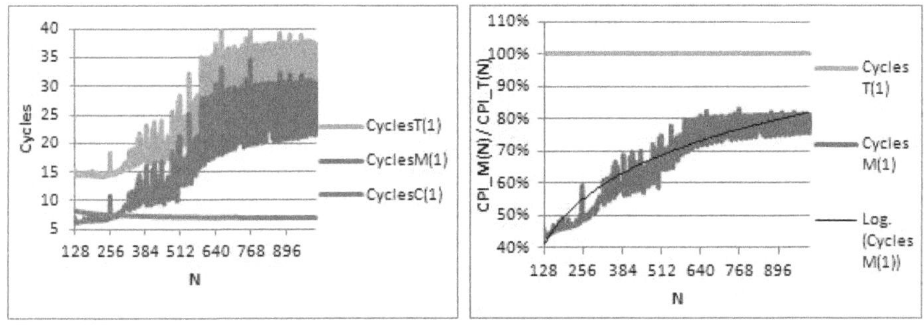

Fig. 6. Decomposed $CPI_T(N)$ for sequential execution on LRU CPU, absolute (left) and relative (right)

$CPI_T(N)$ Comparison. Comparing figures 2 and 5 we can conclude that both infrastructures have similar curves for $CPI_T(N)$ for particular region. The important conclusion is that FIFO CPU needs more cycles per core for each matrix size N regardless of cache region (dedicated or shared). However, the LRU CPU has different features. Sequential execution has the best $CPI_T(N)$ in dedicated per core L_1 and L_2 regions and parallel execution on greater number of cores in shared L_3 and L_4 regions.

$CPI_T(N)$ Decomposition Comparison. Comparing figures 3 and 6 (left) we can conclude that both infrastructures have similar curves for $CPI_T(N)$. The graphs show that $CPI_T(N)$ is greater in LRU than FIFO. However, the real comparison should be the regions $N > 362$ on LRU CPU with $N > 886$ on FIFO CPU as explained in the previous subsection. $CPI_M(N)$ is almost parallel compared to $CPI_T(N)$ for all matrix size N in both infrastructures. Also, the similar result is the fact that $CPI_C(N)$ is almost constant for each matrix size N for both CPUs.

$CPI_M(N)$ Comparison. Comparing figures 3 and 6 (right) we can conclude that $CPI_M(N)$ is relative more closer to $CPI_T(N)$ in LRU than FIFO. However, it is because L_4 region in LRU begins for $N > 362$ and for FIFO CPU for $N > 886$. Therefore the real comparison should be the regions $N > 362$ on LRU CPU with $N > 886$ on FIFO CPU, which are the beginning of L_4 region and the relative values in LRU CPU are better than FIFO CPU. LRU CPU has average of 59.77% in the region of $N = 362$ and FIFO CPU has average 65.84% in the region of $N = 886$.

5 Conclusion and Future Work

In this paper we determine that both cache replacement policies provide similar speed and average cycles per instruction $CPI_T(N)$ for sequential and parallel execution. However, the results show that LRU replacement policy provides best $CPI_T(N)$ for sequential execution in dedicated per core cache memory. Parallel execution provides the best $CPI_T(N)$ in shared memory LRU CPU, i.e. LRU produces greater speedup than FIFO and is more appropriate rather than FIFO cache replacement policy for dense matrix multiplication algorithm.

Our plan for future work is to analyze the performance of other cache replacement policies for sequential and parallel execution, as well as other compute intensive, memory demanding and cache intensive algorithms. With appropriate simulator we can compare different replacement policies with the same cache size, associativity and cache levels.

References

1. Al-Zoubi, H., Milenkovic, A., Milenkovic, M.: Performance evaluation of cache replacement policies for the spec cpu2000 benchmark suite. In: Proceedings of the 42nd Annual Southeast Regional Conference, ACM-SE 42, pp. 267–272. ACM, New York (2004)
2. Duong, N., Cammarota, R., Zhao, D., Kim, T., Veidenbaum, A.: SCORE: A Score-Based Memory Cache Replacement Policy. In: Emer, J. (ed.) JWAC 2010 - 1st JILP Worshop on Computer Architecture Competitions: Cache Replacement Championship, Saint Malo, France (2010)
3. Gupta, R., Tokekar, S.: Proficient pair of replacement algorithms on l1 and l2 cache for merge sort. J. of Computing 2(3), 171–175 (2010)
4. Gusev, M., Ristov, S.: Matrix multiplication performance analysis in virtualized shared memory multiprocessor. In: MIPRO, 2012 Proc. of the 35th International Convention, pp. 264–269. IEEE Conference Publications (2012)
5. Gusev, M., Ristov, S.: Performance gains and drawbacks using set associative cache. Journal of Next Generation Information Technology (JNIT) 3(3), 87–98 (2012)
6. He, L., Sun, Y., Zhang, C.: Adaptive Subset Based Replacement Policy for High Performance Caching. In: Emer, J. (ed.) JWAC 2010 - 1st JILP Worshop on Computer Architecture Competitions: Cache Replacement Championship, Saint Malo, France (2010)

7. Hennessy, J.L., Patterson, D.A.: Computer Architecture, 5th edn. A Quantitative Approach (2012)
8. Intel: Intel smart cache (May 2012), `http://www.intel.com/content/www/us/en/architecture-and-technology/intel-smart-cache.html`
9. Ishii, Y., Inaba, M., Hiraki, K.: Cache Replacement Policy Using Map-based Adaptive Insertion. In: Emer, J. (ed.) JWAC 2010 - 1st JILP Worshop on Computer Architecture Competitions: Cache Replacement Championship, Saint Malo, France (2010)
10. Jaleel, A., Theobald, K.B., Steely Jr., S.C., Emer, J.: High performance cache replacement using re-reference interval prediction (rrip). SIGARCH Comput. Archit. News 38(3), 60–71 (2010)
11. Janapsatya, A., Ignjatović, A., Peddersen, J., Parameswaran, S.: Dueling clock: adaptive cache replacement policy based on the clock algorithm. In: Proceedings of the Conference on Design, Automation and Test in Europe, DATE 2010, pp. 920–925 (2010)
12. Lira, J., Molina, C., González, A.: Lru-pea: a smart replacement policy for non-uniform cache architectures on chip multiprocessors. In: Proceedings of the 2009 IEEE International Conference on Computer Design, ICCD 2009, pp. 275–281. IEEE Press, Piscataway (2009)
13. OpenMP (2012), `https://computing.llnl.gov/tutorials/openMP/`
14. Pimple, M., Sathe, S.: Architecture aware programming on multi-core systems. International Journal of Advanced Computer Science and Applications (IJACSA) 2, 105–111 (2011)
15. Qureshi, M.K., Jaleel, A., Patt, Y.N., Steely, S.C., Emer, J.: Adaptive insertion policies for high performance caching. SIGARCH Comput. Archit. News 35(2), 381–391 (2007)
16. Reineke, J., Grund, D.: Relative competitive analysis of cache replacement policies. Sigplan Not. 43(7), 51–60 (2008)
17. Ristov, S., Gusev, M.: Achieving maximum performance for matrix multiplication using set associative cache. In: 2012 The 8th Int. Conf. on. Computing Technology and Information Management (ICCM 2012), vol. 2, pp. 542–547 (2012)
18. Zhang, K., Wang, Z., Chen, Y., Zhu, H., Sun, X.H.: Pac-plru: A cache replacement policy to salvage discarded predictions from hardware prefetchers. In: Proceedings of the 2011 11th IEEE/ACM International Symposium on Cluster, Cloud and Grid Computing, CCGRID 2011, pp. 265–274. IEEE Computer Society, Washington, DC (2011)

Multimodal Medical Image Retrieval

Ivan Kitanovski, Katarina Trojacanec, Ivica Dimitrovski, and Suzana Loskovska

Faculty of Computer Science and Engineering, Skopje, Macedonia
{ivan.kitanovski,katarina.trojachanec,ivica.dimitrovski,
suzana.loshkovska}@finki.ukim.mk

Abstract. Medical image retrieval is one of the crucial tasks in everyday medical practices. This paper investigates three forms of medical image retrieval: text, visual and multimodal retrieval. We investigate by evaluating different weighting models for text retrieval. In the case of the visual retrieval, we focused on extracting low-level features and examining their performance. For, the multimodal retrieval we used late fusion to combine the best text and visual results. We found that the choice of weighting model for text retrieval dramatically influences the outcome of the multimodal retrieval. The results from the text and visual retrieval are fused using linear combination, which is among the simplest and most frequently used methods. Our results clearly show that the fusion of text and visual retrieval with an appropriate fusion technique improves the retrieval performance.

Keywords: Information Retrieval, Medical Imaging, Content-based Image Retrieval, Medical Image Retrieval.

1 Introduction

The task of medical image retrieval consists of retrieving the most relevant images to a given query from a database of images [1]. Medical image retrieval from medical image databases does not aim to replace the physician by predicting the disease of a particular case but to assist him/her in diagnosis. By consulting the output of a medical image retrieval system, the physician can gain more confidence in his/her decision or even consider other possibilities.

There are two forms of medical image retrieval: text-based (textual) and content-based (visual) [1]. In text-based image retrieval images are usually manually annotated with keywords or a short caption, which describe their content, or in the case of medical images the keywords are related to modality of the image, the present body part, the disease or anomaly depicted. In the latter stage, the user provides textual queries and the retrieval is performed using traditional text retrieval techniques. In visual retrieval the images are represented using descriptors (automatically generated) which describe the visual content of the images. Descriptors are usually numerical by nature and are represented as vectors of numbers [2]. In the retrieval phase, the user provides visual queries (query images) and the retrieval is

S. Markovski and M. Gusev (Eds.): *ICT Innovations 2012*, AISC 207, pp. 81–89.
DOI: 10.1007/978-3-642-37169-1_8

performed by comparing descriptors of the query images to those of all images in the database [3].

Recently, multimodal image retrieval arises as an active research topic [4]. Multimodal image retrieval is the process of using both text-based and visual-based techniques for retrieval. In multimodal retrieval the user provides textual queries and query images and retrieval should provide an ordered set of images related to that complex query. There are many papers on this subject. The authors of [5] use late fusion to combine the results from text-based and visual-based retrieval. For the text-based retrieval, they use a bag-of-words representation on the image captions and DFR-BM25 model for the retrieval. In the visual-based retrieval, they describe the images using a low-level feature called CEDD, and the retrieval is performed using Img(Rummager). Then, using late fusion they combine the results. The most efficient strategy was a linear combination scheme. In [6], the authors use Late Semantic Combination for multimodal retrieval. They represent each image caption with a bag-of-words representation and for the retrieval they compare several models: Dirichlet Smoothed Language (DIR), Log-logistic Information-Based Model (LGD), Smoothed Power Law Information-Based Model (SPL) and Lexical Entailment based IR Model (AX). In the visual retrieval, the images are described with ORH and COL features and they use dot product as similarity measure for the visual retrieval. In [7] the authors use linear late fusion for multimodal retrieval. The text-based retrieval is performed using Lucene and the visual-based using Lira. The multimodal retrieval is performed by linear combination of scores from the text-based and visual-based retrieval and re-ranking. The authors of [8] first perform text-based retrieval and use those results as a filter for the visual-based retrieval. They use low-level texture and color features (CEDD) for the visual-based retrieval. Text-based retrieval is performed using a Lucene.

The paper is organized as follows: Section 2 presents the feature set which is used for the text-based and visual-based retrieval. Multimodal retrieval techniques are presented in section 3. The experimental setup is described in section 4. Section 5 provides the experimental results and a brief discussion. The concluding remarks are given in section 6.

2 Feature Set

Feature selection is a very important part in every information retrieval system, since it directly influences the performance. In this paper we analyze text features for the text-based retrieval and visual features for visual-based retrieval.

2.1 Textual Features

Text-based retrieval is needed when we have text describing the image content i.e. image caption. From the related work we can conclude that regarding the text-based retrieval a traditional bag-of-words representation can be used for the image caption.

The image captions are first pre-processed. Pre-processing includes stemming and stop words removal [5], which is needed so we can extract only the vital information.

The choice of a weighting model may crucially affect the retrieval hence we evaluated the positive and negative sides of different weighting models. We evaluated the following models: PL2 [9], BM25 [9], DFR-BM2 [9], BB2 [9] and one of the most popular TF-IDF [10]. We choose these models as one the most commonly used in real practice.

2.2 Visual Features

Related work shows that low-level features are typically used in content-based image retrieval systems, since they typically deal with large image databases. These features are called low-level because they have little or nothing with human perception. We decided to use the following features:

- Color and Edge Directivity Descriptor (CEDD) combines EHD feature [11] with color histogram information. This descriptor is limited in size to 54 bytes per image, which makes it appropriate for large image databases. Important attribute of the CEDD is the low computational power needed for its extraction, in comparison to the needs of the most MPEG-7 descriptors.
- Fuzzy Color and Texture Histogram (FCTH) is a fuzzy version of CEDD feature which contains fuzzy set of color and texture histogram [12]. FCTH contains results from the combination of three fuzzy systems including histogram, color and texture information. This feature is limited in size to 72 bytes per image and that makes it suitable for use in large image databases.
- The Scalable Fuzzy Brightness and Texture Directionality Histogram (BTDH), was originally created for representing radiology images [13]. BTDH is very similar to FCTH feature. The main difference from FCTH feature is using brightness instead of color histogram. It combines brightness and texture characteristics and their spatial distribution in one compact vector by using a two-unit fuzzy system. This feature does not contain color data, since it was meant for grayscale images.

3 Fusion Techniques for Multimodal Retrieval

Multimodal information retrieval refers to the task of using multiple media to perform a retrieval task. Multimodal retrieval is usually done by fusing multiple modalities. Fusing multiple modalities can improve the overall accuracy in the decision making process [14].

The fusion of multiple modalities can be performed at feature level or decision level. In fusion at feature level, also known as early fusion, various features extracted from the input data are combined in some fashion and then that newly created feature is sent as input to the module that performs the analysis task. In fusion at decision level, also known as late fusion, the analysis units first provide the local decisions that are obtained based on individual features. Afterwards a decision fusion unit combines local decisions to create a new fused decision vector which is analyzed to provide a final decision about the task. To utilize the merits of both approaches, researchers have attempted to create hybrid techniques which are a combination of both feature and decision level techniques.

Related work shows that the late fusion strategy is frequently used. Late fusion has many advantages over early fusion. The decisions usually have the same representation. For instance, the result of both text-based and visual-based retrieval is an ordered list of images. Hence, the implementation of fusion techniques becomes much easier. Furthermore, late fusion allows for modularity and scalability in terms of modalities used in the fusion process, which is quite difficult to achieve with early fusion techniques. Additionally, late fusion allows us to use the optimal analyzing methods for each modality separately which cannot be done with early fusion techniques.

Because of these merits we used late fusion strategy in our experiments. We turned to Linear Weighed Fusion strategy, one of the simplest and most widely used methods. We applied this strategy to results obtained from the separate text-based and visual-based retrievals. Each, retrieval contains an ordered list of images with computed similarity scores. The weighted average function is applied by multiplying each individual similarity with a weight value. The weight assignment to individual scores defines the importance of each modality in the decision making process. If a modality has a high weight it will have significant impact on the final results and vice versa. In this way we can control the influence of individual modalities.

4 Experimental Setup

The experiment consists of three kinds of retrieval: text-based, visual-based and multimodal retrieval. We use the text-based and visual-based retrieval to find the best weighting model and descriptor, which we latter use in the multimodal retrieval. The goal of the experiments is to determine whether multimodal retrieval can improve the overall retrieval performance.

The data for the experiments is provided from the collection of the ImageCLEF 2011 medical task [1]. The collection contains textual and visual information. It consists of 230088 images (Fig.1 shows sample images), each described with a short text (image caption). The queries which we used for testing are the same which were provided for the medical retrieval task. Participants were given a set of 30 textual queries with 2-3 sample images for each query. The queries are classified into textual, visual or mixed (multimodal) based on the data and techniques used. For the text-based retrieval, we only use the text queries. On the other hand, for the visual-based retrieval we use the images provided for each query. Finally, for the multimodal retrieval we use both text and image data provided for every query. It is important to note that the collection also contains ground truth data, so that proper evaluation can be performed.

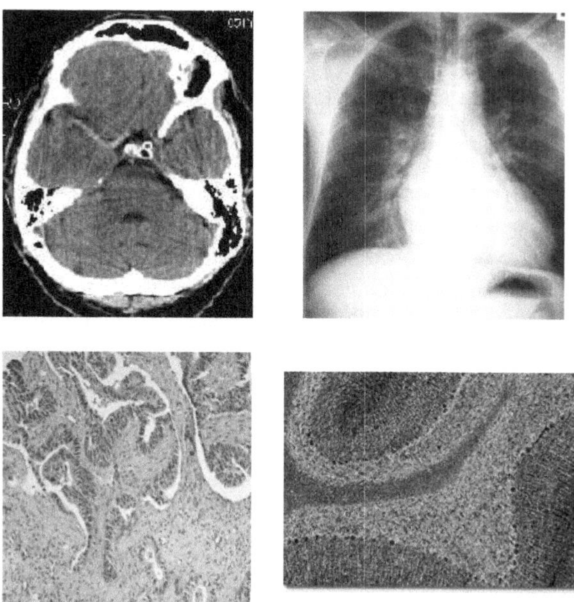

Fig. 1. Sample images from the ImageCLEF 2011 medical images collection

The text-based retrieval was performed using Terrier IR Platform [15], open source search engine written in Java which is developed at School of Computer Science, University of Glasgow. For stemming we used Porter stemmer [16] for English, since the image captions and text queries are in English. Terrier also has a predefined stop words list, which we use in the preprocessing stage. All weighting models which we analyze are integrated in Terrier.

The visual-based retrieval was performed with the aid of the Img(Rummager) application [17], developed in the Automatic Control Systems & Robotics Laboratory at the Democritus University of Thrace-Greece. CEDD, FCTH and BDTH features are implemented in the application. The retrieval stage greatly relies on the distance/similarity function used to quantitatively compare the images. We compute the similarity score based on Tanimoto distance [11], since it is one most frequently used methods for the visual features which we use to describe the images. Since, there were multiple images per query we used an averaging technique in this stage [5]. Here we calculated a mean descriptor from all images in a query, thus creating one new feature vector which will be passed as a query.

The multimodal retrieval is performed using late fusion strategy. In this stage we pick the best weighting model for text-based retrieval and the best performing descriptor for visual-based retrieval. Then, we combine the results from the separate retrievals using linear combination. The formula by which we calculated the score for each image in the retrieval is the following:

$$(text\_score * w_1 + visual\_score * w_2) / 100 = score \quad (1)$$

After comparing different studies [18] and experimenting with various parameters we determined to multiply the text score with 85 and the visual score with 15, thus giving a greater influence to the text-based component.

Before we combine the score we need to normalize them to get more valid and accurate results since different modalities calculate different ranges of values in the similarity score. Here we apply the most common used method for normalization i.e. Min-Max normalization [19]. This normalization ensures that the values of the scores are in the range from 0 to 1. The lowest value is set to 0 and the highest value is set to 1. This allows us to compare values that are measured using different scales. After normalization takes place, we turn to linear combination of the modified retrieval scores.

5 Results and Discussion

To evaluate the retrieval results thoroughly we calculate five different statistics: MAP, P10, P20, Rprec and the total number of retrieved relevant images (documents) [20].

The results from the text-based retrieval are presented in Table 1. The results show that BM25 model has superior MAP of 0.2144 over other models. It is interesting to note that BB2 has a 0.3700 P10 precision which is better by 0.0067 from BM25. We can see that all other statistics show the superiority of BM25. This is to be expected according to the research presented in [21]. Although its advantage over other models is not big, we decided to use this model for the multimodal retrieval experiments.

Table 2 presents the results from the visual-based retrieval. We can see that CEDD feature performs best in terms of MAP with a value of 0.0142. This is expected since previous experience on ImageCLEF shows that CEDD is among the best features for describing medical images, especially in large databases [17]. We can note that the MAP of visual-based retrieval is not satisfactory, which is also expected according to the previous experiences in ImageCLEF [1]. Additionally, we can note the number of retrieved relevant images is three times than in text-based retrieval. This is why we make the influence of the text-based retrieval more significant than the visual-based in the multimodal retrieval.

Multimodal retrieval was performed using the best weighting model in the text-based retrieval and the best performing feature in the visual retrieval, which are BM25 and CEDD, accordingly. Multimodal retrieval at the worst scenario should provide the same results as the text-based retrieval. Table 3 shows the results from the conducted experiments. For the multimodal retrieval, we made three types of experiment to assess the change in retrieval performance. First, we make linear combination of the text-based and visual-based retrieval. The second approach slightly modifies the text-based retrieval with query expansion, since the text-based retrieval has the crucial impact on final result. The third approach uses query expansion and word weighting. This approach assigns weights to special words, in our case image modalities (i.e. MRI, CT, X-RAY etc.). We added a weight of 2.5 to these words using query language of Terrier. The results show that there is an improvement of the retrieval compared to text-based retrieval in every multimodal experiment. The first (mixed) approach has a MAP of 0.2148 which is a slight improvement over the text-based retrieval. The third approach (mixed + ww) provides an additional improvement over text-based retrieval with a MAP of 0.2232 and is the best overall retrieval method.

Table 1. Comparison of weighting models in text-based retrieval

Model	*MAP*	*P10*	*P20*	*Rprec*	*# of rel. docs*
BB2	0.2059	0.3700	0.3100	0.2425	1472
BM25	0.2144	0.3633	0.3200	0.2449	1504
DFR-BM25	0.2054	0.3533	0.2967	0.2426	1494
PL2	0.1970	0.3533	0.2967	0.2413	1474
TF-IDF	0.2048	0.3533	0.3033	0.2398	1482

Table 2. Comparison of features in visual-based retrieval

Feature	*MAP*	*P10*	*P20*	*Rprec*	*# of rel. docs*
CEDD	0.0142	0.0867	0.0733	0.0401	552
FCTH	0.0134	0.0633	0.0483	0.0342	621
BDTH	0.0053	0.0419	0.0372	0.0216	217

Table 3. Results of the multimodal retrieval experiments

Mode	*MAP*	*P10*	*P20*	*Rprec*	*# of rel. docs*
mixed	0.2148	0.3600	0.3233	0.2579	1531
mixed + qe	0.2179	0.3833	0.3433	0.2577	1483
mixed + ww	0.2232	0.3933	0.3467	0.2568	1458

6 Conclusion

In this paper, we investigated the performance of five weighting models for text-based retrieval and three low-level features for visual-based retrieval. The results clearly show that text-based retrieval is superior compared to visual retrieval. The best results in the text-base retrieval were obtained using the BM25 weighting model (0.2144) and in the case of the visual-based retrieval the best performance was achieved in the case of CEDD feature (0.0142).

Additionally, we investigated in late fusion for multimodal retrieval. We used linear combination for late fusion of the text-based and visual-based retrieval results. The obtained results show that by combining the two modalities the overall retrieval performance can be improved, with a MAP of 0.2232.

Medical image retrieval is a crucial task which can aid the work of medical practitioners. It is a very complex which can be improved in many aspects from improving current weighting models to developing or modifying features to describe the image content and creating different techniques to combine these to modalities.

References

1. Kalpathy–Cramer, J., Muller, H., Bedrick, S., Eggel, I., de Herrera, A.G.S., Tsikrika, T.: Overview of the CLEF 2011 medical image classification and retrieval tasks (2011)
2. Sonka, M., Hlavac, V., Boyle, R., et al.: Image processing, analysis, and machine vision, vol. 2. PWS publishing Pacific Grove, CA (1999)
3. Deb, S., Zhang, Y.: An overview of content-based image retrieval techniques. In: 18th International Conference on Advanced Information Networking and Applications, vol. 1, pp. 59–64 (2004)
4. Müller, H., Kalpathy–Cramer, J., Eggel, I., Bedrick, S., Radhouani, S., Bakke, B., Kahn Jr., C.E., Hersh, W.: Overview of the CLEF 2009 medical image retrieval track. In: Peters, C., Caputo, B., Gonzalo, J., Jones, G.J.F., Kalpathy-Cramer, J., Müller, H., Tsikrika, T., et al. (eds.) CLEF 2009. LNCS, vol. 6242, pp. 72–84. Springer, Heidelberg (2010)
5. Alpkocak, A., Ozturkmenoglu, O., Berber, T., Vahid, A.H., Hamed, R.G.: DEMIR at ImageCLEFMed 2011: Evaluation of Fusion Techniques for Multimodal Content-based Medical Image Retrieval. In: 12th Workshop of the Cross-Language Evaluation Forum (CLEF), Amsterdam, Netherlands (2011)
6. Csurka, G., Clinchant, S., Jacquet, G.: XRCE's Participation at Medical Image Modality Classification and Ad-hoc Retrieval Tasks of ImageCLEF (2011)
7. Gkoufas, Y., Morou, A., Kalamboukis, T.: IPL at ImageCLEF 2011 Medical Retrieval Task. Working Notes of CLEF (2011)
8. Castellanos, A., Benavent, X., Benavent, J., Garcia-Serrano, A.: UNED-UV at Medical Retrieval Task of ImageCLEF (2011)
9. Amati, G., Van Rijsbergen, C.J.: Probabilistic models of information retrieval based on measuring the divergence from randomness. ACM Transactions on Information Systems (TOIS) 20, 357–389 (2002)
10. Hiemstra, D.: A probabilistic justification for using tf-idf term weighting in information retrieval. International Journal on Digital Libraries 3, 131–139 (2000)
11. Chatzichristofis, S.A., Boutalis, Y.S.: CEDD: Color and edge directivity descriptor: A compact descriptor for image indexing and retrieval. In: Gasteratos, A., Vincze, M., Tsotsos, J.K. (eds.) ICVS 2008. LNCS, vol. 5008, pp. 312–322. Springer, Heidelberg (2008)
12. Chatzichristofis, S.A., Boutalis, Y.S.: Fcth: Fuzzy color and texture histogram-a low level feature for accurate image retrieval. In: Ninth International Workshop on Image Analysis for Multimedia Interactive Services, pp. 191–196 (2008)
13. Chatzichristofis, S.A., Boutalis, Y.S.: Content based radiology image retrieval using a fuzzy rule based scalable composite descriptor. Multimedia Tools and Applications 46, 493–519 (2010)
14. Atrey, P.K., Hossain, M.A., El Saddik, A., Kankanhalli, M.S.: Multimodal fusion for multimedia analysis: a survey. Multimedia Systems 16, 345–379 (2010)
15. Ounis, I., Amati, G., Plachouras, V., He, B., Macdonald, C., Johnson, D.: Terrier information retrieval platform. In: Losada, D.E., Fernández-Luna, J.M. (eds.) ECIR 2005. LNCS, vol. 3408, pp. 517–519. Springer, Heidelberg (2005)

16. Porter, M.F.: An algorithm for suffix stripping (1980)
17. Chatzichristofis, S.A., Boutalis, Y.S., Lux, M.: Img (rummager): An interactive content based image retrieval system. In: Second International Workshop on Similarity Search and Applications, pp. 151–153 (2009)
18. Croft, W.B.: Combining approaches to information retrieval. In: Advances in Information Retrieval, pp. 1–36 (2002)
19. Jain, A., Nandakumar, K., Ross, A.: Score normalization in multimodal biometric systems. Pattern Recognition 38, 2270–2285 (2005)
20. Manning, C.D., Raghavan, P., Schutze, H.: Introduction to information retrieval. Cambridge University Press, Cambridge (2008)
21. He, B., Ounis, I.: Term frequency normalisation tuning for BM25 and DFR models. In: Losada, D.E., Fernández-Luna, J.M. (eds.) ECIR 2005. LNCS, vol. 3408, pp. 200–214. Springer, Heidelberg (2005)

Modeling Lamarckian Evolution: From Structured Genome to a Brain-Like System

Liljana Bozinovska[1] and Nevena Ackovska[2]

[1] Department of Biological Sciences and Physical Sciences
South Carolina State University, USA
lbozinov@scsu.edu
[2] Faculty of Computer Science and Engineering
University Sts Cyril and Methodius, Macedonia
nevena.ackovska@finki.ukim.mk

Abstract. The paper addresses development of a brain-like system based on Lamarckian view toward evolution. It describes a development of an artificial brain, from an artificial genome, through a neural stem cell. In the presented design a modulon level of genetic hierarchical control is used. In order to evolve such a system, two environments are considered, genetic and behavioral. The genome comes from the genetic environment, evolves into an artificial brain, and then updates the memory through interaction with the behavioral environment. The updated genome can then be sent back to the genetic environment. The memory units of the artificial brain are synaptic weights which in this paper represent achievement motivations of the agent, so they are updated in relation to particular achievement. A simulation of the process of learning by updating achievement motivations in the behavioral environment is also shown.

Keywords: Lamarckian evolution, hierarchical genome, behavioral and genetic environment, neural stem cell, artificial brain, achievement motivation, limbic system.

1 Introduction

Development of evolvable systems and evolutionary biology is of great interest to contemporary research [1-5]. However, evolution of systems is not only a characteristic of the living systems. In the recent years one can observe great interest in the evolution of software systems and artificial intelligent systems [6-10].

In this work we focus a development (morphogenesis) of a *brain-like system* capable of *achievement motivation* [11] learning, starting from a structured genome. This work is a research effort in the area of neuroevolution. Neural networks are currently standard topic in Artificial Intelligence textbooks (e.g. [12], [13]). However, among the new topics in neural networks research is the issue of genetic evolution of neural nets [14]. Examples of works on evolvable systems based on neural networks are [15-17]. Integrated view toward emotion and cognition using neural networks is presented in [7].

S. Markovski and M. Gusev (Eds.): *ICT Innovations 2012*, AISC 207, pp. 91–100.
DOI: 10.1007/978-3-642-37169-1_9

The work presented here follows development of an artificial neural network presented in [18]: First is development from genome to a neural network. The concept of *genetic environment* is used for genomes and genotypes. Second is rewiring of a neural network based on axon growing (e.g. [19]), and its relation to reconfigurable computing systems (e.g. [20]). Final step is adjustment of weights in interaction with the *behavioral environment*. The adjustment of weights has been accepted as dominant approach toward neural plasticity (e.g. [21]).

The neural network that we consider in this research is the Crossbar Adaptive Array (CAA) neural architecture, which introduced emotions in learning mechanisms [18], [22-24]. Although early works on integrating emotions and feelings in building neural networks appeared in early 1980-ties [18] it was not until middle 1990's that emotions and feelings become of major interest in approaches toward artificial intelligence systems [25-28]. The emotions in artificial neural networks were related to the hormonal system [29], [30]. Currently, emotion is widely addressed issue in artificial systems (e.g. [31-33]) and is considered in several influential books [34], [35]. In this paper we view the CAA system as limbic system, since it computes both behaviors and emotions in the same neural structure. In a sense it is a brain which appeared in the *intelligence evolution* before development of a neocortex.

The paper is structured in several sections. The next section gives an overview of the concepts of evolution and heredity. Afterwards, the model of evolvable brain-like system based on a structured genome will be presented. It exhibits three distinct features: 1) it is capable of building neural architecture from a structured genome via a neural stem cell, 2) the architecture is capable of learning using emotion computation principle, and 3) it is capable of exporting a genome back in the genetic environment with changes obtained from the behavioral environment. Next, a simulation using this model will be presented, which shows how such architecture can learn in the environment it lives in. Further we present Darwinian and Lamarckian way of exporting genomes to the environment. The final section is conclusion.

2 Evolution, Heredity and Next Generation

The dominant theory about evolution of the living systems on Earth today is the Darwinian Theory [36]. According to that theory, after a rather random change which appears in a genotype in the genetic environment, a phenotype organism is created and tested for survivability in the behavioral environment, and if passed the test the new organism is established on Earth.

Another, earlier point of view was stated by Lamarck [37]. He is best known for his Theory of Inheritance of Acquired Characteristics, first presented in 1801: "*If an organism changes during life in order to adapt to its environment, those changes are passed on to its offspring*". He believed that the changes to the living system, enforced from the environment, can be passed to the *next generation*. So the change is not random, it is rather learned in some way from the environment.

Today there are applications which show progress in both software and artificial intelligence evolution using Lamarckian approach. Examples are found in computer games [38] and in computing weights in learning mechanism [39].

3 Evolvable System Based on Structured Genome

In this section we will describe the genesis of a brain-like neural structure starting from its genome. Studying genetic systems from the standpoint of flexible manufacturing systems [40], [41] as well as from the standpoint of operating systems [42], [43] we realized that a hierarchical control structure is needed to drive evolution of a neural network. We use three level genetic hierarchy for controlling group of genes: operon, regulon, and modulon. [44].

3.1 The Starting Genome

Here we describe a model of an initial chromosome that is capable of producing an artificial brain, consisting of a crossbar neural network. The genetic structure capable of evolving such architecture is named GALA modulon (modulon for Generic Architecture for Learning Agents) (Fig. 1)

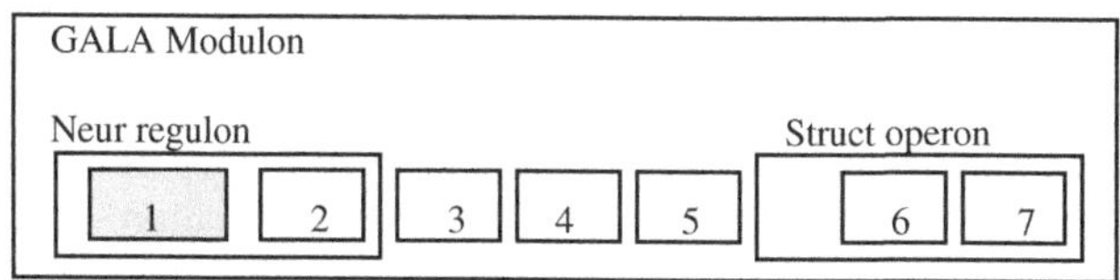

Fig. 1. A single chromosome genome, modulon-level regulated

We consider a single chromosome modulon containing the following co-regulated genetic structures: *Neur*-regulon, *Learn*-gene, *Eval*-gene, *Clone*-gene, and *Struct*-operon. The genes included in this structure could be value encoding genes or function encoding genes.

Neur-regulon consists of an operon and a gene, denoted as *Env* operon (marked as 1 in Fig. 1) and *Equipot* gene (denoted as 2 in Fig. 1). The *Neur*-regulon can be represented as a structure that expresses m+1 genes, m of which are represented by the *Env* operon and encode m synaptic weights of the phenotype neural cell, and a separate gene encoding the neural threshold which is assumed to be a function of the encoded weights. The *Env* operon is the environment encoding operon. *Env*-operon can be denoted as $[w_{01}, w_{02}, .., w_{0j}, .. ,w_{0m}]$. The value encoding genes $\{w_{0j}\}$ of this operon encode a motivation-emotion system of the phenotype agent. The *Equipot* gene can be denoted as $[f(w_{01}, w_{02}, .., w_{0j}, .. ,w_{0m})]$ since it computes a function of the genes in the *Env* operon.

Eval-gene (numerated 3 in Fig. 1) encodes a function, $[\boldsymbol{g}(w_{10}, w_{20}, .. w_{i0}, .. w_{n0})]$. The number n is specified by the GALA modulon, as the number of copies produced by the co-regulated *Neur*-regulon.

Learn-gene (denoted as 4 in Fig. 1) encodes the synaptic plasticity rule for the synaptic weights, can be denoted as $[\boldsymbol{h}(w, u, x, r)]$, where w is a synaptic weight, x is the synaptic input, *u* and *r* are advice input and performance evaluation input, respectively.

It is assumed that the functions ***f( )***, ***g( )***, and ***h( )*** are only specified in the regulon, the regulon does not actually compute them.

Clone-gene (denoted as 5 in Fig. 1) controls the number of copies of the neural stem cells that will be produced the *Neur*-regulon and *Eval*-gene.

Struct-operon contains all the structural information as how to assemble the whole GALA structure. Among others, it contains two genes, *Behav*-gene and *Emot*-gene. The *Behav*-gene (denoted as 6 in Figure 1) encodes a neuron that selects the output behavior, and the *Emot*-gene (denoted as 7 in Figure 1) selects the global emotion of the whole system. The *Struct*-operon contains additional genes that decide, depending on the environment conditions, whether some genes will be activated or not. Not all the genes are expressed in the organisms that could be generated by this genome.

3.2 Obtaining the First Neural Stem Cell

The *Struct*-operon and the *Neur*-regulon produce a neural stem cell in the process of replicating the chromosome. The neural stem cell can replicate itself, but replication produces two different cells: one is clone of itself and the other is a neuronal progenitor cell, which can only differentiate further toward neural cell [45]. This neuronal progenitor cell obtained after the first replication has index 1. The expression of the *Learn*-gene defines the synaptic plasticity rule. Besides the situation input s_j, the neuron receives an advice input, u_1, and another, performance evaluation input r_1. The synaptic plasticity mechanism is defined as

$$w'_{1j} = w_{1j} + r_1 s_j u_1$$

which means that the neuron learns by a three-factor correlation rule. This means that the past performance, the current situation and the advice for future behavior are needed in order a neuron to learn its behavior lesson how to perform later in situation s_j, if it is encountered again.

That cell contains the expressed *m* synaptic weights and has *m* synaptic inputs that are associated with m situations from the behavioral environment. The values of the synaptic weights, and the threshold potential, are available to the future copies of this neuron. While the GALA modulon merely *specifies* the output function, the phenotype neuron actually *computes* it. In computation of the neural output, the neuron uses standard McCulloch–Pitts [46] rule.

3.3 Obtaining a Brain-Like Structure

In order to achieve a brain-like neural architecture, the *Struct*-operon from the initial neural stem cell will enable production of two types of the neural progenitor cells: n times produced by the *Neur* regulon and m times produced by the *Eval*-gene. The obtained neural structure is result of expression of the *Behav* and *Emot* genes of the *Struct* operon. For the sake of clarity, Fig. 2 shows only the first and the *n*-th neuron produced from the *Neur* regulon. The neurons in-between are not shown.

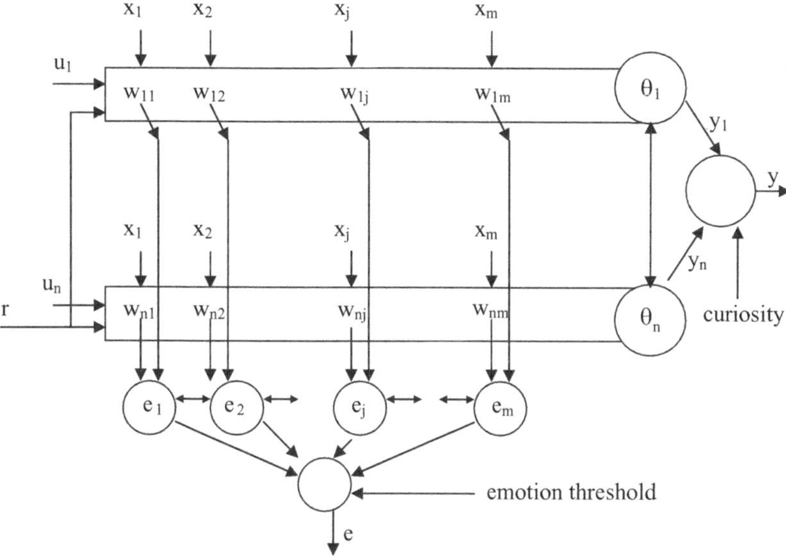

Fig. 2. An artificial brain: it computes both actions and emotions in the same memory structure

The artificial brain shown in Fig. 2 can perform both behavior algebra and emotion algebra. It can perform various functions, examples being summing function, maximum function, and multiplexing function. As example, the behavior computing neuron can enable execution of other, curiosity driven behavior, not the learned one. As another example, the emotion computing neuron can accept influence from other neural systems, before it sends the signal about the emotional state of the system to the outside world and/or back to the learning structure of the agent.

The CAA architecture is a *brain like structure*. It is able to perform functions such as receiving a situation, emotion computation, evaluation of received situation, behavior selection, judging previous behaviors, and updating its memory structure to adjust for a future behavior. The memory elements are synaptic weights, and they are storing achievement motivations. The memory element w_{ij} contains achievement motivation toward i-th behavior provided that j-th situation is encountered.

4 Computer Simulation

Here we describe our simulation system, prepared for Windows environment. In the simulation we used two creatures with different genomes and put them in the same behavioral environment. Fig. 3. shows the result of the initial genome which produces the initial neuron and that neuron is cloned 4 times to produce the artificial brain. Note that at this stage all the four rows (horizontal neurons) have the same weights (the same achievement motivations). They reflect genetically defined predisposition of the creature to the 20 different situations of the behavioral environment.

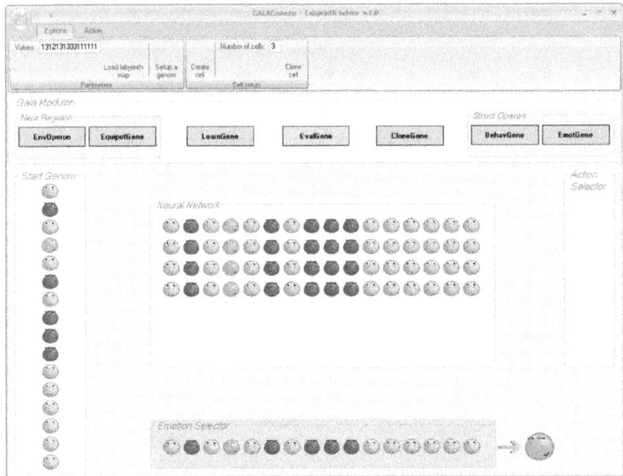

Fig. 3. Generating the neural architecture

The behavioral environment and the changing state of the artificial brain are shown in Fig. 4.

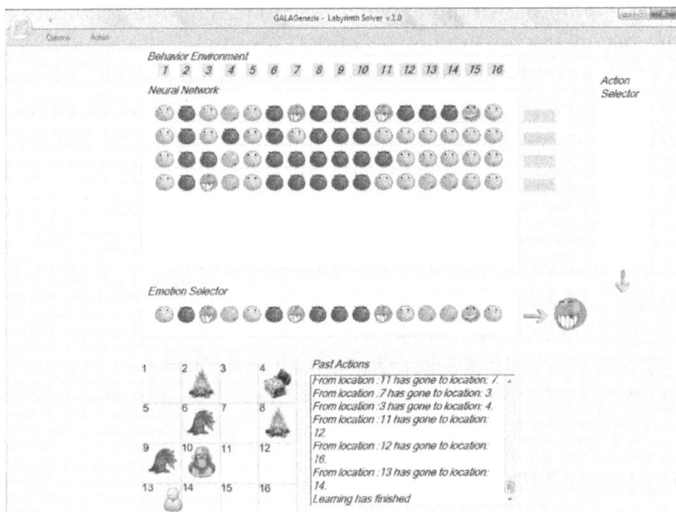

Fig. 4. Simulation of building memory structure to be exported in the genetic environment

The upper part of the screen shows the artificial brain of the considered creature. Each of the weight of the neurons represent achievement motivation toward performing action i in situation j, where i =1,…,4 is the index of the row (actions) and j is index of the considered situation (j=1,2,…,20). The upper row of the screen shows that this behavioral environment has 20 situations. Neural Network shows the artificial brain of the creature after learning. Each emoticon represents an emotion value of achievement motivation toward performing action i (row) in situation j (column). Emotion selector is the result of emotion computation of the whole brain.

5 Exporting Genomes: Lamarckian and Darwinian Way

Here we discuss the process of export to the genetic environment. Consider a species having artificial brain which encoded policy of its behavior in a behavioral environment in its memory elements. At some point of time, let us say that after several iterations and learning in the behavioral environment, the species is considered mature, i.e. capable of exporting its memory into the genetic environment. The memory can be stored in a genome and exported to the genetic environment to create next generation creature. That is consistent with Lamarckian way of evolution.

In our GALA modulon structure, the env-operon (marked as 1 in Fig. 1) enables such an export to the genetic environment. A new genome, possibly consisting of more than one chromosome, will be sent to the genetic environment to build new species that will later appear in the same behavioral environment. In such a case, the env-operon can be encoded as a separate chromosome. Since we have in this case a multichromosome genome, we assume that this particular chromosome with the env-operon is not subject to a crossover process. However, crossover may take place among some other genes of the genome.

To consider the difference between Darwinian and Lamarckian evolution and how it is modeled in our system, assume that the genome that developed the initial species (organism) has the following evaluation of the environment situations.

input_chromosome1(speciesA) = (OO☹OO☺OOO)

Here the species is genetically prepared to meet nine situations in the behavioral environment, and one of them is genetically evaluated as unpleasant (e.g. “it is cold”) and one is evaluated pleasant (e.g. “eating, resupplying energy”). After learning, the species would be able to export, as a part of its genome, the following Lamarckian chromosome:

Lamarckian.output_chromosome1(speciesA) = (O☺☹☺☺☺O☹☺)

To control the Lamarckian-type evolution, some threshold mechanism could be included in the mechanism of genome export. If the exporting algorithm is of Darwinian type, then the export chromosome can be the same as the input Darwinian chromosome:

Darwinian.output_chromosome1(speciesA) = (OO☹OO☺OOO),

and some difference could occur only due to processes like mutation and crossover, not due to a learning process.

6 Conclusion

The work addresses a relevant question in neuroevolution of a brain, natural and artificial as well. The neuroevolution considered in this paper starts from a structured genome. The paper considers three stages of evolving a creature: 1) building neural brain-like architecture from a structured genome via neural stem cells 2) import it into behavioral environment from corresponding genetic environment and perform learning in the behavioral environment and develop policy of proper behavior in the environment, and 3) exporting a genome back in the genetic environment with changes obtained from the behavioral environment, which is modeling the Lamackian evolution. The neural structure we exploited is the CAA architecture, an artificial brain structure with achievement motivation memory elements.

In this work we also developed a simulation of the neuroevolution from a structured genome to a single neuron structure to a multineuron brain like structure. This architecture is self-learning, it computes achievement motivations for each action by learning achievement obtained due to previous execution of that particular action.

We believe that this work contributes toward understanding the evolution of a brain like structure, the one that computes both behaviors and emotions. For stage of evolution of brains, we would say that it models the limbic system, the stage of the brain evolution before appearance of a neocortex.

References

1. Cangelosi, A., Parisi, D., Nolfi, S.: Cell division and migration in a 'genotype' for neural networks. Network 5, 497–515 (1994)
2. Vaario, J., Ogata, N., Shimohara, K.: Synthesis of environment directed and genetic growth. In: Artificial Life V, pp. 207–214. Foundation of Advancement of International Science, Nara (1996)
3. Eggenberger, P.: Creation of neural networks based on developmental and evolutionary principles. In: Gerstner, W., Hasler, M., Germond, A., Nicoud, J.-D. (eds.) ICANN 1997. LNCS, vol. 1327, Springer, Heidelberg (1997)
4. Bull, L.: On the evolution of multicellularity and eusociality. Journal of Artificial Life 5(1), 1–15 (1999)
5. Reil, T.: Dynamics of gene expression in an artificial gene – Implications for biological and artificial ontogeny. In: Floreano, D., Mondada, F. (eds.) ECAL 1999. LNCS, vol. 1674, pp. 457–466. Springer, Heidelberg (1999)
6. Nehaniv, C.L., Hewitt, J., Christiansen, B., Wernick, P.: What Software Evolution and Biological Evolution Don't Have in Common. In: Second International IEEE Workshop on Software Evolvability, SE 2006, pp. 58–65 (2006)
7. Perlovsky, L.I.: Integrated Emotions, Cognition, and Language. In: International Joint Conference on Neural Networks, IJCNN 2006, pp. 1570–1575 (2006)
8. Baluja, S.: Evolution of an artificial neural network based autonomousland vehicle controller. IEEE Trans. Syst. Man Cybern. 26(3), 450–463 (1996)
9. Floreano, D., Kato, T., Marocco, D., Sauser, E.: Coevolution of activevision and feature selection. Biol. Cybern. 90(3), 218–228 (2004)
10. Kohl, N., Stanley, K., Miikkulainen, R., Samples, M., Sherony, R.: Evolving a real-world vehiclewarning system. In: Proc. Genetic Evol. Comput. Conf., pp. 1681–1688 (2006)
11. McLelland, D.: The Achieving Society. Van Nostrand, Princeton (1961)
12. Russell, S., Norvig, P.: Artificial Intelligence: A Modern Approach. Prentice-Hall, Englewood Cliffs (1995)
13. Pfeifer, R., Scheier, C.: Understanding Intelligence. The MIT Press (2000)

14. Jones, T.: Artificial Intelligence – A Systems Approach. Infinity Science Press (2008)
15. Elman, J.: Learning and development in neural networks: The importance of starting small. Cognition 48, 71–99 (1993)
16. Gruau, F., Whitley, D.: Adding learning to the cellular development of neural networks. Evolutionary Computation (1-3), 213–233 (1993)
17. Nolfi, S., Miglino, O., Parisi, D.: Phenotypic plasticity in evolving neural networks. In: Graussier, D., Nicoud, J.-D. (eds.) International Conference from Perception to Action, pp. 146–157. IEEE Computer Society Press, Los Alamitos (1994)
18. Bozinovski, S.: A self-learning system using secondary reinforcement. In: Trappl, R. (ed.) Cybernetics and Systems Research, pp. 397–402. North-Holland (1982)
19. Barinaga, M.: Newborn neurons search for meaning. Science (299), 32–34 (2003)
20. Buell, D., El-Ghazawi, T., Gaj, K., Kindratenko, V.: High performance reconfigurable computing. IEEE Computer, 23–27 (2007)
21. Cooper, L., Nathan, I., Blais, B., Shouval, H.: Theory of Cortical Plasticity. World Scientific (2004)
22. Bozinovski, S.: Crossbar Adaptive Array: The First Connectionist Network that Solved the Delayed Reinforcement Learning Problem. In: Dobnikar, A., Steele, N., Pearson, D., Alberts, R. (eds.) Artificial Neural Networks and Genetic Algorithms, pp. 320–325. Springer (1999)
23. Bozinovski, S., Bozinovska, L.: Self-learning agents: A connectionist theory of emotion, based on crossbar value judgment. Cybernetics and Systems: An International Journal 32, 637–669 (2001)
24. Bozinovski, S., Bozinovska, L.: Evolution of a Cognitive Architecture for Emotional Learning from a Modulon-structured Genome. Journal of Mind and Behavior 29(1-2), 195–216 (2008)
25. Aube', M., Senteni, A.: What are emotions for? Commitments management and regulation within animals/animats encounters. In: Maes, P., Mataric, M., Mayer, J.-A., Pollack, J., Wilson, S. (eds.) From Animals to Animats 4, pp. 246–271. MIT Press (1996)
26. Botelho, L., Coelho, H.: Emotion-based attention shift in autonomous agents. In: Müller, J.P., Wooldridge, M.J., Jennings, N.R. (eds.) ECAI-WS 1996 and ATAL 1996. LNCS, vol. 1193, pp. 277–291. Springer, Heidelberg (1997)
27. Canamero, D., Numaoka, C., Petta, P. (eds.) Grounding Emotions in Adaptive Systems. Workshop Proceedings, Simulation of Adaptive Behavior Conference, Zurich (1998)
28. Petta, P., Trappl, R.: Personalities for synthetic actora: Current issues and some perspectives In R. In: Trappl, R., Petta, P. (eds.) Creating Personalities for Synthetic Actors. LNCS (LNAI), vol. 1195, pp. 209–218. Springer, Heidelberg (1997)
29. Gadanho, S., Hallam, J.: Robot learning driven by emotions. Adaptive Behavior 9(1), 42–64 (2002)
30. Bozinovski, S., Schoel, P.: Emotions and hormones in learning: A biology inspired control architecture for a trainable goalkeeper robot. GMD Report 73. German National Center for Information Technology, Sankt Augustin, Germany (1999)
31. Castelfranchi, C.: Affective appraisal vs. cognitive evaluation in social emotions and interactions. In: Paiva, A.M. (ed.) Affective Interactions. LNCS (LNAI), vol. 1814, pp. 76–106. Springer, Heidelberg (2000)
32. Brave, S., Nass, C.: Emotion in human-computer interaction. In: Jacko, J., Sears, A. (eds.) The Human-Computer Interaction Handbook, pp. 81–96. Lawrence Erlbaum Associates (2003)

33. Yu, C., Xu, L.: An emotion-based approach to decision making and self learning in autonomous robot control. In: Proc. IEEE 8th World Congress on Intelligent Control and Automation, Hangzou, China, pp. 2386–2390 (2004)
34. Fellous, J.-M., Arbib, M.: Who Needs Emotions? The Brain Meets the Robot. Oxford University Press (2005)
35. Minsky, M.: The Emotion Machine. Simon and Schuster (2006)
36. Darwin, C.: On the Origin of Species by Means of Natural Selection. John Murray, London (1859)
37. Lamarck, J.: Philosophie Zoologique. Oxford Univ. Press, Oxford (1809)
38. Parker, M., Bryant, B.: Lamarckian neuroevolution for visual control in the Quake II environment. In: Proc. IEEE Congr. Evol. Comput., pp. 2630–2637 (2009)
39. Gançarski, P., Blansché, A.: Darwinian, Lamarckian, and Baldwinian (Co)Evolutionary Approaches for Feature Weighting in K-MEANS-Based Algorithms. IEEE Transactions on Evolutionary Computation 12(5), 617–629 (2008)
40. Bozinovski, S., Bozinovska, L.: Flexible production lines in genetics: A model of protein biosynthesis process. In: Proc. Int. Conf. on Robotics, pp. 1–4 (1987)
41. Demeester, L., Eichler, K., Loch, C.: Organic production systems: What the biological cell can teach us about manufacturing. Manufacturing and Service Operations management 6(2), 115–132 (2004)
42. Bozinovski, S., Jovancevski, G., Bozinovska, N.: DNA as a real time, database operating system. In: Proc. SCI 2001, Orlando, pp. 65–70 (2001)
43. Danchin, A., Noria, S.: Genome structures, operating systems and the image of a machine. In: Vincente, M., Tamames, J., Valencia, A., Mingorance, J. (eds.) Molecules in Time and Space. Kluwer (2004)
44. Lengeler, J., Mueller, B., di Primio, F.: Cognitive abilities of unicellular mechanisms. GMD Report 57, Institute for Autonomous Intelligent Systems, German National Research Center for Information Technologies, Sankt Augustin (1999) (in German)
45. Sakaguchi, D., Van Hofelen, S., Grozdanic, S., Kwon, Y., Kardon, R., Young, M.: Neural progenitor cell transplants into the developing and mature central nervous system. In: Ourednik, J., Ourednik, V., Sakaguchi, D., Nilsen-Hamilton, M. (eds.) Stem Cell Biology. Annals of the New Your Academy of Sciences, vol. (1049), pp. 118–134 (2005)
46. McCulloch, W., Pitts, W.: A logical calculus of the ideas immanent from nervous activity. Bulletin of Mathematical Biophysics 5, 115–133 (1943)

Numerical Verifications of Theoretical Results about the Weighted $(\mathcal{W}(b);\gamma)-$ Diaphony of the Generalized Van der Corput Sequence

Vesna Dimitrievska Ristovska[1] and Vassil Grozdanov[2]

[1] University "Ss Cyril and Methodius", FINKI,
16, Rugjer Boshkovikj str., 1000 Skopje, Macedonia
vesna.dimitrievska.ristovska@finki.ukim.mk
[2] South-West University "Neophit Rilsky", Department of Mathematics,
66, Ivan Mihailov str., 2700 Blagoevgrad, Bulgaria
vassgrozdanov@yahoo.com

Abstract. The weighted $(\mathcal{W}(b);\gamma)-$diaphony is a new quantitative measure for the irregularity of distribution of sequences. In previous works of the authors it has been found the exact order $\mathcal{O}\left(\frac{1}{N}\right)$ of the weighted $(\mathcal{W}(b);\gamma)-$diaphony of the generalized Van der Corput sequence. Here, we give an upper bound of the weighted $(\mathcal{W}(b);\gamma)-$ diaphony, which is an analogue of the classical Erdös-Turán-Koksma inequality, with respect to this kind of the diaphony. This permits us to make a computational simulations of the weighted $(\mathcal{W}(b);\gamma)-$diaphony of the generalized Van der Corput sequence. Different choices of sequences of permutations of the set $\{0,1,\ldots,b-1\}$ are practically realized and the $(\mathcal{W}(b);\gamma)-$diaphony of the corresponding generalized Van der Corput sequences is numerically calculated and discussed.

Keywords: Uniform distribution of sequences, Diaphony, Generalized Van der Corput sequence

1 Introduction

Let $s \geq 1$ be a fixed integer and will denote the dimension. Following Kuipers and Niederreiter [7] we will recall the concept of uniform distribution of sequences in the $s-$dimensional unit cube $[0,1)^s$. So, let $\xi = (\mathbf{x}_n)_{n\geq 0}$ be an arbitrary sequence of points in the $s-$dimensional unit cube $[0,1)^s$. For an arbitrary subinterval J of $[0,1)^s$ with a volume $\mu(J)$ and each integer $N \geq 1$ we set $A_N(\xi;J) = |\{n : 0 \leq n \leq N-1, \mathbf{x}_n \in J\}|$. The sequence ξ is called uniformly distributed in $[0,1)^s$ if the equality $\lim\limits_{N\to\infty} \dfrac{A_N(\xi;J)}{N} = \mu(J)$ holds for every subinterval J of $[0,1)^s$.

The quantitative theory of the uniformly distribution of sequences $[0,1)^s$ studies the numerical measures for the irregularity of their distribution. The classical diaphony of Zinterhof [9] and some recently introduced $b-$adic versions of the diaphony, (see [5]) are more interesting analytical measures for the irregularity

S. Markovski and M. Gusev (Eds.): *ICT Innovations 2012*, AISC 207, pp. 101–110.
DOI: 10.1007/978-3-642-37169-1_10

of the distribution of sequences. This fact is a consequence of the numerous applications of the uniformly distributed sequences and of the diaphony, as a basic their characteristic, especially to the quasi-Monte Carlo methods. Let $b \geq 2$ be an arbitrary fixed integer. The process of the synchronization between the construction of digitally $b-$adic nets and sequences, and the $b-$adic versions of the diaphony, as a tool of investigation of these objects, is quite natural.

We will recall the concept of the Walsh functions in base b. Following Chrestenson [1], for a non-negative integer k and a real $x \in [0,1)$, with the $b-$adic representations $k = \sum_{i=0}^{\nu} k_i b^i$ and $x = \sum_{i=0}^{\infty} x_i b^{-i-1}$, where for $i \geq 0$ $x_i, k_i \in \{0,1,\ldots,b-1\}$, $k_\nu \neq 0$ and for infinitely many values of i $x_i \neq b-1$, the $k-$th function of Walsh ${}_b wal_k : [0,1) \to \mathbb{C}$ is defined as

$$ {}_b wal_k(x) = e^{\frac{2\pi \mathbf{i}}{b}(k_0 x_0 + k_1 x_1 + \ldots + k_\nu x_\nu)}. $$

The set $\mathcal{W}(b) = \{ {}_b wal_k(x) : k = 0,1,\ldots; x \in [0,1)\}$ is called Walsh functional system in base b.

Let $\mathbb{N}$ be the set of the positive integer numbers and $\mathbb{N}_0 = \mathbb{N} \cup \{0\}$. For an arbitrary vector $\mathbf{k} = (k_1, \ldots, k_s) \in \mathbb{N}_0^s$ the $\mathbf{k}-$th function of Walsh in base b is defined as

$$ {}_b wal_{\mathbf{k}}(\mathbf{x}) = \prod_{h=1}^{s} {}_b wal_{k_h}(x_h), \quad \mathbf{x} = (x_1, \ldots, x_s) \in [0,1)^s. $$

In [2] the authors introduced a new version of the weighted $b-$adic diaphony, which is the following:

Definition 1. *Let $\gamma = (\gamma_1, \gamma_2, \ldots, \gamma_s)$, where $\gamma_1 \geq \gamma_2 \geq \ldots \geq \gamma_s > 0$ be an arbitrary vector of weights. For each integer $N \geq 1$ the weighted $(\mathcal{W}(b);\gamma)-$diaphony $F_N(\mathcal{W}(b);\gamma;\xi)$ of the first N elements of the sequence $\xi = (\mathbf{x}_n)_{n \geq 0}$ of points in $[0,1)^s$ is defined as*

$$ F_N(\mathcal{W}(b);\gamma;\xi) = \left(C^{-1}(b;s;\gamma) \sum_{\mathbf{k} \in \mathbb{N}_0^s,\ \mathbf{k} \neq \mathbf{0}} R(b;\gamma;\mathbf{k}) \left| \frac{1}{N} \sum_{n=0}^{N-1} {}_b wal_{\mathbf{k}}(\mathbf{x}_n) \right|^2 \right)^{\frac{1}{2}}, $$

for each vector $\mathbf{k} = (k_1, \ldots, k_s) \in \mathbb{N}_0^s$ the coefficient $R(b;\gamma;\mathbf{k}) = \prod_{h=1}^{s} \rho(b;\gamma_h;k_h)$ and for a real $\gamma > 0$ and an arbitrary integer $k \geq 0$, the coefficient $\rho(b;\gamma;k)$ is given as

$$ \rho(b;\gamma;k) $$

$$ = \begin{cases} 1, & \text{if } k = 0, \\ \gamma b^{-4a}, & \text{if } k = k_{a-1} b^{a-1}, \ k_{a-1} \in \{1,\ldots,b-1\}, \ a \geq 1, \\ \gamma b^{-2(a+j)} + \gamma b^{-4a}, & \text{if } k = k_{a-1} b^{a-1} + k_{j-1} b^{j-1} + k_{j-2} b^{j-2} + \ldots + k_0, \\ & k_{a-1}, \ k_{j-1} \in \{1,\ldots,b-1\}, \ 1 \leq j \leq a-1, \ a \geq 2, \end{cases} $$

with constant $C(b;s;\gamma) = \prod_{h=1}^{s} \left[1 + \frac{\gamma_h (b+2)}{b(b+1)(b^2+b+1)} \right] - 1.$

It is shown the fact that the weighted $(\mathcal{W}(b);\gamma)-$diaphony is a numerical measure for uniform distribution of sequences, in sense that the sequence $\xi = (\mathbf{x}_n)_{n\geq 0}$ is uniformly distributed in $[0,1)^s$ if and only if, the equality $\lim_{N\to\infty} F_N(\mathcal{W}(b);\gamma;\xi) = 0$ holds for an arbitrary vector γ with non-increasing positive weights.

The purposes of this paper are:

- to obtain an upper bound of the $(\mathcal{W}(b);\gamma)-$ diaphony of an arbitrary one-dimensional sequence in the terms of finite sum and an error term, which is an analogue of the classical inequality of Erdös-Turán-Koksma;
- to give a computational simulation of the $(\mathcal{W}(b);\gamma)-$ diaphony of the generalized Van der Corput sequences.

2 On the Weighted $(\mathcal{W}(b);\gamma)-$Diaphony of the Generalized Van der Corput Sequence

With the next definition we will remind the construction of a large class of sequences with very good distribution in the interval $[0,1)$.

Definition 2. *Let $\Sigma = (\sigma_i)_{i\geq 0}$ be a sequence of permutations of the set $\{0,1,\ldots,b-1\}$. If an arbitrary non-negative integer n has the $b-$adic representation $n = \sum_{i=0}^{\infty} a_i(n)b^i$, then we replace $S_b^{\Sigma}(n) = \sum_{i=0}^{\infty} \sigma_i(a_i(n))b^{-i-1}$. The sequence $S_b^{\Sigma} = \{S_b^{\Sigma}(n)\}_{n\geq 0}$ is called a generalized Van der Corput sequence.*

The sequence S_b^{Σ} has been defined by Faure [4], and it is an example of a class of sequences, which have very well distribution and an effective constructive algorithm. If $\Sigma = I-$ the sequence of identities $\sigma_i(a) = a$ holds for each integer $a \in \{0,1,\ldots,b-1\}$ and $i \geq 0$, the obtained sequence S_b^I is the well known sequence of Halton [6]. When $b = 2$, the sequence S_2^I is the original Van der Corput [8] sequence.

In the next Theorem we will expose estimations of the $(\mathcal{W}(b);\gamma)-$ diaphony of the generalized Van der Corput sequence, which have been obtained in [3].

Theorem 1. *Let $S_b^{\Sigma} = (S_b^{\Sigma}(n))_{n\geq 0}$ be an arbitrary generalized Van der Corput sequence. Then, the following inequalities hold:*

(i) There is a positive constant $C(b)$, depending only on the base b, such that for each integer $N \geq 1$, the weighted $(\mathcal{W}(b);\gamma)-$diaphony $F_N(\mathcal{W}(b);\gamma;S_b^{\Sigma})$ of the sequence S_b^{Σ} satisfies the inequality

$$F_N(\mathcal{W}(b);\gamma;S_b^{\Sigma}) \leq \frac{C(b)}{N};$$

(ii) There is a positive constant $C_1(b)$, depending only on the base b, such that the inequality

$$F_N(\mathcal{W}(b);\gamma;S_b^{\Sigma}) \geq \frac{C_1(b)}{N}$$

holds, for infinitely many values of N of the form $N = b^{\nu}+1$, where $\nu > 0$ is an arbitrary integer.

Theorem 1 shows that the $(\mathcal{W}(b);\gamma)-$ diaphony of an arbitrary generalized Van der Corput sequence has an exact order $\mathcal{O}\left(\frac{1}{N}\right)$.

3 A Theoretical Base for Numerical Approximations

In the next Theorem we will expose an upper estimation of the $(\mathcal{W}(b);\gamma)-$diaphony, which is an analogue of the inequality of Erdös-Turán-Koksma.

Theorem 2. *Let $\xi = (x_n)_{n\geq 0}$ be an arbitrary one-dimensional sequence. There is a constant $C_2(b)$ depending only on the base b, such that for each integer $M > 0$ the inequality*

$$(F_N(\mathcal{W}(b);\gamma;\xi))^2 \leq$$

$$C^{-1}(b;1;\gamma)\left[\sum_{k=1}^{M-1}\rho(b;\gamma;k)\left|\frac{1}{N}\sum_{n=0}^{N-1}{}_bwal_k(x_n)\right|^2 + C_2(b)\cdot\frac{1}{M^2}\right]$$

holds.

Proof. Let $M > 0$ be an arbitrary integer. We will use the following presentation

$$C(b;1;\gamma).(F_N(\mathcal{W}(b);\gamma;\xi))^2 = \sum_{k=1}^{\infty}\rho(b;\gamma;k)\left|\frac{1}{N}\sum_{n=0}^{N-1}{}_bwal_k(x_n)\right|^2$$

$$= \sum_{k=1}^{M-1}\rho(b;\gamma;k)\left|\frac{1}{N}\sum_{n=0}^{N-1}{}_bwal_k(x_n)\right|^2 + \sum_{k=M}^{\infty}\rho(b;\gamma;k)\left|\frac{1}{N}\sum_{n=0}^{N-1}{}_bwal_k(x_n)\right|^2. \quad (1)$$

For each integer $k \geq M$ we use the estimation $\left|\frac{1}{N}\sum_{n=0}^{N-1}{}_bwal_k(S_b^{\Sigma}(n))\right| \leq 1$, and from (1) we obtain the inequality

$$C(b;1;\gamma).(F_N(\mathcal{W}(b);\gamma;\xi))^2$$

$$\leq \sum_{k=1}^{M-1}\rho(b;\gamma;k)\left|\frac{1}{N}\sum_{n=0}^{N-1}{}_bwal_k(x_n)\right|^2 + \sum_{k=M}^{\infty}\rho(b;\gamma;k). \quad (2)$$

For the fixed integer $M > 0$ there exists an integer $p > 0$ such that $b^p \leq M < b^{p+1}$. We will realize the following calculations

$$\sum_{k=M}^{\infty}\rho(b;\gamma;k) \leq \sum_{k=b^p}^{\infty}\rho(b;\gamma;k) = \sum_{a=p+1}^{\infty}\sum_{k_{a-1}=1}^{b-1}\rho(b;\gamma;k_{a-1}b^{a-1})$$

$$+\sum_{a=p+1}^{\infty}\sum_{k_{a-1}=1}^{b-1}\sum_{j=1}^{a-1}\sum_{k_{j-1}=1}^{b-1}\sum_{k=k_{a-1}b^{a-1}+k_{j-1}b^{j-1}}^{k_{a-1}b^{a-1}+(k_{j-1}+1)b^{j-1}-1}\rho(b;\gamma;k)$$

$$= \gamma(b-1)\sum_{a=p+1}^{\infty}b^{-4a} + \sum_{a=p+1}^{\infty}\sum_{k_{a-1}=1}^{b-1}\sum_{j=1}^{a-1}\sum_{k_{j-1}=1}^{b-1}\sum_{k=k_{a-1}b^{a-1}+k_{j-1}b^{j-1}}^{k_{a-1}b^{a-1}+(k_{j-1}+1)b^{j-1}-1}(\gamma b^{-2(a+j)} + \gamma b^{-4a})$$

$$= \gamma(b-1)\sum_{a=p+1}^{\infty} b^{-4a} + \gamma \sum_{a=p+1}^{\infty} b^{-2a} \sum_{k_{a-1}=1}^{b-1}\sum_{j=1}^{a-1} b^{-2j} \sum_{k_{j-1}=1}^{b-1} \sum_{k=k_{a-1}b^{a-1}+k_{j-1}b^{j-1}}^{k_{a-1}b^{a-1}+(k_{j-1}+1)b^{j-1}-1} 1$$

$$+\gamma \sum_{a=p+1}^{\infty} b^{-4a} \sum_{k_{a-1}=1}^{b-1}\sum_{j=1}^{a-1}\sum_{k_{j-1}=1}^{b-1} \sum_{k=k_{a-1}b^{a-1}+k_{j-1}b^{j-1}}^{k_{a-1}b^{a-1}+(k_{j-1}+1)b^{j-1}-1} 1$$

$$= \gamma(b-1)\sum_{a=p+1}^{\infty} b^{-4a} + \frac{\gamma(b-1)^2}{b}\left[\sum_{a=p+1}^{\infty} b^{-2a}\sum_{j=1}^{a-1} b^{-j} + \sum_{a=p+1}^{\infty} b^{-4a}\sum_{j=1}^{a-1} b^{j}\right]$$

$$= \gamma(b-1)\left[\sum_{a=p+1}^{\infty} b^{-4a} + \frac{1}{b}\sum_{a=p+1}^{\infty} b^{-2a}\left(1-\frac{b}{b^a}\right) + \sum_{a=p+1}^{\infty} b^{-4a}\left(\frac{b^a}{b}-1\right)\right]$$

$$= \gamma(b-1)\left[\frac{1}{b}\sum_{a=p+1}^{\infty} b^{-2a} - \frac{b-1}{b}\sum_{a=p+1}^{\infty} b^{-3a} + \sum_{a=p+1}^{\infty} b^{-4a} - \sum_{a=p+1}^{\infty} b^{-4a}\right]$$

$$< \frac{\gamma(b-1)}{b}\sum_{a=p+1}^{\infty} b^{-2a} = \frac{\gamma b}{b+1}\cdot\frac{1}{b^{2(p+1)}} < \frac{\gamma b}{b+1}\cdot\frac{1}{M^2}. \quad (3)$$

From (2) and (3) we obtain the inequality

$$C(b;1;\gamma).(F_N(\mathcal{W}(b);\gamma;\xi))^2$$

$$\leq \sum_{k=1}^{M-1} \rho(b;\gamma;k)\left|\frac{1}{N}\sum_{n=0}^{N-1} {}_b wal_k(x_n)\right|^2 + \frac{\gamma b}{b+1}\cdot\frac{1}{M^2}.$$

We remind the fact that $C(b;1;\gamma) = \frac{\gamma(b+2)}{b(b+1)(b^2+b+1)}$ and from the above inequality finally obtain that

$$(F_N(\mathcal{W}(b);\gamma;\xi))^2$$

$$\leq C^{-1}(b;1;\gamma)\sum_{k=1}^{M-1} \rho(b;\gamma;k)\left|\frac{1}{N}\sum_{n=0}^{N-1} {}_b wal_k(x_n)\right|^2 + \frac{b^2(b^2+b+1)}{b+2}\cdot\frac{1}{M^2}.$$

Theorem 2 guarantees the fact, that the error which is obtained when we approximate the infinite sum on k in the definition of $(F_N(\mathcal{W}(b);\gamma;\xi))^2$ with finite sum on k ($1 \leq k \leq M$) has an order $\mathcal{O}\left(\frac{1}{M^2}\right)$.

4 Graphical Numerical Results

In this section numerical results obtained for different values of base b and for different values of sequence Σ of permutations are graphically shown. The sequence of permutations Σ was taken by a random choice and all of computations were made with working precision of 30 decimals. Here, the parameter γ is chosen to be equal to 1. According to Theorem 2, if we set $M = N^2$, the error

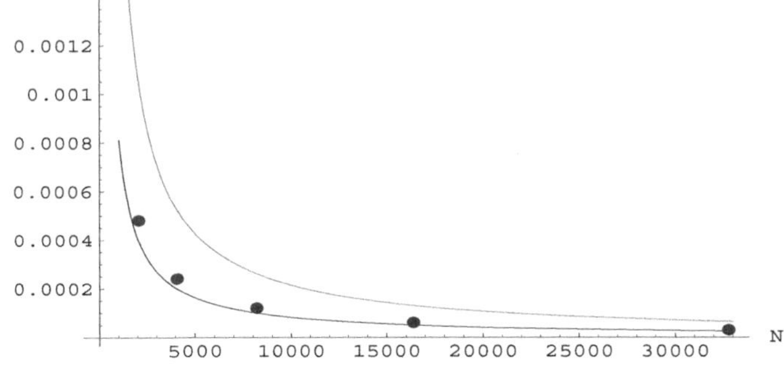

Fig. 1. $b = 2, N = b^{\nu} + 1$

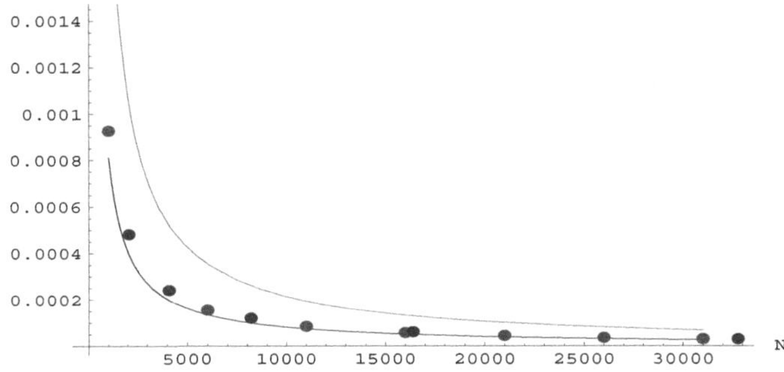

Fig. 2. b=2

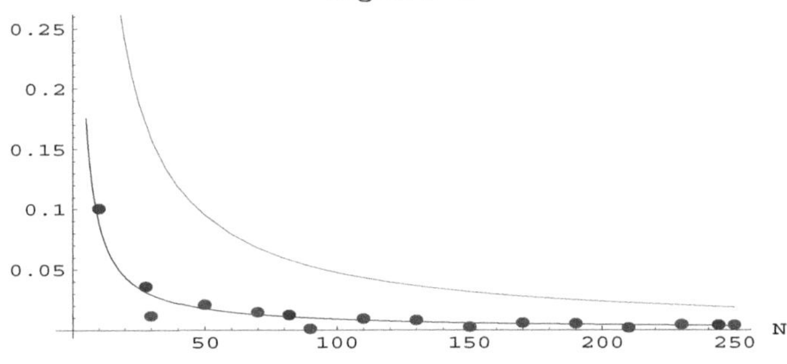

Fig. 3. b=3

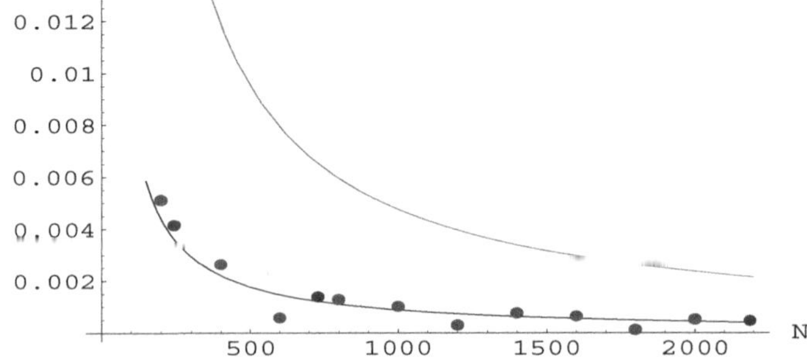

Fig. 4. b=3

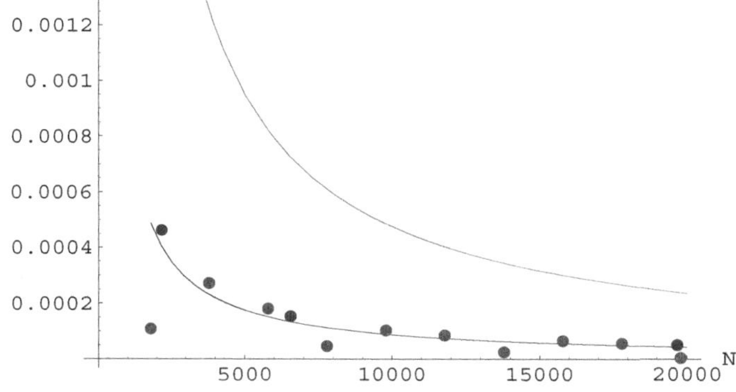

Fig. 5. b=3

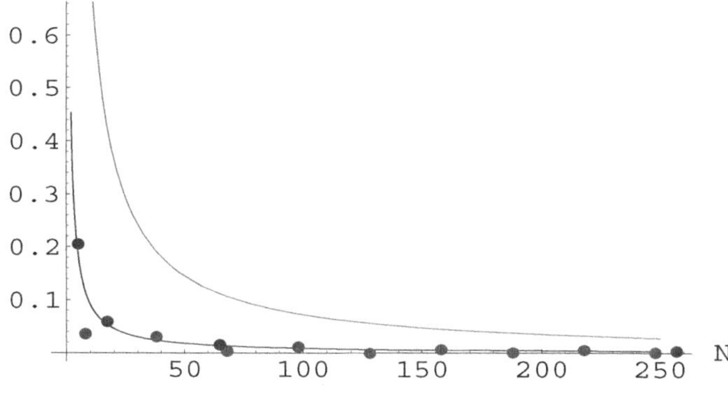

Fig. 6. b=4

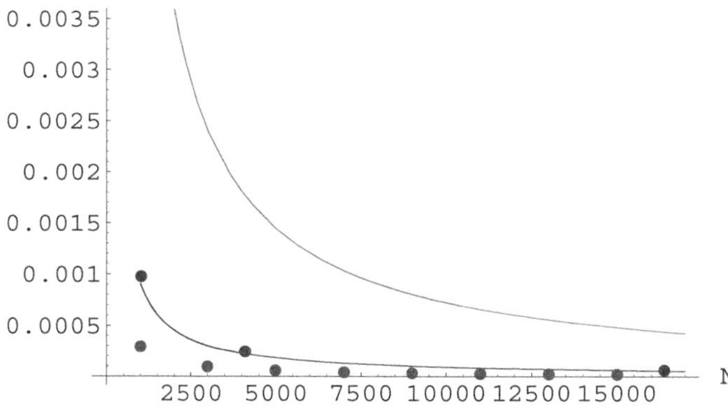

Fig. 7. b=4

which is obtained when we approximate the infinite sum on k in the definition of $F_N(\mathcal{W}(b);\gamma;\xi)$ with finite sum on k $(1 \leq k \leq M)$ has an order $\mathcal{O}\left(\frac{1}{M}\right) = \mathcal{O}\left(\frac{1}{N^2}\right)$.

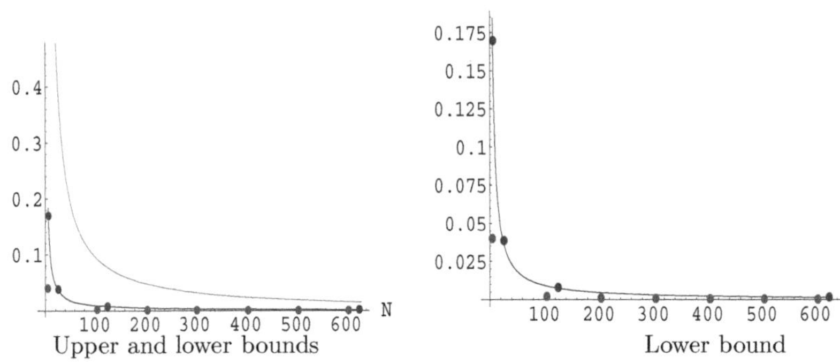

Fig. 8. b=5

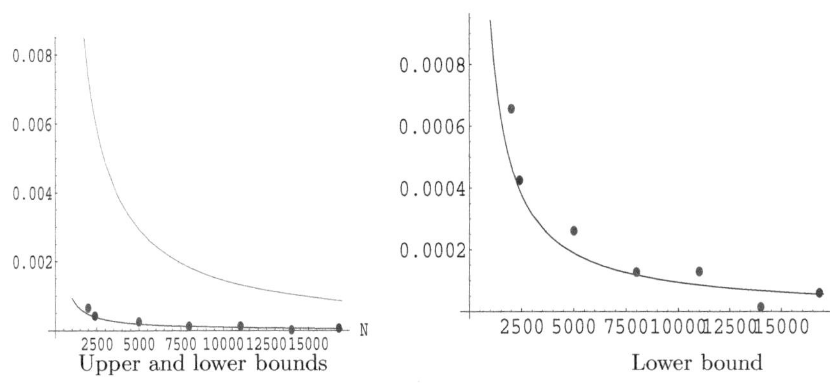

Fig. 9. b=7

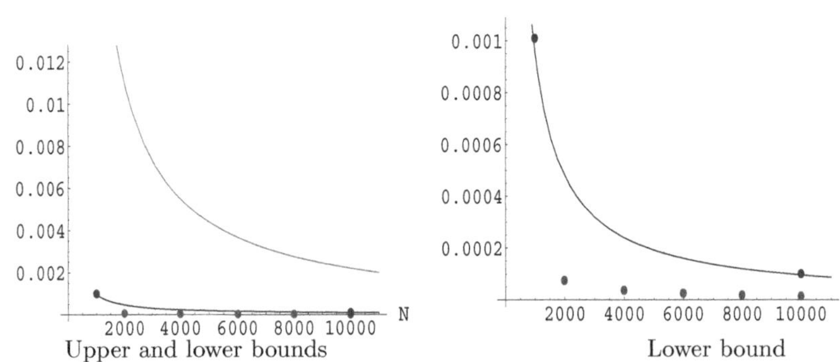

Fig. 10. b=10

The number N of summands in these numerical computations is at least 1000, and computations with $N = 10^8$ were made but they didn't make an important influence on the final results (see Figure 5 and Figure 7).

Table 1. First 15 elements of sequence Σ

4	3	2	0	1
0	3	4	1	2
4	0	2	3	1
2	0	3	4	1
3	0	2	1	4
1	2	4	3	0
3	2	0	4	1
3	0	2	1	4
0	4	2	1	3
3	4	0	1	2
3	2	0	4	1
0	3	1	4	2
0	4	2	3	1
4	1	3	2	0
4	2	1	0	3

$b = 5$

1	8	0	5	3	4	7	2	9	6
2	3	4	7	9	5	6	1	0	8
2	1	4	0	6	7	9	8	5	3
3	5	8	6	4	9	0	2	1	7
7	1	3	0	5	9	8	6	2	4
5	2	3	8	1	6	7	4	9	0
8	1	2	0	6	3	4	5	9	7
0	3	5	9	7	6	2	4	1	8
1	4	2	7	3	9	0	5	8	6
6	8	7	5	3	4	1	0	9	2
5	7	9	2	0	8	3	4	1	6
2	7	9	8	6	3	4	1	0	5
8	0	2	7	1	9	4	5	3	6
2	4	1	6	5	9	3	8	0	7
2	8	3	9	1	0	7	6	4	5

$b = 10$

All of numerical simulations have verified theoretical upper and lower estimations presented in Theorem 1, i.e. obtained approximated values of the $(\mathcal{W}(b);\gamma)-$diaphony of an arbitrary generalized Van der Corput sequence S_b^{Σ} are under the upper bound and some of them are over the lower bound. The upper bound is presented with green line, the lower bound is presented with blue line, the values of the $F_N(\mathcal{W}(b);1;S_b^{\Sigma})$ diaphony which are obtained for arbitrary N are with red points and the values of the $F_N(\mathcal{W}(b);1;S_b^{\Sigma})$ diaphony which are obtained for $N = b^{\nu}+1, \nu \in \mathbb{N}$ are with violet points. The upper bound is satisfied for arbitrary N. The lower bound is satisfied for all N of the form $N = b^{\nu}+1, \nu \in \mathbb{N}$ (see Figure 1, Figure 2, Figure 8 and Figure 10) and for some integers N which are not of this form (see Figure 3, Figure 4, Figure 6 and Figure 9). Two examples of sequence Σ of random permutations are given in Table 1.

In this paper computer simulations are being used to verify obtained theoretical estimations in one-dimensional case. The next our step will be to make software simulations of the $(\mathcal{W}(b);\gamma)-$diaphony of some classes of s-dimensional sequences. This will be motivation for us, to make investigations in the direction to prove analogue theorem in s- dimensional case. These results will have applications in quasi-Monte Carlo methods for numerical integration. We will stress the fact that the techniques and the methods, which was used to compute the order of the $(\mathcal{W}(b);\gamma)-$diaphony of the generalized Van der Corput sequense have applications for construction of some classes of pseudo- random numbers by using for example linear, quadratic, or more general, polynomial recursive procedures for construction of low-discrepancy sequences and nets. It will be another motivation for us to continue our investigations in that field.

References

1. Chrestenson, H.E.: A class of generalized Walsh functions. Pacific J. Math. 5, 17–31 (1955)
2. Dimitrievska Ristovska, V., Grozdanov, V., Kusakatov, V., Stoilova, S.: Computing complexity of a new type of the diaphony. In: The 9th International Conference for Informatics and Information Technology, CIIT, Bitola (2012) (inprint)
3. Dimitrievska Ristovska, V., Grozdanov, V., Mavrodieva, D., Stoilova, S.: On the weighted $(\mathcal{W}(b);\gamma)-$diaphony of the generalized Van der Corput sequence and the Zaremba-Halton net (in submitting)
4. Faure, H.: Discrepances de suite associées à un systèm de numération (en dimension un). Bull. Soc. Math. France 109, 143–182 (1981)
5. Grozdanov, V., Stoilova, S.: The $b-$adic diaphony. Rendiconti di Matematica, Serie VII 22, 203–221 (2002)
6. Halton, J.H.: On the efficiency of certain quasi-random sequences of points in evaluating multi-dimensional integrals. Numer. Math. 2, 84–90 (1960)
7. Kuipers, L., Niederreiter, H.: Uniform distribution of sequences. John Wiley & Sons, N. Y. (1974)
8. Van der Corput, J.G.: Verteilungsfunktionen. Proc. Kon. Ned. Akad. Wetensch. 38, 813–821 (1935)
9. Zinterhof, P.: Über einige Abschätzungen bei der Approximation von Funktionen mit Gleichverteilungsmethoden. S. B. Akad. Wiss., Math.-Naturw. Klasse. Abt. II 185, 121–132 (1976)

Stability Analysis of Impulsive Stochastic Cohen-Grossberg Neural Networks with Mixed Delays

Biljana Tojtovska

Faculty of Computer Science and Engineering,
Ss. Cyril and Methodius University, Skopje, Macedonia
biljana.tojtovska@finki.ukim.mk

Abstract. For impulsive stochastic Cohen-Grossberg neural networks with mixed time delays, we study in the present paper pth moment ($p \geq 2$) stability on a general decay rate. Using the theory of Lyapunov function, M-matrix technique and some famous inequalities we generalize and improve some known results referring to the exponential stability. The presented theory allows us to study the pth moment stability even if the exponential stability cannot be shown. Some examples are presented to support and illustrate the theory.

Keywords: Impulsive stochastic neural networks, moment stability analysis, general decay function.

1 Introduction

Neural networks have proved to be useful in many applications but failed to give an accurate description of biological neural networks, partly due to their deterministic nature. Many authors have already emphasized the stochastic properties of the human brain [1–3]. For more realistic description time delays in the synaptic transmission of the neurons should be assumed and in the artificial neural networks this is justified with the existence of delays in communication channels. The impulsive effects in the network should also be incorporated in the model. They may come in the form of various stimuli received by the receptors [4] or as sudden perturbations coming from frequency changes, additional noise etc. This brings us to the model of impulsive stochastic neural network with time delays.

All such models are essentially based on complex stochastic differential equations (SDEs) of the Itô's type. Very often they can not be exactly solved and thus it is important to know whether the model has a unique solution and which are its properties. The stability of the solution of an SDE indicates that the system is not sensitive to small changes of the initial state or parameters of the model. In the case of neural networks stability means existence of thermal equilibrium which is essential for the learning process. Using different techniques many authors have studied the stability properties of different models of stochastic neural

S. Markovski and M. Gusev (Eds.): *ICT Innovations 2012*, AISC 207, pp. 111–120.
DOI: 10.1007/978-3-642-37169-1_11

networks. The literature on this topic is extensive and here we give only few references [5–8]. The stability with respect to a general decay function is also of interest in many cases and although some results do exist (ex. [9, 10]), to the best of our knowledge only few of them address the models of stochastic neural networks (ex. [11, 12]). The goal of this paper is to give sufficient conditions on the pth moment stability with respect to a general decay function, for impulsive stochastic Cohen-Grossberg neural network with mixed time delays. The study in this paper is motivated by the results presented in [7], where the authors study the pth moment exponential stability. We extend the usual notion on a general decay function, which allows us to study the pth moment stability even when the exponential stability cannot be shown by applying the results from paper [7].

This paper is organized as follows: In Section 2, we introduce some notions and basic assumptions for the model. In Section 3, we give our main result, new criteria on a general decay pth moment stability of our model. A numerical example is presented in Section 4 to support and illustrate our theory.

For more details on the theory of neural networks, stochastic and stability analysis, M-matrix theory and numerical simulations of SDEs used in this paper we refer to [13–15].

2 Preliminaries

Throughout the paper, we assume that all random variables and processes are defined on a complete probability space $(\Omega, \mathcal{F}, \mathbb{P})$ with a natural filtration $\{\mathcal{F}_t\}_{t\geq t_0}$ generated by a standard n-dimensional Brownian motion $w = \{w(t), t \geq t_0\}$.

We consider $\mathbb{R}^n$ with the norm $||x|| = \left(\sum_{i=1}^n |x_i|^p\right)^{\frac{1}{p}}$ for $p \geq 1$. Let $t_0, \tau \geq 0$ be some constants and $t_0 < t_1 < t_2 < ...$, be a sequence such that $\lim_{n->\infty} t_n = \infty$. Denote that $C\big((-\infty, t_0]; \mathbb{R}^n\big)$ is the space of continuous functions φ defined on $(-\infty, t_0]$ into $\mathbb{R}^n$, equipped with the norm $||\varphi|| = \left(\sup_{s\in(-\infty,t_0]} ||\varphi(s)||^p\right)^{\frac{1}{p}}$, while $\mathcal{L}^p_{\mathcal{F}_{t_0}}((-\infty, t_0]; \mathbb{R}^n)$ is the family of all $\mathcal{F}_{t_0}$-adapted $C\big((-\infty, t_0]; \mathbb{R}^n\big)$-valued random variables ϕ satisfying $\mathbb{E}||\phi||^p < \infty$. For any interval $J \subseteq \mathbb{R}$ we define:

$$PC[J; \mathbb{R}^n] = \{x(t) : J \to \mathbb{R}^n | x(t) \text{ is continuous on } J \setminus J_0, \text{ where } J_0 \text{ is at most countable set and for } s \in J \setminus J_0 \quad x(s^+), x(s^-) \quad \text{exist and} \quad x(s^+) = x(s)\}.$$

For any function $\phi \in C[J, \mathbb{R}^n]$ or $\phi \in PC[J; \mathbb{R}^n]$ we will assume that it is bounded and define $||\phi|| = (\sup_J |\phi(s)|^p)^{\frac{1}{p}}$. Note also that for $x(t) \in PC[\mathbb{R}, \mathbb{R}^n]$, $||x_t|| = \left(\sum_{i=1}^n \sup_{s\in[t-\tau,t]} |x_i(s)|^p\right)^{\frac{1}{p}}$, $t \geq t_0$. For $\lambda(t) : \mathbb{R}^+ \to \mathbb{R}$ we define:

$$\wp(\lambda(t)) = \{\phi(t) : \mathbb{R}^+ \to \mathbb{R} \,|\, \phi(t) \text{ is piecewise continuous and there is a constant } \gamma_0 > 0 \quad \text{such that} \quad \int_0^{+\infty} |\phi(s)| \lambda^{\gamma_0}(s) ds < \infty\}. \tag{1}$$

In the sequel we also use the following notations: $N = \{1, 2, ..., n\}$; for $A = (a_{ij})_{m\times n} \in \mathbb{R}^{m\times n}$ we say that A is a non-negative matrix i.e. $A \geq 0$ iff for all $i, j \in N$, $a_{ij} \geq 0$; $z \in \mathbb{R}^n$ is a positive vector iff $z > 0$.

In this paper we study a model of impulsive stochastic Cohen-Grossberg neural network with mixed time delays, given by the following system of SDEs: For all $i \in N$, $t \geq t_0$, $t \neq t_k$,

$$dx_i(t) = -h_i(x_i(t))\Bigg[c_i(t, x_i(t)) - \sum_{j=1}^{n} a_{ij}(t) f_j(x_j(t)) - \sum_{j=1}^{n} b_{ij}(t) g_j(x_{ij,t})$$
$$- \sum_{j=1}^{n} d_{ij}(t) \int_{-\infty}^{t} l_{ij}(t-s) k_j(x_j(s)) ds\Bigg] dt + \sum_{j=1}^{n} \sigma_{ij}\left(t, x_j(t), x_{ij,t}\right) dw_j(t), \quad (2)$$

and for $t = t_k$, $k \in \mathbb{N}$

$$x_i(t) = p_{ik}(x_1(t^-), ..., x_n(t^-)) + q_{ik}(x_1(t^- - \tau_{i1}(t^-)), ..., x_n(t^- - \tau_{in}(t^-))),$$
$$x_i(t_0 + s) = \phi_i(s), \quad s \in (-\infty, t_0].$$

Let us denote that $h_i(x(t))$ are amplification functions, $c_i(t, x_i(t))$ are appropriately behaved functions dependent on t and on the state process $x(t)$. $a_{ij}(t)$, $b_{ij}(t)$ and $d_{ij}(t)$ are interconnection functions, while $f_i(x_i(t))$, $g_i(x_{i,t})$ and $k_i(x_i(t))$ are activation functions. $l_{ij}(t) \in \wp(\lambda(t))$ are the delay kernel functions and the term $\sigma(t, x(t), x_t) = \left[\sigma_{ij}(t, x_j(t), x_{ij,t})\right]_{n\times n}$ is a diffusion-coefficient matrix. $\tau_{ij}(t)$ represents the transmission delay from the ith to the jth unit, $0 \leq \tau_{ij}(t) \leq \tau$ and $x_{ij,t} = x_j(t - \tau_{ij}(t))$ are delayed processes.[1]

The following definition on pth moment stability with a certain decay rate $\lambda(t)$ will be used in the sequel.

Definition 1. *Let the function $\lambda \in C(\mathbb{R}; \mathbb{R}^+)$ be a non-decreasing function on $(-\infty, t_0 - \tau)$, strictly increasing on $[t_0 - \tau, +\infty)$ with $\lim_{t\to\infty} \lambda(t) = \infty$. Let it be differentiable on $[t_0, +\infty)$ s.t. $\lambda'(t) \leq \lambda(t)$ and also let $\lambda(s+t) \leq \lambda(s)\lambda(t)$ for all s,t in its domain. Then, Eq. (2) is said to be pth moment stable with a decay $\lambda(t)$ of order γ if there exists a pair of constants $\gamma > 0$ and $c(\phi) > 0$ such that for any $\phi \in \mathcal{L}^p_{\mathcal{F}_{t_0}}((-\infty, t_0]; \mathbb{R}^n)$ it holds*

$$\mathbb{E}||x(t, \phi)||^p \leq c(\phi)\lambda^{-\gamma}(t), \quad t \geq t_0$$

In this paper we combine the stochastic version of the Lyapunov direct method with M-matrix theory and some famous inequalities. Next we give definition of an M-matrix.

Definition 2. *Let $D = (d_{ij}) \in \mathbb{R}^n$, $d_{ij} \leq 0 < d_{ii}$ $(i \neq j, i, j \in N)$. D is an M-matrix if there is $z \in \mathbb{R}^n, z > 0$ such that $Dz > 0$. For a non-singular M-matrix we define the set $\Omega_M(D) = \{z \in \mathbb{R}^n | z > 0, Dz > 0\}$ which is closed under linear combinations over $\mathbb{R}^+$.*

[1] We assume that the conditions of existence and uniqueness of a solution of an impulsive stochastic neural networks are satisfied (see for ex. [16]).

In order to prove the main results, we introduce some additional assumptions for the coefficients of Eq. (2). For $\lambda(t)$ a decay function, $i,j \in N$, we suppose that:

($\mathbf{H_1}$) There exist positive constants $\underline{h}_i, \overline{h}_i$ and $\underline{c}_i$ such that

$$0 < \underline{h}_i \leq h_i(x) \leq \overline{h}_i \quad x \in \mathbb{R}, \qquad x c_i(t,x) \geq \underline{c}_i x^2 \quad (t,x) \in [t_0,\infty) \times \mathbb{R}.$$

($\mathbf{H_2}$) The activation functions and the functions $c_i(t,x)$ are Lipshitz continuous, that is, there exist positive constants $\beta_i, \delta_i, k_i, \overline{c}_i$ such that for all $i \in N$, $t \in [t_0,\infty)$ and $x,y \in \mathbb{R}$ it holds

$$|f_i(x) - f_i(y)| \leq \beta_i\,|x-y|, \qquad |g_i(x) - g_i(y)| \leq \delta_i\,|x-y|,$$

$$|k_i(x) - k_i(y)| \leq k_i\,|x-y|, \qquad |c_i(t,x) - c_i(t,y)| \leq \overline{c}_i|x-y|.$$

($\mathbf{H_3}$) The functions $a_{ij}(t), b_{ij}(t), d_{ij}(t)$ are bounded, that is, there exist non-negative constants $\overline{a}_{ij}, \overline{b}_{ij}, \overline{d}_{ij}$ such that

$$|a_{ij}(t)| \leq \overline{a}_{ij}, \quad |b_{ij}(t)| \leq \overline{b}_{ij}, \quad |d_{ij}(t)| \leq \overline{d}_{ij}.$$

($\mathbf{H_4}$) The diffusion functions $\sigma_{ij}(t,x,y)$ are globally Lipshitz continuous, uniformly in t, and there exist non-negative constants μ_{ij} and ν_{ij} such that

$$\sigma_{ij}^2(t,x,y) \leq \mu_{ij}\,x^2 + \nu_{ij}\,y^2, \quad (t,x,y) \in [t_0,\infty) \times \mathbb{R}^2.$$

($\mathbf{H_5}$) There exist non-negative matrices $P_k = (p_{ij}^{(k)})_{n\times n}$ and $Q_k = (q_{ij}^{(k)})_{n\times n}$ such that for all $x = (x_1,...,x_n)^T \in \mathbb{R}^n$ and $y = (y_1,...,y_n)^T \in \mathbb{R}^n$, $i \in N$, $k \in \mathbb{N}$ it holds

$$|p_{ik}(x) - p_{ik}(y)| \leq \sum_{j=1}^n p_{ij}^{(k)}|x_j - y_j|, \quad |q_{ik}(x) - q_{ik}(y)| \leq \sum_{j=1}^n q_{ij}^{(k)}|x_j - y_j|.$$

($\mathbf{H_6}$)[2] There exist $z = (z_1,...,z_n)^T \in \Omega_M(D)$ and a constant $\gamma \in (0,\gamma_0]$ such that for any $k \in \mathbb{N}$ it holds

$$2^{p-1}\left[\left(\sum_{j=1}^n (p_{ij}^{(k)})^{\frac{p}{p-1}}\right)^{p-1} + \left(\sum_{j=1}^n (q_{ij}^{(k)})^{\frac{p}{p-1}}\right)^{p-1}\lambda^\gamma(\tau)\right]\sum_{j=1}^n z_j \leq \min_{i\in N} z_i.$$

3 Main Results

In this section we study the general decay pth moment stability of the impulsive stochastic neural network given by Eq. (2). First we present two lemmas which will help us prove the main result.

[2] Condition ($\mathbf{H_6}$) substitutes conditions ($\mathbf{A_7}$) and ($\mathbf{A_8}$) in [7].

Lemma 1. *Let* $\tilde{P} = (\tilde{p}_{ij})_{n\times n}$, $\tilde{p}_{ij} \geq 0$ *for* $i \neq j$, $\tilde{W} = (\tilde{w}_{ij})_{n\times n} \geq 0$, $\tilde{Q}(t) = (\tilde{q}_{ij}(t))_{n\times n} \geq 0$ *and* $\tilde{q}_{ij}(t) \in \wp(\lambda(t))$, $\tilde{Q} = (\tilde{q}_{ij})_{n\times n} = (\int_0^\infty \tilde{q}_{ij}(s)ds)_{n\times n}$ *and let* $D = -(\tilde{P}+\tilde{W}+\tilde{Q})$ *be a non-singular* M*-matrix. Let* $u(t) = (u_1(t), ..., u_n(t))^T \in PC[\mathbb{R};\mathbb{R}^n]$ *satisfy the following inequality*

$$D^+u(t) \leq \tilde{P}u(t) + \tilde{W} \sup_{s\in[t-\tau,t]} u(s) + \int_0^\infty \tilde{Q}(s)u(t-s)ds \quad t \geq t_0 \tag{3}$$

with an initial condition $u_{t_0}(s) \in PC, s \in (-\infty, t_0]$. *Let* $u(t) \leq z\frac{\lambda^{-\gamma}(t)}{\lambda^{-\gamma}(t_0)}$ *for* $t \in (-\infty, t_0]$ *and some decay function* $\lambda(t)$, *where* $z = (z_1, z_2, ..., z_n)^T \in \Omega_M(D)$ *and* $\gamma < \gamma_0$ *is a positive constant determined by*

$$\left[\gamma I + \tilde{P} + \tilde{W}\lambda^\gamma(\tau) + \int_0^{+\infty} \tilde{Q}(s)\lambda^\gamma(s)ds\right] z \equiv T(\gamma)z < 0. \tag{4}$$

Then $u(t) \leq z\frac{\lambda^{-\gamma}(t)}{\lambda^{-\gamma}(t_0)}$ *for* $t \geq t_0$.

Proof. The proof is similar to the proof of Theorem 3.1 in [17]. Since $D = -(\tilde{P}+\tilde{W}+\tilde{Q}) = -T(0)$ is a non-singular M-matrix, there is $z \in \Omega_M(D)$ such that $T(0)z < 0$. From the continuity of $T(\gamma)$ it follows that there is $\gamma > 0$ such that $T(\gamma)z < 0$. Let the initial condition be satisfied and let $\varepsilon > 0$. We will prove that $u_i(t) < (1+\varepsilon)z_i\frac{\lambda^{-\gamma}(t)}{\lambda^{-\gamma}(t_0)} = y_i(t)$ for $t \geq t_0, i \in N$. If we assume the contrary, then there exists $t^* > t_0$ and $k \in N$ s.t. $u_i(t) \leq y_i(t)$ for $t \in (-\infty, t^*)$, $i \in N$, $u_k(t^*) = y_k(t^*)$ and $D^+u_k(t^*) \geq y_k'(t^*)$.Then

$$D^+u_k(t^*) \leq \sum_{j=1}^n \left[\tilde{p}_{kj}u_j(t^*) + \tilde{w}_{kj} \sup_{s\in[t^*-\tau,t^*]} u_j(s) + \int_0^\infty \tilde{q}_{kj}(s)u_j(t^*-s)ds\right]$$

$$\leq \sum_{j=1}^n \left[\tilde{p}_{kj}\lambda^{-\gamma}(t^*) + \tilde{w}_{kj} \sup_{s\in[t^*-\tau,t^*]} \lambda^{-\gamma}(s) + \int_0^\infty \tilde{q}_{kj}(s)\lambda^{-\gamma}(t^*-s)ds\right] \frac{(1+\varepsilon)z_j}{\lambda^{-\gamma}(t_0)}$$

$$\leq \sum_{j=1}^n \left[\tilde{p}_{kj} + \tilde{w}_{kj}\lambda^\gamma(\tau) + \int_0^\infty \tilde{q}_{kj}(s)\lambda^\gamma(s)ds\right] (1+\varepsilon)z_j\frac{\lambda^{-\gamma}(t^*)}{\lambda^{-\gamma}(t_0)}$$

$$< -\gamma(1+\varepsilon)z_k\frac{\lambda^{-\gamma}(t^*)}{\lambda^{-\gamma}(t_0)}\frac{\lambda'(t*)}{\lambda(t^*)} = y_k'(t^*),$$

which is a contradiction. ◇

Lemma 2. *Let* $\tilde{P}, \tilde{W}, \tilde{Q}(t), \tilde{Q}$ *and* D *be defined as in Lemma 1. Let* $V_i(x) = C^2[\mathbb{R}^n;\mathbb{R}^+]$, $i \in N$ *be functions such that for* $t \geq t_0$

$$LV_i(x(t)) \leq \sum_{j=1}^n \tilde{p}_{ij}V_j(x(t)) + \sum_{j=1}^n \tilde{w}_{ij}[V_j(x(t))]_\tau + \sum_{j=1}^n \int_0^\infty \tilde{q}_{ij}(s)V_j(x(t-s))ds$$

where $x(t)$ *is the solution of the system (2). Let for some* $k \in \mathbb{N}$ *hold* $\mathbb{E}V_i(x(t)) \leq z_i\frac{\lambda^{-\gamma}(t)}{\lambda^{-\gamma}(t_0)}$, $t \in (-\infty, t_k]$, $i \in N$ *where* $z = (z_1, ..., z_n)^T \in \Omega_M(D)$ *and* $\gamma \in (0, \gamma_0)$ *satisfies the inequality (4) from Lemma 1.*

Then
$$\mathbb{E}V_i(x(t)) \le z_i \frac{\lambda^{-\gamma}(t)}{\lambda^{-\gamma}(t_0)} \quad t \in [t_k, t_{k+1}). \tag{5}$$

Proof. This lemma is generalisation of Theorem 3.1 in [7]. The proof is similar and can be obtained with the help of Itô's formula, the upper-right Dini derivative and direct application of Lemma 1. ◇

The following theorem, which is the main result of this paper, gives sufficient conditions under which the impulsive stochastic neural network (2) is pth moment stable with a decay $\lambda(t)$ of order γ. In the proof, besides Lyapunov operator we use the famous Jensen's, Young's and Hölder's inequality.

Theorem 1. *Let $\phi \in \mathcal{L}^p_{\mathcal{F}_{t_0}}((-\infty, t_0]; \mathbb{R}^n)$, $\lambda(t)$ be a decay function, γ be defined as before and let the conditions (H1)-(H7) be satisfied for Eq.(2). Also, let $\tilde{P} = (\tilde{p}_{ij})_{n\times n}$, $\tilde{W} = (\tilde{w}_{ij})_{n\times n}$ and $\tilde{Q} = (\tilde{q}_{ij})_{n\times n} = \left(\int_0^\infty \tilde{q}_{ij}(s)ds\right)_{n\times n}$ be matrices defined as follows*

$$\begin{aligned}
\tilde{p}_{ij} &= \frac{m_i}{\min_{j\in N} m_j}\left(\overline{h}_i \beta_j \overline{a}_{ij} + (p-1)\mu_{ij}\right), \quad i \neq j \\
\tilde{p}_{ii} &= -p\underline{h}_i\underline{c}_i + (p-1)\overline{h}_i \sum_{j=1}^n \left[\beta_j \overline{a}_{ij} + \delta_j \overline{b}_{ij} + k_j \overline{d}_{ij} \int_0^\infty |l_{ij}(s)|ds\right] \\
&\quad + \overline{h}_i \beta_i \overline{a}_{ii} + (p-1)\mu_{ii} + \frac{1}{2}(p-1)(p-2)\sum_{j=1}^n (\mu_{ij} + \nu_{ij}), \qquad (6) \\
\tilde{w}_{ij} &= \frac{m_i}{\min_{j\in N} m_j}\left(\overline{h}_i \delta_j \overline{b}_{ij} + (p-1)\nu_{ij}\right), \quad \tilde{q}_{ij}(t) = \frac{m_i}{\min_{j\in N} m_j}\overline{h}_i \overline{d}_{ij} k_j |l_{ij}(t)|.
\end{aligned}$$

If $D = -(\tilde{P}+\tilde{W}+\tilde{Q})$ is a nonsingular M-matrix for some choice of the constants m_i, then the equilibrium solution $x(t) \equiv 0$ of Eq.(2) is pth moment stable with a decay function $\lambda(t)$ of order γ.

Proof. Let $V_i(x) = m_i|x_i|^p$ be Lyapunov functions, where $x \in \mathbb{R}^n$, $m_i > 0$, $i \in N$, and let $x(t)$ be the solution to Eq. (2).

$$\begin{aligned}
LV_i(x(t)) &= -pm_i|x_i(t)|^{p-2}x_i(t)h_i(x_i(t))c_i(t, x_i(t)) \\
&\quad + pm_i|x_i(t)|^{p-2}x_i(t)h_i(x_i(t))\sum_{j=1}^n a_{ij}(t)f_j(x_j(t)) \\
&\quad + pm_i|x_i(t)|^{p-2}x_i(t)h_i(x_i(t))\sum_{j=1}^n b_{ij}(t)g_j(x_{j,t}) \\
&\quad + pm_i|x_i(t)|^{p-2}x_i(t)h_i(x_i(t))\sum_{j=1}^n d_{ij}(t)\int_{-\infty}^t l_{ij}(t-s)k_j(x_j(s))ds \\
&\quad + \frac{p(p-1)}{2}m_i|x_i(t)|^{p-2}\sum_{j=1}^n \sigma_{ij}^2(t, x_j(t), x_{j,t}) \equiv \sum_{k=1}^5 I_k(t).
\end{aligned}$$

By applying $(\mathbf{H_1}) - (\mathbf{H_4})$ and Young's inequality we get the following estimates

$$I_1(t) \le -pm_i\underline{h}_i\underline{c}_i|x_i(t)|^p, \tag{7}$$

$$I_2(t) \le (p-1)m_i\overline{h}_i|x_i(t)|^p\sum_{j=1}^n\beta_j\overline{a}_{ij} + \frac{m_i}{\min_{j\in N}m_j}\overline{h}_i\sum_{j=1}^n m_j\beta_j\overline{a}_{ij}|x_j(t)|^p, \tag{8}$$

$$I_3(t) \le (p-1)m_i\overline{h}_i|x_i(t)|^p\sum_{j=1}^n\delta_j\overline{b}_{ij} + \frac{m_i}{\min_{j\in N}m_j}\overline{h}_i\sum_{j=1}^n m_j\delta_j\overline{b}_{ij}|x_j(t)|^p_\tau, \tag{9}$$

$$I_4(t) \le (p-1)m_i\overline{h}_i|x_i(t)|^p\sum_{j=1}^n\overline{d}_{ij}k_j\int_0^\infty|l_{ij}(s)|ds$$
$$+\frac{m_i}{\min_{j\in N}m_j}\overline{h}_i\sum_{j=1}^n\overline{d}_{ij}k_j\int_0^\infty m_j|l_{ij}(s)||x_j(t-s)|^p ds, \tag{10}$$

$$I_5(t) \le \frac{(p-1)(p-2)}{2}m_i|x_i(t)|^p\sum_{j=1}^n(\mu_{ij}+\nu_{ij})$$
$$+(p-1)\frac{m_i}{\min_{j\in N}m_j}\Big(\sum_{j=1}^n\mu_{ij}m_j|x_j(t)|^p + \sum_{j=1}^n\nu_{ij}m_j|x_j(t)|^p_\tau\Big). \tag{11}$$

Combining the inequalities (7)-(11) we get

$$LV_i(x(t)) \le \sum_{j=1}^n\tilde{p}_{ij}V_j(x(t)) + \sum_{j=1}^n\tilde{w}_{ij}V_j(x)_\tau + \sum_{j=1}^n\int_0^\infty\tilde{q}_{ij}(s)V_j(x(t-s))ds, \tag{12}$$

where $\tilde{p}_{ii}, \tilde{p}_{ij}, \tilde{w}_{ij}, \tilde{q}_{ij}(t)$ are given by (6).

Since $\phi \in \mathcal{L}^p_{\mathcal{F}_{t_0}}((-\infty,t_0];\mathbb{R}^n)$, $D = -(\tilde{P}+\tilde{W}+\tilde{Q})$ is an M-matrix and $\lambda(t)$ is a non-decreasing function on $(-\infty,t_0]$, we get that for $t\le t_0$, $z^{(0)}\in\Omega_M(D)$ and $\gamma<\gamma_0$ it holds $\mathbb{E}V_i(x(t)) \le \max_{i\in N}m_i\frac{z_i^{(0)}}{\min_{i\in N}z_i^{(0)}}||\phi||^p\frac{\lambda^{-\gamma}(t)}{\lambda^{-\gamma}(t_0)} = dz_i^{(0)}\frac{\lambda^{-\gamma}(t)}{\lambda^{-\gamma}(t_0)}$. Let now z_0 be chosen such that $(\mathbf{H_7})$ holds. From $dz^{(0)} = z \in \Omega_M(D)$ we conclude that the conditions from Lemma 2 are fulfilled and this together with (12) gives $\mathbb{E}V_i(x(t)) \le z_i\frac{\lambda^{-\gamma}(t)}{\lambda^{-\gamma}(t_0)}$ for $t\in[t_0,t_1)$. Let us suppose that for $m = 1,..,k$ it holds $\mathbb{E}V_i(x(t)) \le z_i\frac{\lambda^{-\gamma}(t)}{\lambda^{-\gamma}(t_0)}$ for $t\in[t_{m-1},t_m)$. Then, for $x = t_k$ by $(\mathbf{H_6})$ and applying Jensen's and Hölder's inequality we get

$$\mathbb{E}V_i(x(t_k)) \le 2^{p-1}\mathbb{E}\Big[\Big(\sum_{j=1}^n p_{ij}^{(k)}|x_j(t_k^-)|\Big)^p + \Big(\sum_{j=1}^n q_{ij}^{(k)}|x_{j,t_k^-}|\Big)^p\Big]$$
$$\le 2^{p-1}\left[\Big(\sum_{j=1}^n(p_{ij}^{(k)})^{\frac{p}{p-1}}\Big)^{p-1}\sum_{j=1}^n\mathbb{E}|x_j(t_k^-)|^p + \Big(\sum_{j=1}^n(q_{ij}^{(k)})^{\frac{p}{p-1}}\Big)^{p-1}\sum_{j=1}^n\mathbb{E}|x_{j,t_k^-}|^p\right]$$
$$\le 2^{p-1}\left[\Big(\sum_{j=1}^n(p_{ij}^{(k)})^{\frac{p}{p-1}}\Big)^{p-1} + \Big(\sum_{j=1}^n(q_{ij}^{(k)})^{\frac{p}{p-1}}\Big)^{p-1}\lambda^\gamma(\tau)\right]\sum_{j=1}^n z_j\frac{\lambda^{-\gamma}(t_k)}{\lambda^{-\gamma}(t_0)} \le z_i\frac{\lambda^{-\gamma}(t_k)}{\lambda^{-\gamma}(t_0)}.$$

Now Lemma 2 implies that for $t \in [t_k, t_{k+1})$ and $i \in N$ $\mathbb{E}V_i(x(t)) \le z_i \frac{\lambda^{-\gamma}(t)}{\lambda^{-\gamma}(t_0)}$ and by the principal of mathematical induction the last inequality holds for any $t \ge t_0$. Finally, for the pth moment we have

$$\mathbb{E}||x(t)||^p \le \frac{1}{\min\limits_{i\in N} m_i} \sum_{i=1}^{n} \mathbb{E}V_i(x(t)) \le \frac{1}{\min\limits_{i\in N} m_i} (\sum_{i=1}^{n} z_i) \frac{\lambda^{-\gamma}(t)}{\lambda^{-\gamma}(t_0)} = c(\phi)\lambda^{-\gamma}(t). \ \diamond$$

4 Numerical Example

We consider the 4th moment general decay stability for the impulsive stochastic neural network represented by Eq. (2) where $x_0 = \phi \in \mathcal{L}^4_{\mathcal{F}_0}((\infty, 2]; \mathbb{R}^2)$, $w(t)$ is a two-dimensional Brownian motion, delay functions are $\tau_{11}(t) = \frac{1}{4}(1 + \sin t)$, $\tau_{12}(t) = 0$, $\tau_{21}(t) = 0$, $\tau_{22}(t) = \frac{1}{4}(1 + \cos t)$, $\tau = 0.5$, $x = (x_1, x_2)^T, y = (y_1, y_2)^T \in \mathbb{R}^2$, $t \ge 2$, the impulsive moments are $t_k = 2^{k+1}$ for $k \in \mathbb{N}$ and

$$H(x) = \begin{bmatrix} \frac{3}{1+2|\sin x_1|} & 0 \\ 0 & \frac{3}{1+2e^{-|x_2|}} \end{bmatrix}, \quad C(t,x) = \begin{bmatrix} (23 + \frac{1}{1+t}) x_1 \\ (23 + \frac{2}{1+t}) x_2 \end{bmatrix},$$

$$A(t) = \begin{bmatrix} \frac{1}{3(1+t)} & -\frac{1}{1+t} \\ -\frac{2}{1+t} & \frac{2}{3(1+t)} \end{bmatrix}, \quad B(t) = \begin{bmatrix} -1 - \frac{3}{1+t} & 0 \\ -1 & \frac{3}{1+t} \end{bmatrix},$$

$$F(x) = \begin{bmatrix} x_1 e^{-|x_1|} \\ x_2 e^{-|x_2|} \end{bmatrix}, \quad G(y) = \begin{bmatrix} \sin y_1 \\ \sin y_2 \end{bmatrix}, \quad D(t) = \begin{bmatrix} \sin t & e^{2-t} \\ \cos t & 1 \end{bmatrix}, \quad K(x) = \begin{bmatrix} x_1 \\ x_2 \end{bmatrix},$$

$$L(t) = \begin{bmatrix} \frac{4}{\pi(1+t^2)^2} & \frac{4}{\pi(1+t^2)^2} \\ \frac{4}{\pi(1+t^2)^2} & \frac{4}{\pi(1+t^2)^2} \end{bmatrix}, \quad \sigma(t,x,y) = \begin{bmatrix} \sqrt{\frac{6}{1+t}} y_1 & -\sqrt{\frac{6}{1+t}} x_2 \\ -\sqrt{\frac{3}{1+t}} x_1 & \sqrt{\frac{3}{1+t}} y_2 \end{bmatrix}.$$

Also for $k \in \mathbb{N}$, $p_{1k}(x) = \frac{1}{4}x_1$, $p_{2k}(x) = \frac{1}{4}x_2$, $q_{1k}(y) = \frac{1}{2^{k+2}}x_1$, $q_{2k}(y) = \frac{1}{2^{k+2}}y_2$. We can easily calculate that:

$$c_1 = 23 + \frac{1}{3}, \ c_2 = 23 + \frac{2}{3}, \ \underline{h}_i = 1, \overline{h}_i = 3, \quad \beta_i = \delta_i = k_i = d_{ij} = 1, \quad i,j = 1,2,$$
$$a_{11} = \frac{1}{9}, \ a_{12} = \frac{1}{3}, \ a_{21} = \frac{2}{3}, \ a_{22} = \frac{2}{9}, \quad b_{11} = 2, \ b_{12} = 0, \ b_{21} = 1, \ b_{22} = 1,$$
$$\mu_{11} = \mu_{22} = 0, \ \mu_{12} = 2, \ \mu_{21} = 1, \quad \nu_{12} = \nu_{21} = 0, \ \nu_{11} = 2, \ \mu_{22} = 1,$$
$$p_{11}^{(k)} = \frac{1}{4}, \ p_{12}^{(k)} = 0, \ p_{21}^{(k)} = 0, \ p_{22}^{(k)} = \frac{1}{4}, \quad q_{ij}^{(k)} = \frac{1}{2^{k+2}}, \quad i = 1,2, \ k \in \mathbb{N}.$$

We also observe that $l_{ij} \notin \wp(e^t)$ since $\int_0^\infty \frac{e^{\gamma t}}{(1+t^2)^2} dt$ does not converge for any $\gamma > 0$. Thus we can not say anything about the pth moment exponential stability using the theory presented in [7]. We will show that the neural network is pth moment stable for the general decay function $\lambda(t)$ defined with $\lambda(t) = 1$ for $t \le 0$ $\lambda(t) = 1 + t$ for $t \ge 0$. First we observe that $l_{ij} \in \wp(\lambda(t))$ and $\int_0^\infty \frac{(1+t)^\gamma}{(1+t^2)^2} dt < \infty$ for $\gamma < 3$. Next, for $p = 4$ we calculate the parameters in Eq.

(6) : $\tilde{P} = \begin{bmatrix} -41 & 7 \\ 5 & -44 \end{bmatrix}$, $\tilde{W} = \begin{bmatrix} 12 & 0 \\ 3 & 6 \end{bmatrix}$, $\tilde{Q} \begin{bmatrix} 3 & 3 \\ 3 & 3 \end{bmatrix}$, $D = \begin{bmatrix} 26 & -10 \\ -11 & 35 \end{bmatrix}$. D is an M-matrix and $\Omega_M(D) = \{z > 0 \mid \frac{11}{35} z_1 < z_2 < 2.6 z_1\}$. For $z = (1,1)^T$ and $\gamma = 1$ the condition ($\mathbf{H_6}$) can be verified. All the conditions in Th. 1 are fulfilled and we can conclude that the model is 4th moment stable with a decay $\lambda(t)$ of order $\gamma = 1$. Fig. 1 below shows one simulation of the neural network with an initial condition $\phi = 1$. The 4th moment $\mathbb{E}||x(t)||^4$ is plotted on Fig. 2.

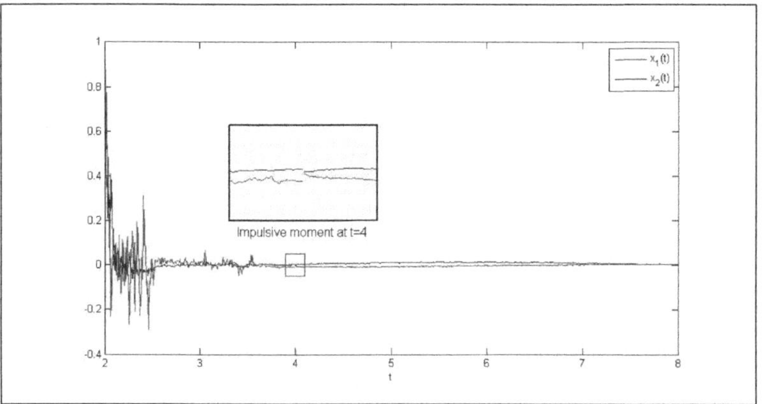

Fig. 1. Simulation of the neural network on the interval $[2, 8)$ for $\phi = 1$

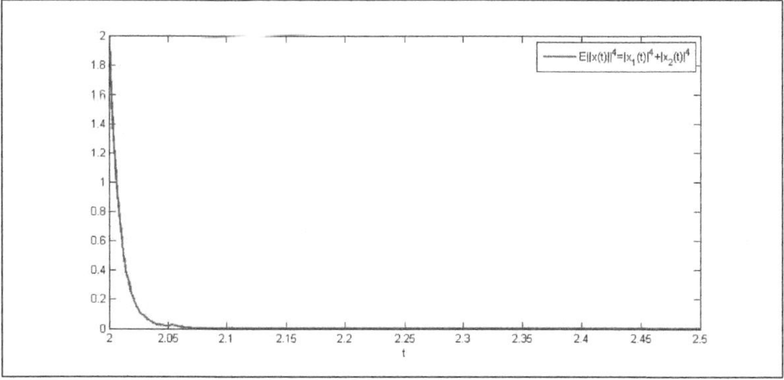

Fig. 2. 4th moment $\mathbb{E}||x(t)||^4$ for the given example

5 Conclusions

In this paper we generalise the results on pth moment exponential stability of impulsive stochastic Cohen-Grossberg neural networks presented in ([7]). By generalising the model and the definition of a decay function and by substituting some of the model assumptions, we give a new result on the pth moment stability with a general decay. We support our theory with a numerical example which can not be discussed with the theory presented in ([7]).

References

1. von Neumann, J.: The computer and the brain. Yale University Press, New Haven (1958)
2. Taylor, J.G.: Spontaneous behaviour in neural networks. J. Theor. Biol. 36(3), 513–528 (1972)
3. Sejnowski, T.J.: Skeleton filters in the brain. In: Hinton, G.E., Anderson, J.A. (eds.) Parallel Models of Associative Memory, Hillsdale LEE, New Jersey, pp. 189–208 (1981)
4. Arbib, M.A.: Brains, Machines, and Mathematics. Springer, New York (1987)
5. Huang, C., He, Y., Chen, P.: Dynamic analysis of stochastic recurrent neural networks. Neural Process. Lett. 27, 267–276 (2008)
6. Huang, C., Cao, J.: Stochastic dynamics of nonautonomous - Grossberg neural networks. Abstract Appl. Anal. Article ID 297147 (2011)
7. Wang, X., Guo, Q., Xu, D.: Exponential p-stability of impulsive stochastic Cohen-Grossberg neural networks with mixed delays. Math. Comput. Simul. 79(5), 1698–1710 (2009)
8. Peng, S., Zhang, Y.: Razumikhin-type theorems on pth moment exponential stability of impulsive stochastic delay differential equations. IEEE Transactions on Automatic Control 55(8), 1917–1922 (2010)
9. Randjelović, J., Janković, S.: On the pth moment exponential stability criteria of neutral stochastic functional differential equations. J. Math. Anal. Appl. 326, 266–280 (2007)
10. Pavlović, G., Janković, S.: Razumikhin-type theorems on general decay stability of stochastic functional differential equations with infinite delay. J. Comp. Appl. Math. 236, 1679–1690 (2012)
11. Hu, Y., Huang, C.: Lasalle method and general decay stability of stochastic neural networks with mixed delays. J. Appl. Math. Computing 38(1-2), 257–278 (2012)
12. Tojtovska, B., Jankovič, S.: On a general decay stability of stochastic Cohen-Grossberg neural networks with time-varying delays. Appl. Math. Comput. 219 (2012)
13. Yegnanarayana, B.: Artificial neural networks. PHI Learning Pvt. Ltd. (2004)
14. Mao, X.: Stochastic differential equations and applications, 2nd edn. Horwood Publishing, Chichester (2007)
15. Horn, R.A., Johnson, C.R.: Matrix Analysis. Cambridge University Press (1990)
16. Song, Q., Wang, Z.: Stability analysis of impulsive stochastic Cohen-Grossberg neural networks with mixed time delays. Physica A 387, 3314–3326 (2008)
17. Xu, D., Yang, Z.: Impulsive delay differential inequality and stability of neural networks. J. Math. Anal. Appl. 305(1), 107–120 (2005)

How Lightweight Is the Hardware Implementation of Quasigroup S-Boxes

Hristina Mihajloska[1], Tolga Yalcin[2], and Danilo Gligoroski[3]

[1] Faculty of Computer Science and Engineering, UKIM, Skopje, Macedonia
hristina.mihajloska@finki.ukim.mk
[2] Embedded Security Group, HGI, Ruhr-University Bochum, Germany
tolga.yalcin@rub.de
[3] Department of Telematics, NTNU, Trondheim, Norway
danilog@item.ntnu.no

Abstract. In this paper, we present a novel method for realizing S-boxes using non-associative algebraic structures - quasigroups, which - in certain cases - leads to more optimized hardware implementations. We aim to give cryptographers an iterative tool for designing cryptographically strong S-boxes (which we denote as Q-S-boxes) with additional flexibility for hardware implementation. Existence of the set of cryptographically strong 4-bit Q-S-boxes depends on the non-linear quasigroups of order 4 and quasigroup string transformations. The Q-S-boxes offer the option to not only iteratively reuse the same circuit to implement several different strong 4-bit S-boxes, but they can also be serialized down to bit level, leading to S-box implementations below 10 GEs. With Q-S-boxes we can achieve over 40% area reduction with respect to a lookup table based implementation, and also over 16% area reduction in a parallel implementation of Present. We plan to generalize our approach to S-boxes of any size in the future.

Keywords: lightweight cryptography, S-boxes, ASIC implementation, quasigroup S-boxes.

1 Introduction

Today, we live in a time where computing reaches its third phase, in which pervasive computing takes over important roles in daily life. Huge deployment of pervasive devices (lightweight solutions), on one hand, promises many benefits that makes our lives better, but on the other hand, opens huge number of questions related to the security and privacy of these devices. In this sense, implementing security into lightweight solutions requires special cryptographic techniques that should be applied in resource-constrained environment. These can be done via lightweight cryptography. The term "lightweight" does not necessary imply weak cryptography. On the contrary, its main goal is to achieve uncompromised security by means of new secure algorithms whose implementation requires as lightweight hardware and software area as possible [1].

S. Markovski and M. Gusev (Eds.): *ICT Innovations 2012*, AISC 207, pp. 121–128.
DOI: 10.1007/978-3-642-37169-1_12

Example for lightweight devices can be RFID tags or smart cards which are used in many industries. They are immensely used in electronic payment, product tracking, transportation and logistics, access control and identification systems. RFID tags use radio-frequency electromagnetic fields to transfer data from a tag attached to an object. In general, they can be divided into passive and active according to the power source that they use. Active RFID tags provide their own power supply in form of battery, whereas passive are powered by the electromagnetic fields used to read them [2]. They have to be extremely low-cost and low-power devices, meaning that they have to be implemented in minimal chip area. Therefore, in what follows we limit our focus to passive RFID tags. We note that in the rest of this paper talking about RFID tags, we refer to the passive tags.

Because of the fact that RFID tags are deployed into untreated environment where enemy has full control over the device and physical access to it, greater attention must be paid to their security. To provide security of these devices, new cryptographic protocols have been proposed in the last few years. These protocols basically used symmetric cryptography, which can be successfully implemented in constrained hardware environment. In most cases of RFID tags, low cost and low power require symmetric cryptographic primitives with low gate count to be used. In order to achieve better results for low gate count in RFID tags, researchers have focused on block ciphers [3] [4] and hash functions [5] [6].

The structure of the paper is the following. In Section 2, we give a brief overview of the lightweight implementation of S-boxes of proposed lightweight block ciphers. In Section 3, we present an iterative tool for designing cryptographically strong S-boxes with the help of quasigroups of order 4. In section 4, hardware implementation results for Q-S-boxes are provided and the results are compared with previous work. Finally, the paper is concluded with future directions in Section 5.

2 Previous Work

From cryptographic point of view, unarguably, the most critical components in lightweight symmetric cipher design are substitution boxes, S-boxes. They are responsible for confusing the input data via their highly non-linear properties. On the other hand, from a lightweight implementation point of view, they are the highest area occupying modules within the cipher and basically determine the overall cipher area. Therefore, implementing lightweight solutions for block ciphers depends on how lightweight can be presented their S-boxes.

In what follows, we are focused on lightweight S-box implementations. It is very common practice to state the area of a digital block implemented on silicon in term of gate equivalents (GE). One GE is equivalent to the area of the two-input NAND gate with the lowest driving strength of the corresponding technology [4].

mCRYPTON is the first special design of block cipher targeted for tiny pervasive devices. It is designed by following the overall architecture of Crypton, but with a little bit simplification of the building blocks to fit the block/key sizes. The nonlinear substitution building block uses four 4×4-bit S-boxes, each of which occupies 27 GE [7].

In [8], a lightweight hardware implementation of the known cipher DES is presented by Leander *et al.* at FSE 2007. DESXL is a modified variant of DES, where the eight S-boxes are substituted by a single cryptographically stronger S-box which is repeated eight times. This 6×4-bit S-box is implemented in a serialized ASIC design, which requires 32 GE for a single 4-bit S-box.

In the same year, Bogdanov *et al.* at CHES conference presented a new ultra-lightweight block cipher, PRESENT [9]. The design of PRESENT is extremely hardware efficient, and the serial implementation of it with 80-bit key length requires as low as 1000 GE [4]. The PRESENT S-box is a 4-bit S-Box which requires 28 GE.

In 2009, Hummingbird is proposed as a new ultra-lightweight crypto algorithm for RFID tags by Engels and Smith *et al.* [10]. It has a hybrid structure of block cipher and stream cipher and was developed with a minimal hardware footprint in mind. Hummingbird uses four Serpent type 4-bit S-boxes which require hardware area in the rage of 20 to 40 GE [11].

In [12], a new lightweight cipher proposal is presented at CHES 2011. This is a 64-bit block cipher, LED, which is as small as PRESENT and faster in software, but slower in hardware. It uses the same PRESENT type S-box in its non-linear layer.

Also in 2011, TWINE, another of the family of lightweight block ciphers [13], was presented. It is the first cipher that combines features like no bit permutation, no Galois-Field matrix and generalized Feistel network. The components that are used are only one 4-bit S-box and 4-bit XOR. The ASIC hardware implementation of TWINE S-box is about 30 GE.

In all these cases, all research made on the 4-bit S-boxes has focused on lookup table based S-boxes. While, they are very well established and easy to analyze, their design space is very limited with a minimum achievable area of about 20 GE, without compromising security. They cannot be optimized via folding, serializing, or any other means, without additional area overhead. Our approach defers from the existing via its specific methodology, which offers a more optimized hardware implementation of S-boxes suitable for lightweight ciphers in RFID tags.

3 Our Approach

In this paper, we present a novel method for realizing S-boxes using non-associative algebraic structures - quasigroups, which - in certain cases - leads to more optimized hardware implementations. We aim to give cryptographers an iterative tool for designing cryptographically strong S-boxes (which we denote as Q-S-boxes) with additional flexibility for hardware implementation. Existence of the set of cryptographically strong 4×4-bit Q-S-boxes depends on the non-linear quasigroups of order 4 and quasigroup string transformations.

Quasigroups of order 4, themselves are 4×2-bit S-boxes. If they are presented as vector valued Boolean functions [14] and more precisely in their ANF, it can be seen that their maximal algebraic degree is 2. Quasigroup string transformations have an advantage to transform a given string with length n bijectively to output string with the same length n, and also, to raise the algebraic degree of the final bijection output. Therefore, quasigroups of order 4 with algebraic degree 2 (called non-linear quasigroups) and quasigroup string transformation (called e-transformation) with adequately chosen leaders generate 4×4-bit Q-S-boxes. An algorithm for generating Q-S-boxes is given in our previous paper [15]. The minimum number of rounds (iterations) for this methodology is 4. Hence, with this methodology we can generate Q-S-boxes in different ways depending on the number of rounds and the number of leaders that we have chosen. In the previous paper [15] we defined an optimal S-box as a Boolean map $S : \mathbb{F}_2^4 \rightarrow \mathbb{F}_2^4$ which is a bijection, has algebraic degree of 3 on all the output bits, and has linearity of 1/4 and differential potential of 1/4. Also, we presented results with 2, 4 and 8 different leaders and 4 and 8 rounds, respectively, and we listed all the Q-S-boxes that fulfill the predetermined criteria to be an optimal.

3.1 Sample Q-S-Box

We apply the methodology for generating cryptographically strong Q-S-boxes to generate a sample Q-S-box with a given quasigroup of order 4 (chosen from the class of 432 non-linear quasigroups) and two different leaders.

Example 1. Let $(Q, *)$ be one non-linear quasigroup and l_1 and l_2 (leaders) be elements from the set $Q = \{0, 1, 2, 3\}$. We present leaders and quasigroup with two-digit binary representation, like:

$l_1 = 1 \rightarrow 01$
$l_2 = 3 \rightarrow 11$

*	0	1	2	3
0	0	2	1	3
1	2	1	3	0
2	1	3	0	2
3	3	0	2	1

$\Longrightarrow$

*	00	01	10	11
00	00	10	01	11
01	10	01	11	00
10	01	11	00	10
11	11	00	10	01

We get the first input block of 4 bits in lexicographic ordering (0000), and then the method for producing the output is shown graphically in Figure 1.

Afterwards, we repeat this procedure on all possible 4-bit input values in lexicographic order, and we obtain permutation of order 16, which is our Q-S-box. The corresponding Q-S-box from this example given in its hexadecimal notation is shown in Table 1.

4 Hardware Implementation of Q-S-Boxes

We start our hardware implementation by mapping the algorithm for generating optimal Q-S-boxes directly to hardware. Like in the Example 1, we implement

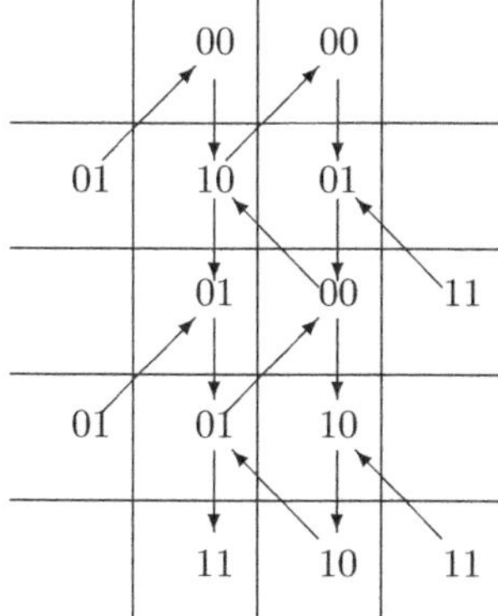

Fig. 1. Four e-transformations that bijectively transforms 4 bits into 4 bits by a quasigroup of order 4

Table 1. Sample of Q-S-box

x	0	1	2	3	4	5	6	7	8	9	A	B	C	D	E	F
S(x)	E	6	C	B	0	1	8	2	D	3	A	F	9	5	4	7

a Q-S-Box of $4 - th$ degree, i.e. with 4 layers of non-linear mappings. This requires 4 × 2-bit lookup tables and multiplexers as shown in Figure 2. The resulting Q-S-box has been synthesized with non-linear quasigroup and several different leaders. In all the cases, it has been found to occupy between 38-46 GE. This number is far above the gate counts of existing ciphers. However, the S-box has 4 layers of non-linear mappings inside, which means that it should be possible to implement the S-box in a round-based approach. This is equivalent to implementing only 2 or 1 of the non-linear mapping layers inside the Q-S-box (which we refer to as Q-S-box rounds) and executing it twice or 4 times, respectively, for each substitution layer operation. One might argue that such a solution would require additional registers for the storage of temporary state of Q-S-box outputs. However this is not the case, since the registers are already integral part of the cipher because of the round based operation. Therefore, it can be realized with minimum overhead with respect to a single-round-operation S-box. Figures 3 and 4 show the multi-round Q-S-box (with a single non-linear layer and 4-round operation) alone and with a generic SP-network cipher.

Gate count of such a Q-S-box heavily depends on the overall cipher structure. Its standalone gate count would be huge due to the extra registers and multiplexers required for the multi-round operation. However, the actual (or effective) gate count depends on the reduction it causes on the overall area of the cipher it is part of. Therefore, we chose the PRESENT block cipher as our target platform and implemented it first with its standard S-boxes in order to determine the overall cipher area and the area of the S-boxes. In the next step, we replaced the PRESENT lookup table based S-box with several different 4-round Q-S-boxes that have the same cryptographic properties as the original PRESENT S-Box.

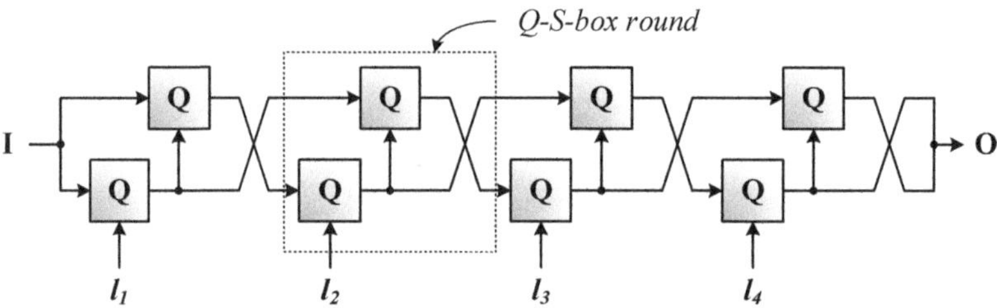

Fig. 2. Direct Q-S-box implementation

As a result, we have observed that the effective area of the 4-round Q-S-box to be between 15.5 to 19.75 GE. This corresponds to a $30 - 45\%$ reduction on the S-box area and $12 - 18\%$ on the overall cipher area, depending on the choice of quasigroup and leaders.

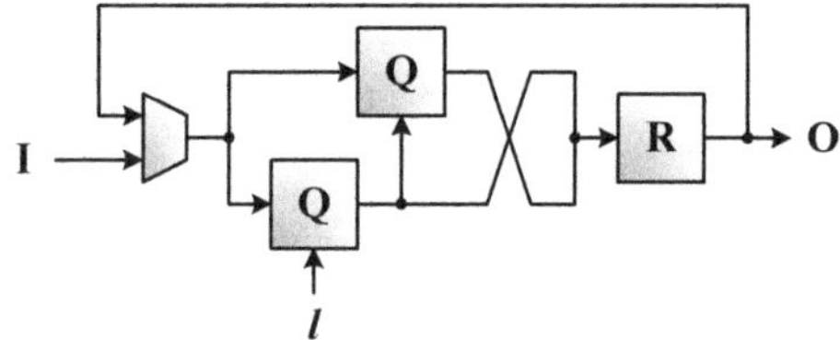

Fig. 3. Multi-round Q-S-box implementation

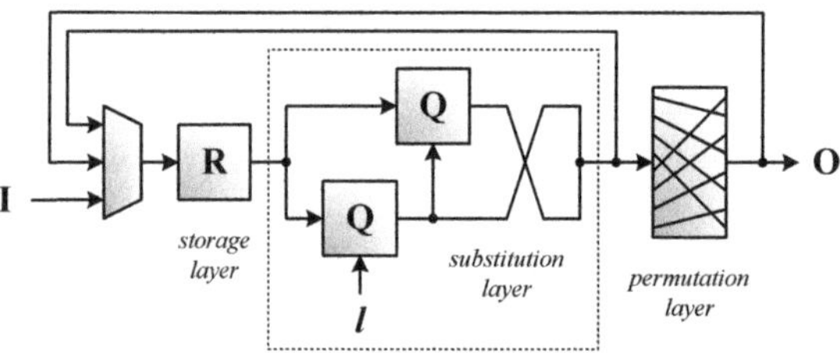

Fig. 4. Multi-round Q-S-box based PRESENT-like cipher implementation

The resultant cipher has 25% of the data throughput of the original cipher. However, compared to a serialized implementation, where the throughput drops to $1/16 - th$, it is still an acceptable compromise. In terms of time-area product, the Q-S-box parallel cipher solution is 3.2 times worse than the original parallel cipher, whereas a serialized implementation is 10 times worse.

We repeated the same procedure to a 2-round Q-S-box based cipher, in the hopes of enhancing the time-area product. However, it turned out that the resultant Q-S-box can achieve only the gate count of the original PRESENT S-box, in the very best case, with zero overall improvement in the cipher area. In similar way, we implemented Q-S-boxes with only a single 4×2 lookup table, with which we could reduce the area by another $2-3\%$ w.r.t. a full 4×4 lookup table based approach, in certain cases. But we deem this much reduction not practical considering the further throughput reduction by a factor 2.

Another argument is that the resultant cipher does not any more hold the cryptographic properties of the original PRESENT cipher. With a completely different substitution layer, all of its cryptanalysis have to be done from scratch. This, in fact, is true. However, it should be noted that we have chosen PRESENT only as the development and proof-of-concept platform for our Q-S-box proposal. In a completely new cipher, designed from scratch with properties and hardware structures of Q-S-boxes taken into consideration, it will be possible to achieve the same security levels as the PRESENT cipher, and possibly even more reduction in gate count. As a matter of fact, applying the Q-S-boxes in another existing cipher might result in completely different area reduction figures. Another active research we still work on is the ability to represent any given S-box as a Q-S-boxes, which would solve all the above mentioned problems.

5 Conclusion and Future Work

Q-S-boxes offer the option to not only iteratively reuse the same circuit to implement several different strong 4-bit S-boxes, but they can also be serialized down to bit level, leading to S-box implementations below 10 GE. With Q-S-boxes we can get over the 40% of area reduction w.r.t. a lookup table based implementation, and also over the 16% of area reduction w.r.t. a parallel implementation of PRESENT.

As the future work, we will focus on two different directions: The first is to extend our current approach to S-boxes of any size, while the second, also the harder, is to express any given S-box as a Q-S-box. This might require application of different techniques such as application of Gröbner bases and brute-force search (possibly with hardware acceleration), all of which will help to extend the application of quasigroups into different fields of research.

References

1. Eisenbarth, T., Kumar, S., Paar, C., Poschmann, A., Uhsadel, L.: A Survey of Lightweight-Cryptography Implementations. IEEE Des. Test 24(6), 522–533 (2007)
2. Finkenzeller, K.: RFID Handbook. John Wiley, Chichester (2003)
3. Feldhofer, M., Wolkerstorfer, J., Rijmen, V.: AES Implementation on a Grain of Sand. Information Security IEEE Proc. 152(1), 13–20 (2005)
4. Rolfes, C., Poschmann, A., Leander, G., Paar, C.: Ultra-Lightweight Implementations for Smart Devices – Security for 1000 Gate Equivalents. In: Grimaud, G., Standaert, F.-X. (eds.) CARDIS 2008. LNCS, vol. 5189, pp. 89–103. Springer, Heidelberg (2008)

5. ONeill, M.: Low-Cost SHA-1 Hash Function Architecture for RFID Tags. In: Proceedings of RFIDSec (2008)
6. Yoshida, H., Watanabe, D., Okeya, K., Kitahara, J., Wu, H., Küçük, Ö., Preneel, B.: MAME: A compression function with reduced hardware requirements. In: Paillier, P., Verbauwhede, I. (eds.) CHES 2007. LNCS, vol. 4727, pp. 148–165. Springer, Heidelberg (2007)
7. Lim, C.H., Korkishko, T.: mCrypton – A Lightweight Block Cipher for Security of Low-Cost RFID Tags and Sensors. In: Song, J., Kwon, T., Yung, M. (eds.) WISA 2005. LNCS, vol. 3786, pp. 243–258. Springer, Heidelberg (2006)
8. Leander, G., Paar, C., Poschmann, A., Schramm, K.: New Lightweight DES Variants. In: Biryukov, A. (ed.) FSE 2007. LNCS, vol. 4593, pp. 196–210. Springer, Heidelberg (2007)
9. Bogdanov, A., Knudsen, L.R., Leander, G., Paar, C., Poschmann, A., Robshaw, M.J.B., Seurin, Y., Vikkelsoe, C.: PRESENT: An Ultra-Lightweight Block Cipher. In: Paillier, P., Verbauwhede, I. (eds.) CHES 2007. LNCS, vol. 4727, pp. 450–466. Springer, Heidelberg (2007)
10. Engels, D., Fan, X., Gong, G., Hu, H., Smith, E.: Ultra-Lightweight Cryptography for Low-Cost RFID Tags: Hummingbird Algorithm and Protocol. Technical report, `http://cacr.uwaterloo.ca/techreports/2009/cacr2009-29.pdf`
11. Leander, G., Poschmann, A.: On the Classification of 4 Bit S-Boxes. In: Carlet, C., Sunar, B. (eds.) WAIFI 2007. LNCS, vol. 4547, pp. 159–176. Springer, Heidelberg (2007)
12. Guo, J., Peyrin, T., Poschmann, A., Robshaw, M.: The LED Block Cipher. In: Preneel, B., Takagi, T. (eds.) CHES 2011. LNCS, vol. 6917, pp. 326–341. Springer, Heidelberg (2011)
13. Suzaki, T., Minematsu, K., Morioka, S., Kobayashi, E.: TWINE: A Lightweight, Versatile Block Cipher. In: ECRYPT Workshop on Lightweight Cryptography 2011 (2011)
14. Gligoroski, D., Dimitrova, V., Markovski, S.: Quasigroups as Boolean Functions, Their Equation Systems and Gröbner Bases. In: Sala, M., Mora, T., Perret, L., Sakata, S., Traverso, C. (eds.) Gröbner Bases, Coding, and Cryptography, pp. 415–420. Springer, Heidelberg (2009)
15. Mihajloska, H., Gligoroski, D.: Construction of Optimal 4-bit S-boxes by Quasigroups of Order 4. In: The Sixth International Conference on Emerging Security Information, Systems and Technologies, SECURWARE 2012, Rome, Italy (2012) (Best paper award)
16. Chabaud, F., Vaudenay, S.: Links between Differential and Linear Cryptanalysis. In: De Santis, A. (ed.) EUROCRYPT 1994. LNCS, vol. 950, pp. 356–365. Springer, Heidelberg (1995)

Comparison of Models for Recognition of Old Slavic Letters

Mimoza Klekovska[1], Cveta Martinovska[2], Igor Nedelkovski[1], and Dragan Kaevski[3]

[1] Faculty of Technical Sciences, University St. Kliment Ohridski, Bitola, R. Macedonia
mimiklek@yahoo.com, igor.nedelkovski@uklo.edu.mk
[2] Computer Science Faculty, University Goce Delcev, Stip, R. Macedonia
cveta.martinovska@ugd.edu.mk
[3] Faculty of Electrical Engineering and Information Technologies, University St. Cyril and Methodius, Skopje, R. Macedonia
d.kaevski@gmail.com

Abstract. This paper compares two methods for classification of Old Slavic letters. Traditional letter recognition programs cannot be applied on the Old Slavic Cyrillic manuscripts because these letters have unique characteristics. The first classification method is based on a decision tree and the second one uses fuzzy techniques. Both methods use the same set of features extracted from the letter bitmaps. Results from the conducted research reveal that discriminative features for recognition of Church Slavic Letters are number and position of spots in the outer segments, presence and position of vertical and horizontal lines, compactness and symmetry. The efficiency of the implemented classifiers is tested experimentally.

Keywords: Handwritten Letter Recognition, Fuzzy Decision, Decision Tree, Feature Extraction, Recognition Accuracy and Precision.

1 Introduction

This paper describes an analysis of digital samples extracted from old manuscripts written in Cyrillic alphabet with constitutional script. The analysis is not related to the literary form or content which is usually of interest in Slavic research studies but investigates the construction of the letters. The subject of the analysis is the contour of a particular grapheme which is extracted from the digital records of the manuscripts. The letter is analyzed as an object with multiple parameters: geometric-mathematical, topological, graphical-aesthetic, construction elements and emanated symbolism. It is observed as an integrated union of information.

The knowledge obtained from the letters analysis consists of style descriptions, harmonic proportions as well as recognition features. This innovative approach uses elements that are closer to human perception of form through visual sensation. The research contributes to the digitization efforts of Church Slavic Cyrillic manuscripts

S. Markovski and M. Gusev (Eds.): *ICT Innovations 2012*, AISC 207, pp. 129–139.
DOI: 10.1007/978-3-642-37169-1_13

found in Macedonian churches and monasteries and is applicable in the work of the Slavic Institutes, libraries and archives.

The existing computer software for letter recognition is not appropriate for these handwritten historical collections due to their specific properties. There are attempts to establish databases of historical Slavic handwritten manuscripts for bibliographical and research purposes [1].

In OCR (ICR) research different approaches are used for letter recognition [2] [3]. Some of the systems incorporate stochastic, neural network, structural or syntactic techniques [4] [5]. Structural pattern recognition is based on a decomposition of patterns in simpler forms while syntactic approach describes patterns with a set of strings generated by formal or fuzzy grammars. While stochastic approaches use the probability that the letter belongs to a class, neural networks use learning techniques in order to recognize different letter styles.

In the last several years a number of handwriting recognition systems have been proposed that are based on segmentation [6] or segmentation-free [7] [8] approaches. The segmentation-free approach identifies the whole words while the segmentation approach separates words into letters. There are projects dealing with the recognition of Old Slavic Glagolitic Script [7] but none so far describes recognition of Old Slavic Cyrillic Script.

Thirty seven letters are used in the recognition system, although different dialects include several other letters. These 37 letters are the oldest letters present from the very beginning of the Cyrillic alphabet. Besides Latin and Greek languages, Old Slavic language was the fourth encoded language in Europe: Gothic in 4th, English-Saxon in 7th -9th, Old German in 7th and Old Slavic in 9th century [9]. Figure 1 shows two Old Slavic Cyrillic manuscripts: Minej 936, from 15th century and Sinai palimpsests, from the end of 12th or early 13th century.

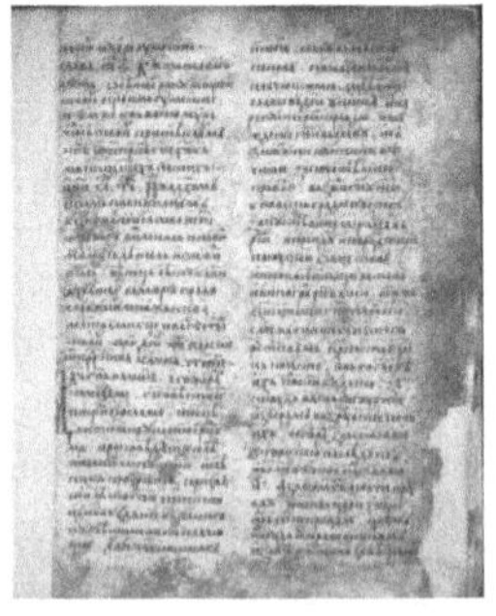

a) Minej 936, 15th century

b) Sinai palimpsests, end of 12th early 13th century

Fig. 1. Manuscripts from different centuries

During a long historical period covering 11th to 18th century, several styles appeared in Europe, like Roman, Gothic, Baroque and Rococo. They had strong influence on writing style of letter graphemes. In the same time the Cyrillic Church Slavic graphemes were unaffected in their stylish appearance [9-10]. This creates a problem in determining the period when a particular manuscript was written. Expert predictions for the period differ even for a century.

Especially the manuscripts used for church liturgical purposes are unaffected by style changes. They are written in Constitutional Script (Fig. 2). The Constitutional Script is characterized with artistic design, disjoined letters which are visually different from the letters of the Semi-Constitutional and the Cursive scripts. The last scripts are newer and were also used in civil purposes, outside the church.

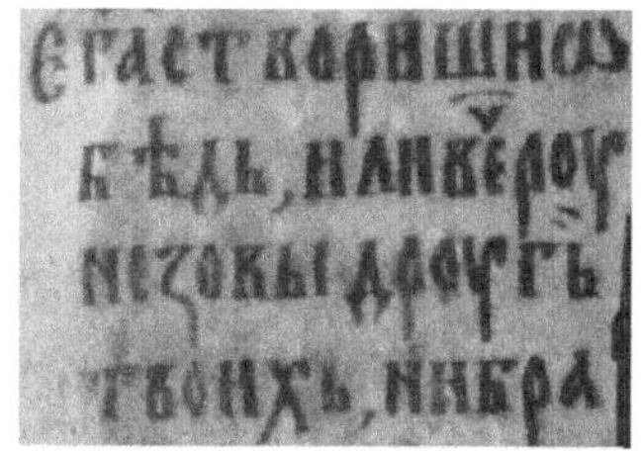

Fig. 2. Manuscript written with Constitutional script

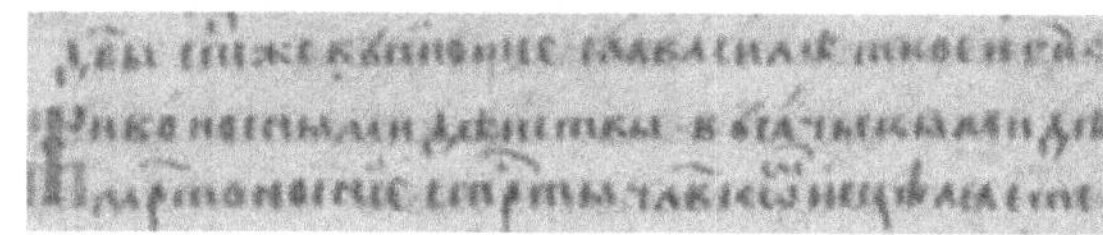

Fig. 3. Conjoint writing (scripta continua) principle

Our research is focused only on manuscripts written with Constitutional Script. This Script looks like printed text and letter contour lines can be easily extracted. Detecting and recognizing words is not a subject of analysis of this paper. Specific property of Church Slavic manuscripts is conjoint writing or so called scripta continua principle (Fig 3). The manuscripts used for recognition purposes are taken from the anthology of written monuments by eminent Macedonian linguists [11] and digital review prepared by Russian linguists [12].

In the next section we describe the process of letter analysis and feature extraction. Then we present a novel methodology for building fuzzy prototypes of each letter for the recognition purposes. Letter recognition is generally based on detecting spots on characteristic segments, existence of vertical and horizontal lines, as well as on determining compactness and symmetry of the whole letter. After that, the performance of the fuzzy classifier is evaluated through experimental results. Then this classifier is compared to a designed decision tree classifier based on the same feature set. The paper ends with a discussion of the results of the performed experiments.

2 Feature Extraction

Different features were considered for the process of letter recognition, such as dimensions of the bitmap image, height to width ratio, harmonic relationship of the height and width, black to white pixels ratio and black to total number of pixels ratio, percentage of pixels symmetric to x and y axes, length of the outer contour expressed in pixels and outer contour length to area occupied ratio.

Some features that are not meaningful for the recognition process have role in distinguishing the style related features of the letter [13]. For example, height to width ratio is used to determine whether the letter belongs to one of the basic styles: Roman or Gothic. The length of the contour line provides information about the curvature of the letter and the presence of decorative elements.

Fig. 4. Contour line for the decorative letter

Simpler forms have smaller contour line for standard matrix area compared to the decorative letters (Fig. 4). The percentage of the black pixels vs. the total matrix area shows whether the letter is light, regular or bold.

The features of the letters are extracted from the bitmaps created as prototypes for each letter (Fig. 4). The prototypes are obtained by applying logical operators on the samples of the digitalized original manuscripts. They are normalized in such a manner that its height is 24 units and the width is proportionaly determined to preserve the original shape. Each bitmap is segmented with two vertical and two horizontal intersections. Segments are topologically inspected to determine the number of spots and the presence of vertical and horizontal lines. The features compactness and symmetry are determined for the non-segmented bitmap.

This approach of extracting meaningful features is inspired by the methods for letter recognition which use histograms and contour profiles [14]. Projected histograms are mappings from two dimensional to one dimensional function, with values that represent sum of the pixels along one direction, horizontal or vertical. Contour prifiles represent remaining part from the border of the matrix to the contour line of the letter. Histograms and contour profiles are illustrated in Figure 5.

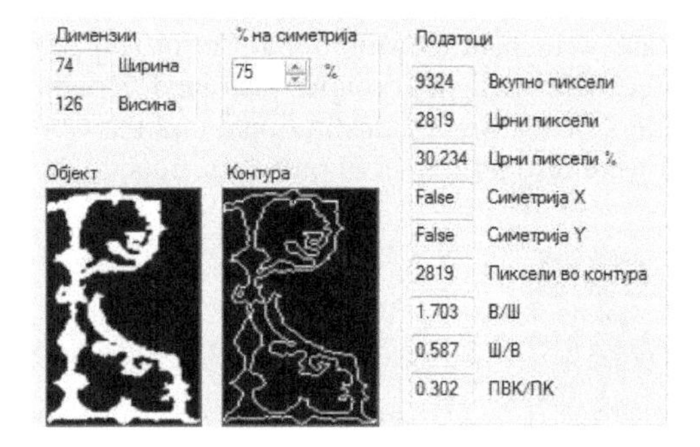

Fig. 5. Construction of histograms and contour profiles

Some of the features extracted from the normalized black and white bitmaps with height and width set proportionally, divide the total set of letters in several subsets, such as letters with emphasized vertical left line, right line or both, then subsets with horizontal line at the top, or at the bottom, or both, middle horizontal or vertical line, etc. (Fig. 6).

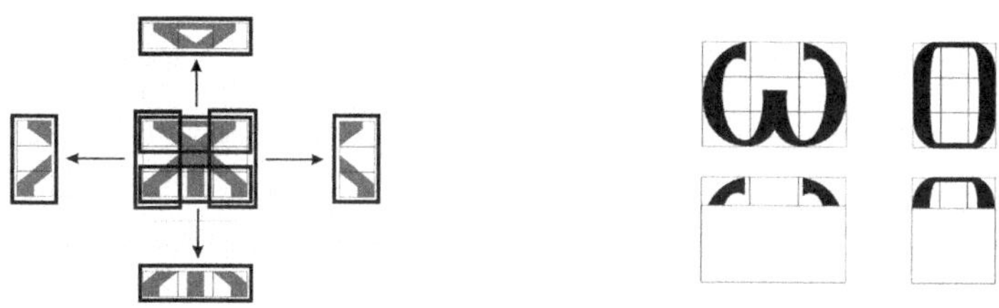

Fig. 6. Vertical and horizontal lines of letters

Fig. 7. Intersections of the letters that form the top, bottom, left and right segments

Fig. 8. Compact and airy letter

The number of spots (one, two or three) in outer segments is another feature that is important for the classification of the letters (Fig 7). The term compactness is used to characterize letters drawn by only one contour line without holes. If the letter has one or more holes it is considered as airy. For example, in Figure 8 letter ω is compact while letter O is airy, with one hole.

3 Fuzzy Classifier

Fuzzy classification of Old Slavic Cyrillic letters is based on prototypes that consist of fuzzy linguistic rules [15]. Fuzziness in letter descriptions emerges from the fact that manuscripts are written by different individuals with different writing styles and that they originate from different historical periods.

Important issue of letter classification is to select meaningful features which characterize the letters more precisely. Some of the extracted features from input data are geometrical (lines, loops, spots) while others are position related. Two letters with same geometrical features might have different positions of these features. For example, letters Φ and B both have two loops but while loops in Φ are positioned left and right, in B they are positioned at the top and at the bottom.

The goal of the first processing module in fuzzy classification is to create a matrix of combined features from the complete set of features I. Thus forming feature associations the number of features that have to be calculated for each letter is reduced.

To create a matrix of combined features a hierarchical fuzzy aggregation based on the operators (1) and (2) is performed. The first aggregation uses Yager union operator [16]

$$U(a_1, a_2, \ldots, a_N) = \min\left\{1, \left(\sum_{i=1}^{N} (a_i)^{\alpha}\right)^{\frac{1}{\alpha}}\right\} \quad (1)$$

where α is a real non-zero number. The value that can be obtained as a result of the union ranges between 1 and $\min(a_1, a_2, \ldots, a_N)$.

The second aggregation is a weighted median aggregation defined by the following formula [8]

$$\mathrm{Med}(a_1, \ldots, a_N, w_1, \ldots, w_N) = \left(\sum_{i=1}^{N} (w_i a_i)^{\alpha}\right)^{\frac{1}{\alpha}} \quad (2)$$

where $w_1, w_2, \ldots w_N$ are weights representing the importance given to input data A_1, A_2, …, A_N, such that $\sum_{i=1}^{N} w_i = 1$ and α is a real non-zero number with values between $\max(a_1, a_2, \ldots, a_N)$ and $\min(a_1, a_2, \ldots, a_N)$.

The precision of the letter recognition system to a certain extent depends on the appropriate selection of features. It is necessary to extract the most representative features from the entire set and to arrange the features by the degree of importance. This is done by calculating the overall measure for the features applying the fuzzy aggregation techniques.

The segmentation process decomposes the letter into S (in our case 6) segments. These segments are obtained with two vertical and two horizontal intersections. First, G global features (symmetry, compactness) related to the whole letter and L local features (structural or position related) for each segment are isolated. Then complex features are created for the segments associating the structural features (lines, spots) with their position or size. Using weight matrix the importance of each feature in the aggregation process can be increased or decreased. The number of features that have to be calculated in the recognition process is reduced by the aggregation process performed with the operators (1) and (2).

In the first step of the aggregation process L x S matrix of linguistic features for S segments is obtained:

$$\bar{I}_s = \{i_{sl} | l = [1, L]\}, s = [1, S] \quad (3)$$

where l are local features for each segment s. Also, the global feature set $\bar{I}_g$ is created.

Then the local features of each segment are combined with position and size related features and combined feature vectors V for each segment are formed:

$$\bar{V}_s = \{\bar{v}_{sc} | i = [1, C]\}, s = [1, S] \quad (4)$$

where C is the number of combined features for each segment. Combined feature vectors $\bar{v}_{sc}$ have to contain only the most significant associations of features for the segments and are formed using estimation function E:

$$\bar{v}_{sc} = E(\bar{I}_{sj}) . \tag{5}$$

The number of combined features C is less than or equal to the number of combination of L+G choose P elements. The combined feature vectors are formed from the local and global features.

The corresponding weight matrix $\bar{W}_s$ is computed during the learning process through statistic evaluation of the prototype samples:

$$\bar{W}_s = \{\bar{w}_{s1}, \dots, \bar{w}_{sC}\} \tag{6}$$

Using the weighted median aggregation (2) the feature vectors for each segment is computed:

$$\bar{\mu}_s = Med(\bar{w}_{sc}, \bar{v}_{sc}) \tag{7}$$

In the last step of this process the best features from the previously generated feature list are selected using Yager's union connective (1):

$$\{\mu_p\} = \min\left\{1, \left(\sum \mu_{ps}\right)\right\} \tag{8}$$

Fuzzy description for each letter is created from the computed subset $\{\mu_p\}$ of meaningful features. For example, let the subset of meaningful features for letter g consists of two combined features and with the aggregation process is obtained $\{\mu_p\} = \begin{bmatrix} 0.8 \\ 0.9 \end{bmatrix}$ then its fuzzy description is „left vertical line" with 80% possibility and „horizontal line at the top" with 90% possibility.

4 Decision Tree Classifier

The set of features which is used for decision tree classifier (Fig. 9) is the same as the one used for fuzzy classifier. As discriminative features are considered compactness, number of holes (one or more), number of spots and their location (one, two or three, on the left or right side, at the top or the bottom), vertical lines (left, middle or right), horizontal lines (top, middle or bottom) and symmetry by x or y-axis.

More precise descriptions of the letters are obtained combining the features. For example, letter "ⰴ ", is described by the following features: a hole, one spot at the left segment, one spot at the upper segment and one spot at the right segment. Less important features for this letter are: one or two spots at the lower segment.

During the learning process when the decision tree is created several measures are tested to determine the best splitting method. These measures are based on the rate of impurity of the children nodes. From the tested splitting measures gain ratio, GINI index and information gain, the best results are obtained by the information gain. The rules extracted from the created decision tree are afterwards corrected according to the measures accuracy and coverage. Corrections include reordering the rules, combining the conditions of the if-then rules or dropping some of the conditions.

As the decision tree in Figure 9 shows, the most discriminative features for letter classification are the number of spots in the outer segments and compactness, i.e. the number of holes. These features as more informative for letter recognition are placed close to the root of the decision tree. Some features might not be explicitly expressed. The features that are less informative for the procedure of classification are placed under the discriminative features in the decision tree.

5 Comparison of the Classifiers

We performed several experiments to test the performance of both classifiers [17]. Table 1 shows the recall and the precision measures for each letter. Recall (R) is computed as a fraction of the number of retrieved correct letters divided by the total number of relevant letters R=TP/(TP+FN). Precision (P) is computed as a fraction of the number of retrieved correct letters, divided with the number of retrieved letters P=TP/(TP+FP). TP (True Positive) is the number of correctly predicted examples and FP (False Positive) is the number of negative examples wrongly predicted as positive.

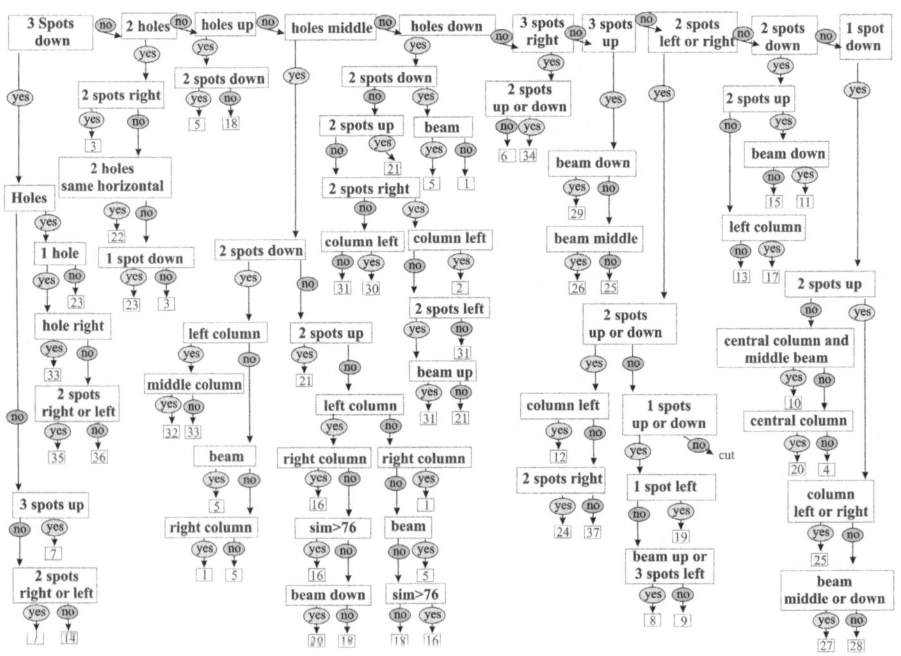

Fig. 9. Decision tree used as a classifier for recognition of Old Slavic handwritten letters

FN (False Negative) is the number of positive examples wrongly predicted as negative. The sum of precision and recall i.e. F1 metric is computed as F1=2RP/(R+P).

The proposed fuzzy classifier recognizes the letters with an average recall of 0.71, average precision of 0.78 and an overall average measure of precision and recall F1 of 0.74. The decision tree classifier recognizes the letters with an average recall of 0.73, average precision of 0.76 and F1 of 0.74.

Figure 10 shows a screen-shot of the application that is created for recognition of Old Slavic handwritten letters. The fuzzy classifier is positioned above the decision tree classifier.

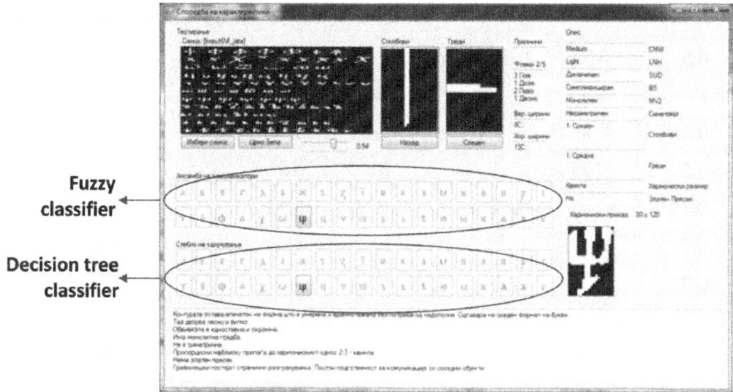

Fig. 10. Screen-shot of the application presenting the results of both classifiers

6 Conclusions

This paper describes two approaches for recognition of Old Slavic Cyrillic letters, one based on a decision tree and the other based on fuzzy techniques. Both methodologies are novel and use the same set of discriminative features, such as position and number of spots in the outer segments, presence and position of vertical and horizontal lines, compactness, symmetry, etc. The meaningfulness of the features for recognition purposes is determined computing the feature variance of the samples. This analysis distributes samples with small feature variance in the same class while samples with significant feature variances are placed in different classes. In this way 23 meaningful features are selected instead of initial 40 statistical and structural features.

Original digitalized manuscripts written with Cyrillic alphabet in the period from 12th till 16th century are used for the experiments in this paper. Training and testing examples are randomly chosen from this database of letters.

According to the experimental results for both classifiers shown in Table 1, the precision and recall measures differ significantly for some letters. The average precision and recall of the classifiers for the whole set of letters are more or less the same. Experiments show that the proposed methodologies give acceptable results.

These considerations lead to an idea of combining the two classifiers to improve the recognition accuracy and precision of the Old Slavic Cyrillic letters.

Table 1. Precision and recall of classifiers (FC-Fuzzy Classifier, DTC-Decision Tree Classifier)

Letters	FC recall	FC precision	DTC recall	DTC precision
а	0.2	1	0.5	0.63
б	0.83	0.71	0.67	1
в	0.75	0.86	0.88	0.64
г	1	0.54	1	0.5
д	0.75	0.5	0.75	0.75
е	0.33	1	0.56	1
ж	0.8	0.66	0.8	1
ѕ	0.75	0.5	0.25	0.33
ꙁ	0.2	0.25	1	0.42
і	0.86	0.7	0.71	1
н	1	0.7	1	0.82
к	1	1	1	0.2
л	0.56	1	0.89	1
м	0.7	1	0.8	1
и	1	0.7	1	0.82
о	1	0.41	0.86	0.86
п	1	0.63	1	1
р	0.33	1	0.5	0.75
с	1	0.75	1	0.82

Letters	FC recall	FC precision	DTC recall	DTC precision
т	1	0.67	0.86	0.58
ȣ	1	0.8	1	0.8
ф	1	0.78	1	0.86
ѳ	0.57	0.8	0.43	0.75
х	0.44	0.67	0.56	0.71
ѡ	0.33	1	0.67	0.2
щ	0.71	1	1	1
ц	0.86	0.55	0.43	0.75
ч	0.43	1	0.43	1
ш	1	1	0.89	1
ь	0.79	0.85	0.71	0.71
ъ	0.64	0.75	0.79	0.65
ю	0.89	0.73	0.89	0.57
ꙗ	0.33	1	0.33	1
ѥ	0.4	1	0.4	1
ѫ	1	0.67	0.5	0.33
ѧ	0.2	1	0.4	1
у	0.67	0.67	0.44	0.67
Total	**0.71**	**0.78**	**0.73**	**0.76**

References

1. Scripta & e-Scripta. The Journal of Interdisciplinary Mediaeval Studies, vol. 6. Institute of Literature, Bulgarian Academy of Sciences. “Boyan Penev” Publishing Center (2008)
2. Mori, S., Suen, C.Y., Yamamoto, K.: Historical Review of OCR Research and Development. Proceedings of the IEEE 80, 1029–1058 (1992)
3. Vinciarelli, A.: A Survey on Off-line Cursive Word Recognition. Pattern Recognition 35, 1433–1446 (2002)
4. Kavallieratou, E., Fakotakis, N., Kokkinakis, G.: Handwritten Character Recognition Based on Structural Characteristics. In: 16th International Conference on Pattern Recognition, pp. 139–142 (2002)
5. Malaviya, A., Peters, L.: Fuzzy Handwriting Description Language: FOHDEL. Pattern Recognition 33, 119–131 (2000)

6. Eastwood, B., Jennings, A., Harvey, A.: A Feature Based Neural Network Segmenter for Handwritten Words. In: International Conference on Computational Intelligence and Multimedia Applications (ICCIMA 1997), Australia, pp. 286–290 (1997)
7. Ntzios, K., Gatos, B., Pratikakis, I., Perantonis, S.J.: An Old Greek Handwritten OCR System based on an Efficient Segmentation-free Approach. Int. Journal on Document Analysis and Recognition 9(2), 179–192 (2007)
8. Chen, C.H., Curtins, J.: Word Recognition in a Segmentation-free Approach to OCR. In: Second International Conference on Document Analysis and Recognition (ICDAR 1993), pp. 573–576 (2003)
9. Antic, V.: Macedonian Medieval Literature. Institute for Macedonian Literature. Skopje, Macedonia (1997) (in Macedonian)
10. Atanasova, S.: Linguistic Analysis of Bitola's Liturgical Book, Institute of Macedonian Language - Skopje, Macedonia (1990) (in Macedonian)
11. Velev, I., Makarijoska, L., Crvenkovska, E.: Macedonian Monuments with Glagolitic and Cyrillic Handwriting, Stip, Macedonia (August 2, 2008) (in Macedonian)
12. Russian Review of Cyrillic Manuscripts, http://xlt.narod.ru/pg/alpha.html
13. Klekovska, M., Nedelkovski, I., Stojcevska-Antic, V., Mihajlov, D.: Automatic Letter Style Recognition of Church Slavic Manuscripts. In: Proc. of 44th Int. Scientific Conf. on Information, Communication and Energy Systems and Technologies, Veliko Tarnovo, Bulgaria, pp. 221–224 (2009)
14. Cheriet, M., Kharma, N., Liu, C.L., Suen, C.Y.: Character Recognition Systems, A Guide for Students and Practioners. John Wiley and Sons, New Jersey (2007)
15. Martinovska, C., Nedelkovski, I., Klekovska, M., Kaevski, D.: Fuzzy Classifier for Church Slavic Handwritten Characters. In: Proc. of 14th Int. Conf. on Enterprise Information Systems, Wroclaw, Poland, pp. 310–313 (2012)
16. Yager, R.: On the Representation of Multi-Agent Aggregation using Fuzzy Logic. Cybernetics and Systems 21, 575–590 (1990)
17. Martinovska, C., Nedelkovski, I., Klekovska, M., Kaevski, D.: Recognition of Old Cyrillic Slavic Letters: Decion Tree versus Fuzzy Classifier Experiments. In: Yager, R.R., Sgurev, V., Hadjiski, M. (eds.) Proc. 6th IEEE Int. Conf. on Intelligent Systems 2012, Sofia, Bulgaria, vol. I, pp. 48–53 (2012)

Emotion-Aware Recommender Systems – A Framework and a Case Study*

Marko Tkalčič, Urban Burnik, Ante Odić, Andrej Košir, and Jurij Tasič

University of Ljubljana Faculty of Electrical Engineering,
Tržaška 25, Ljubljana, Slovenia
{marko.tkalcic,urban.burnik,ante.odic,
andrej.kosir,jurij.tasic}@fe.uni-lj.si

Abstract. Recent work has shown an increase of accuracy in recommender systems that use emotive labels. In this paper we propose a framework for emotion-aware recommender systems and present a survey of the results in such recommender systems. We present a consumption-chain-based framework and we compare three labeling methods within a recommender system for images: (i) generic labeling, (ii) explicit affective labeling and (iii) implicit affective labeling.

Keywords: recommender systems, emotion detection, multimedia consumption chain.

1 Introduction

In the pursuit of increasing the accuracy of recommender systems, researchers started to turn to more user-centric content descriptors in recent years. The advances made in affective computing, especially in automatic emotion detection techniques, paved the way for the exploitation of emotions and personality as descriptors that account for a larger part of variance in user preferences than the generic descriptors (e.g. genre) used so far.

However, these research efforts have been conducted independently, stretched among the two major research areas, *recommender systems* and *affective computing*. In this paper we (i) provide a unifying framework that will allow the members of the research community to identify the position of their activities and to benefit from each other's work and (ii) present the outcomes of a comparative study with affective labeling in an image recommender system.

2 The Unifying Framework

When using applications with recommender systems the user is constantly receiving various stimuli (e.g. visual, auditory etc.) that induce emotive states.

* This work was partially funded by the European Commission within the FP6 IST grant number FP6-27312 and partially by the Slovenian Research Agency ARRS under the grant P2-0246.

S. Markovski and M. Gusev (Eds.): *ICT Innovations 2012*, AISC 207, pp. 141–150.
DOI: 10.1007/978-3-642-37169-1_14

These emotions influence, at least partially (according to the bounded rationality model [9]) the user's decisions on which content to choose. Thus it is important for the recommender system application to detect and make good use of emotive information.

2.1 Describing Emotions

There are two main approaches to describe the emotive state of a user: (i) the *universal emotions model* and (ii) the *dimensional model*. The universal emotions model assumes there is a limited set of distinct emotional categories. There is no unanimity as to which are the universal emotions, however, the categories proposed by Ekman [5] (i.e. happiness, anger, sadness, fear, disgust and surprise) appear to be very popular. The dimensional model, on the contrary, describes each emotion as a point in a continuous multidimensional space where each dimension represents a quality of the emotion. The dimensions that are used most frequently are valence, arousal and dominance (thus the VAD acronym) although some authors refer to these dimensions with different names (e.g. pleasure instead of valence in [13] or activation instead of arousal in [7]). The circumplex model, proposed by Posner et al. [15], maps the basic emotions into the VAD space (as depicted in Fig. 1)

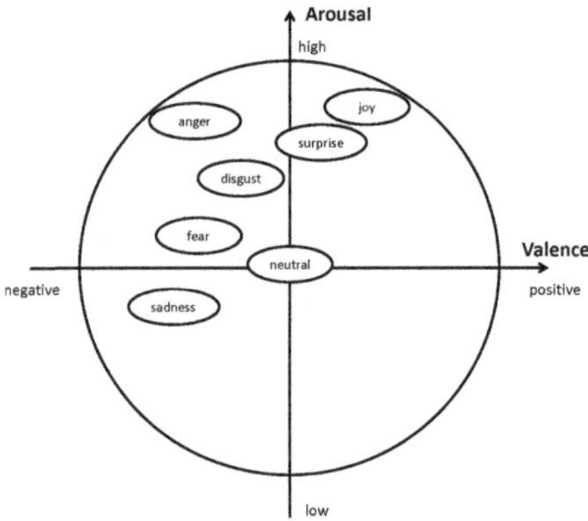

Fig. 1. Basic emotions in the valence-arousal plane of the dimensional model

2.2 The Role of Emotions in the Consumption Chain

During the user interaction with a recommender system and the content consumption that follows, emotions play different roles in different stages of the process. We divided the user interaction process in three stages, based on the role that emotions play (as shown in Fig. 2): (i) *the entry stage*, (ii) *the consumption stage* and (iii) *the exit stage*.

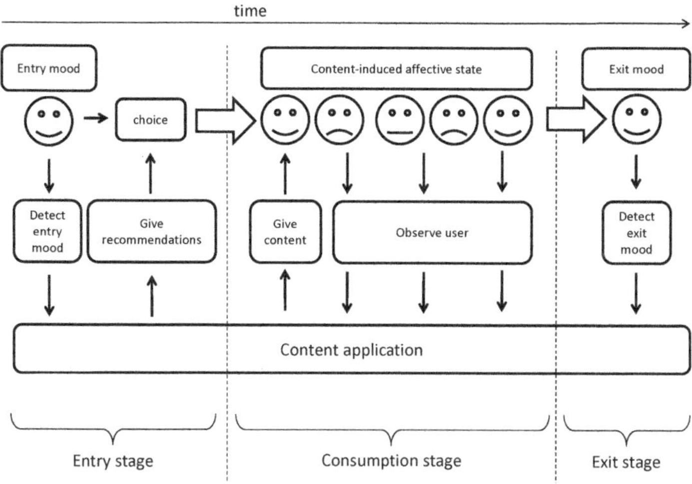

Fig. 2. The unifying framework: the role of emotions in user interaction with a recommender system

2.3 Detecting Affective States

Affective states of end users (in any stage of the proposed interaction chain) can be detected in two ways: (i) explicitly or (ii) implicitly. The implicit detection of emotions is more accurate but it's an intrusive process that breaks the interaction. The implicit approach is less accurate but it's well suited for user interaction purposes since the user is not aware of it. Furthermore, Pantić et al. [14] argued that explicit acquisition of users' affect has further negative properties as users may have side-interests that drive their explicit affective labeling process (egoistic tagging, reputation-driven tagging or asocial tagging).

The most commonly used procedure for the explicit asessment of emotions is the Self Assessment Manikin (SAM) developed by [3]. It is a questionnaire where users assess their emotional state in the three dimensions: valence, arousal and dominance.

The implicit acquisition of emotions is usually done through a variety of modalities and sensors: video cameras, speech, EEG, ECG etc. These sensors measure various changes of the human body (e.g. facial changes, posture changes, changes in the skin conductance etc.) that are known to be related to specific emotions. For example, the Facial Action Coding System (FACS), proposed by Ekman [4], maps emotions to changes of facial characteristic poionts. There are excellent surveys on the topic of multimodal emotion detection: [20,14,8]. In general, raw data is acquired from one or more sensors during the user interaction. These signals are processed to extract some low level features (e.g. Gabor based features are popular in the processing of facial expression video signals). Then some kind of classification or regression technique is applied to yield distinct emotional classes or continuous values. The accuracy of emotion detection ranges from over 90% on posed datasets (like the Kanade-Cohn dataset [10]) to slightly better than coin tossing on spontaneous datasets (like the LDOS-PerAff-1 dataset [18]) [17,1].

2.4 Entry Stage

The first part of the proposed framework (see Fig. 2) is the entry stage. When a user starts to use a recommender system, she is in an affective state, the *entry mood.* The entry mood is caused by some previous user's activities, unknown to the system. When the recommender system suggests a limited amount of content items to the user, the entry mood influences the user's choice. In fact, the user's decision making process depends on two types of cognitive processes, the rational and the intuitive, the latter being strongly influenced by the emotive state of the user, as explained by the bounded rationality paradigm [9]. For example, a user might want to consume a different type of content when she is happy than when she is sad. In order to adapt the list of recommended items to the user's entry mood the system must be able to detect the mood and to use it in the content filtering algorithm as contextual information.

2.5 Consumption Stage

The second part of the proposed framework is the consumption stage (see Fig. 2). After the user starts with the consumption of the content she experiences affective responses that are induced by the content. Depending on the type of content, these responses can be (i) single values (e.g. the emotive response to watching an image) or (ii) a vector of emotions that change over time (e.g. while watching a movie or a sequence of images). Figure 3 shows how emotions change over time in the consumption stage. The automatic detection of emotions can help building emotive profiles of users and content items that can be exploited for content-based recommender algorithms.

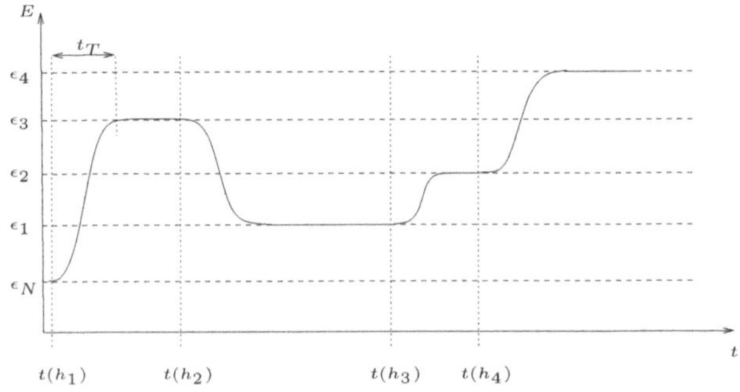

Fig. 3. The user's emotional state ϵ is continuously changing as the time sequence of the visual stimuli $h_i \in H$ induce different emotions

2.6 Exit Stage

After the user has finished with the content consumption she is in what we call the exit mood. The main difference between the consumption stage and the exit stage is that the exit mood will influence the user's next actions, thus having an active part, while in the consumption stage the induced emotions did not influence any actions but were a passive response to the stimuli. In case that the user continues to use the recommender system the exit mood for the content just consumed is the entry mood for the next content to be consumed.

The automatic detection of the exit mood can be useful as an indicator of the user's satisfaction with the content. Thus the detection of the exit mood can be seen as an unobtrusive feedback collection technique.

3 Affective Labeling

In this section we present the outcomes of a comparative study of the performance of an image recommender system that uses (i) generic metadata (e.g. the genre), (ii) explicitly acquired affective metadata and (iii) implicitly acquired metadata.

3.1 Affective Modeling in a CBR System

We use the valence-arousal-dominance (VAD) emotive space for describing the users' emotive reactions to images. In the VAD space each emotive state is described by three parameters, namely valence, arousal and dominance. A single user $u \in U$ consumes one or more content items (images) $h \in H$. As a consequence of the image h being a visual stimulus, the user u experiences an emotive response which we denote as $er(u,h) = (v,a,d)$ where v, a and d are scalar values that represent the valence, arousal and dominance dimensions of the emotive response er. The set of users that have watched a single item h are denoted with U_h. The emotive responses of all users U_h, that have watched the item h form the set $ER_h = \{er(u,h) : u \in U_h\}$. We model the image h with the item profile that is composed of the first two statistical moments of the VAD values from the emotive responses ER_h which yields the six tuple

$$\mathcal{V} = (\bar{v}, \sigma_v, \bar{a}, \sigma_a, \bar{d}, \sigma_d) \tag{1}$$

where $\bar{v}$, $\bar{a}$ and $\bar{d}$ represent the average VAD values and σ_v, σ_a and σ_d represent the standard deviations of the VAD values for the observed content item h. An example of the affective item profile is shown in Tab. 1.

The preferences of the user are modeled based on the explicit ratings that she/he has given to the consumed items. The observed user u rates each viewed item either as relevant or non-relevant. A machine learning (ML) algorithm is trained to separate relevant from non-relevant items using the affective metadata in the item profiles as features and the binary ratings (relevant/non-relevant) as classes. The user profile $up(u)$ of the observed user u is thus an ML algorithm dependent data structure.

Table 1. Example of an affective item profile $\mathcal{V}$ (first two statistical moments of the induced emotion values v, a and d)

Metadata field	Value
$\bar{v}$	3.12
σ_v	1.13
$\bar{a}$	4.76
σ_a	0.34
$\bar{d}$	6.28
σ_d	1.31

4 Experiment

We used our implementation of an emotion detection algorithm (see [17]) for implicit affective labeling and we compared the performance of the CBR system that uses explicit vs. implicit affective labels. We used the LDOS-PerAff-1 dataset [18].

The emotion detection procedure used to give affective labels to the content images involved three stages: (i) pre-processing, (ii) low level feature extraction and (iii) emotion detection. We formalized the procedure with the mappings

$$I \rightarrow \Psi \rightarrow E \tag{2}$$

where I represents the frame from the video stream, Ψ represents the low level features corresponding to the frame I and E represents the emotion corresponding to the frame I.

In the pre-processing stage we extracted and registered the faces from the video frames to allow precise low level feature extraction. We used the eye tracker developed by [19] to extract the locations of the eyes. The detection of emotions from frames in a video stream was performed by comparing the current video frame I_t of the user's face to a neutral face expression. As the LDOS-PerAff-1 database is an ongoing video stream of users consuming different images we averaged all the frames to get the neutral frame. This method is applicable when we have a non supervised video stream of a user with different face expressions.

The low level features used in the proposed method were drawn from the images filtered by a Gabor filter bank. We used a bank of Gabor filters of 6 different orientation and 4 different spatial sub-bands which yielded a total of 24 Gabor filtered images per frame. The final feature vector had the total length of 240 elements.

The emotion detection was done by a k-NN algorithm after performing dimensionality reduction using the principal component analysis (PCA).

Each frame from the LDOS-PerAff-1 dataset was labeled with a six tuple of the induced emotion $\mathcal{V}$. The six tuple was composed of scalar values representing the first two statistical moments in the VAD space. However, for our purposes we opted for a coarser set of emotional classes $\epsilon \in E$. We divided the whole VAD space into 8 subspaces by thresholding each of the three first statistical moments $\bar{v}$, $\bar{a}$ and $\bar{d}$. We thus gained 8 rough classes. Among these, only 6 classes actually contained at least one item so we reduced the emotion detection problem to a classification into 6 distinct classes problem.

4.1 Overview of the CBR Procedure

Figure 4 shows the overview of the CBR experimental setup. After we collected the ratings and calculated the affective labels for the item profiles, we trained the user profiles with four different machine learning algorithms: the SVM, NaiveBayes, AdaBoost and C4.5. We split the dataset in the train and test sets using the ten-fold cross validation technique. We then performed ten training-classifying iterations which yielded the confusion matrices that we used to assess the performance of the CBR system.

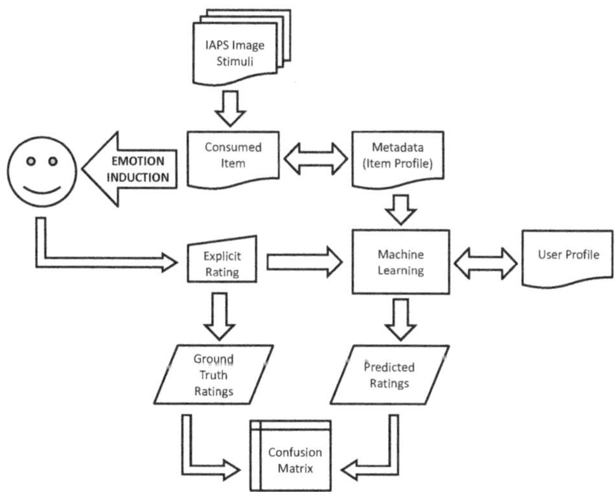

Fig. 4. Overview of the CBR experiment

The set of images $h \in H$ that the users were consuming, had a twofold meaning: (i) they were used as content items and (ii) they were used as emotion induction stimuli for the affective labeling algorithm. We used a subset of 70 images from the IAPS dataset [11]. The IAPS dataset of images is annotated with the mean and standard deviations of the emotion responses in the VAD space which was useful as the ground truth in the affective labeling part of the experiment. We had 52 users taking part in our experiment (mean = 18.3 years, 15 males).

4.2 Affective CBR System Evaluation Methodology

The results of the CBR system were the confusion matrices of the classification procedure that mapped the images H into one of the two possible classes: relevant or non-relevant class. From the confusion matrices we calculated the recall, precision and F measure as defined in [6].

We also compared the performances of the CBR system with three types of metadata: (i) generic metadata (genre and watching time as done by [16]), (ii) affective metadata given explicitly and (iii) affective metadata acquired implicitly with the proposed emotion detection algorithm. For that purpose we transferred the statistical testing of the confusion matrices into the testing for the equivalence of two estimated discrete probability distributions [12]. To test the equivalence of the underlying distributions we used the Pearson χ^2 test. In case of significant differences we used the scalar measures precision, recall and F measure to see which approach was significantly better.

5 Results

We compared the performance of the classification of items into relevant or non relevant through the confusion matrices in the following way: (i) Explicitly acquired affective metadata vs Implicitly acquired metadata, (ii) explicitly acquired metadata vs. generic metadata and (iii) implicitly acquired metadata vs. generic metadata. In all three cases the p value was $p < 0.01$. Table 2 shows the scalar measures precision, recall and F measures for all three approaches.

Table 2. The scalar measures P, R, F for the CBR system

metadata/labeling method	classifier	P	R	F
implicit affective labeling	AdaBoost	0.61	0.57	0.59
	C4.5	0.58	0.50	0.53
	NaiveBayes	0.56	0.62	0.59
	SVM	0.64	0.47	0.54
explicit affective labeling	AdaBoost	0.64	0.56	0.60
	C4.5	0.62	0.54	0.58
	NaiveBayes	0.56	0.59	0.58
	SVM	0.68	0.54	0.60
generic metadata	AdaBoost	0.57	0.41	0.48
	C4.5	0.60	0.45	0.51
	NaiveBayes	0.58	0.57	0.58
	SVM	0.61	0.55	0.58

6 Discussion and Conclusion

As we already reported in [17], the application of the emotion detection algorithm on spontaneous face expression videos has a low performance. We identified three main reasons for that: (i) weak supervision in learning, (ii) non-optimal video acquisition and (iii) non-extreme facial expressions.

In supervised learning techniques there is ground truth reference data to which we compare our model. In the induced emotion experiment the ground truth data is weak because we did not verify whether the emotive response of the user equals to the predicted induced emotive response.

Second, the acquisition of video of users' expressions in real applications takes place in less controlled environments. The users change their position during the session. This results in head orientation changes, size of the face changes and changes of camera focus.

The third reason why the accuracy drops is the fact that face expressions in spontaneous videos are less extreme than in posed videos. As a consequence the changes on the faces are less visible and are hidden in the overall noise of the face changes. The dynamics of face expressions depend on the emotion amplitude as well as on the subjects' individual differences.

The comparison of the performance of the CBR with explicit vs. implicit affective labeling shows significant differences regardless of the ML technique employed to predict the ratings. The explicit labeling yields superior CBR performance than the implicit labeling. However, another comparison, that between the implicitly acquired affective labels and generic metadata (genre and watching time) shows that the CBR with implicit affective labels is significantly better than the CBR with generic metadata only. Although not as good as explicit labeling, the presented implicit labeling technique brings additional value to the CBR system used.

The usage of affective labels is not present in state-of-the-art commercial recommender systems, to the best of the authors' knowledge. The presented approach allows to upgrade an existing CBR system by adding the unobtrusive video acquisition of users' emotive responses. The results showed that the in-clusion of affective metadata, although acquired with a not-so-perfect emotion detection algorithm, significantly improves the quality of the selection of recom-mended items. In other words, although there is a lot of noise in the affective labels acquired with the proposed method, these labels still describe more variance in users' preferences than the generic metadata used in state-of-the-art recom-mender systems.

References

1. Bartlett, M.S., Littlewort, G.C., Frank, M.G., Lainscsek, C., Fasel, I.R., Movellan, J.R.: Automatic Recognition of Facial Actions in Spontaneous Expressions. Journal of Multimedia 1(6), 22–35 (2006)
2. Eisenbarth, T., Kumar, S., Paar, C., Poschmann, A., Uhsadel, L.: A Survey of Lightweight-Cryptography Implementations. IEEE Des. Test 24(6), 522–533 (2007)
3. Bradley, M.M., Lang, P.J.: Measuring emotion: the self-assessment manikin and the semantic differential. Journal of Behavior Therapy and Experimental Psychiatry 25(1), 49–59 (1994)
4. Ekman, P.: Facial expression and emotion. American Psychologist 48(4), 384 (1993)

5. Ekman, P.: Basic Emotions. In: Handbook of Cognition and Emotion, pp. 45–60 (1999)
6. Herlocker, J.L., Konstan, J.A., Terveen, L., Riedl, J.A.: Evaluating collaborative filtering recommender systems. ACM Trans. Inf. Syst. 22(1), 5–53 (2004)
7. Ioannou, S.V., Raouzaiou, A.T., Tzouvaras, V., Mailis, T.P., Karpouzis, K.C., Kollias, S.D.: Emotion recognition through facial expression analysis based on a neurofuzzy network. Neural Networks: The Official Journal of the International Neural Network Society 18(4), 423–435 (2005)
8. Jaimes, A., Sebe, N.: Multimodal human computer interaction: A survey. Computer Vision and Image Understanding 108(1-2), 116–134 (2007)
9. Kahneman, D.: A perspective on judgment and choice: mapping bounded rationality. The American Psychologist 58(9), 697–720 (2003)
10. Kanade, T., Cohn, J.F., Tian, Y.: Comprehensive database for facial expression analysis. In: Proceedings of the Fourth IEEE International Conference on Automatic Face and Gesture Recognition, pp. 46–53 (2000)
11. Lang, P.J., Bradley, M.M., Cuthbert, B.N.: International affective picture system (IAPS): Affective ratings of pictures and instruction manual. Technical report, University of Florida (2005)
12. Lehmann, E.L., Romano, J.P.: Testing Statistical Hypotheses. Springer Texts in Statistics. Springer, New York (2005)
13. Mehrabian, A.: Pleasure-arousal-dominance: A general framework for describing and measuring individual differences in Temperament. Current Psychology 14(4), 261–292 (1996)
14. Pantic, M., Vinciarelli, A.: Implicit human-centered tagging Social Sciences. IEEE Signal Processing Magazine 26(6), 173–180 (2009)
15. Posner, J., Russell, J., Peterson, B.S.: The circumplex model of affect: an integrative approach to affective neuroscience, cognitive development, and psychopathology. Development and Psychopathology 17(3), 715–734 (2005)
16. Tkalčič, M., Burnik, U., Košir, A.: Using affective parameters in a content-based recommender system for images. User Modeling and User-Adapted Interaction 20(4), 279–311 (2010)
17. Tkalčič, M., Odić, A., Košir, A., Tasič, J.: Comparison of an Emotion Detection Technique on Posed and Spontaneous Datasets. In: Proceedings of the 19th ERK Conference, Portorož (2010)
18. Tkalčič, M., Tasič, J., Košir, A.: The LDOS-PerAff-1 Corpus of Face Video Clips with Affective and Personality Metadata. In: Proceedings of Multimodal Corpora: Advances in Capturing, Coding and Analyzing Multimodality LREC, p. 111 (2009)
19. Valenti, R., Yucel, Z., Gevers, Z.: Robustifying eye center localization by head pose cues. In: IEEE Conference on Computer Vision and Pattern Recognition, pp. 612–618 (2009)
20. Zeng, Z., Pantic, M., Roisman, G.I., Huang, T.S.: A Survey of Affect Recognition Methods: Audio, Visual, and Spontaneous Expressions. IEEE Trans. Pattern Analysis & Machine Intelligence 31, 39–58 (2009)

OGSA-DAI Extension for Executing External Jobs in Workflows

Ğorgi Kakaševski[1], Anastas Mishev[2], Armey Krause[3], and Solza Grčeva[1]

[1] Faculty of Informatics, FON University, bul. Vojvodina b.b. 1000 Skopje, R. Macedonia
{gorgik,solza.grceva}@fon.edu.mk

[2] Faculty of Computer Science and Engineering, Ss. Cyril and Methodius University, 1000 Skopje, Republic of Macedonia
anastas.mishev@finki.ukim.mk

[3] EPCC, University of Edinburgh, James Clerk Maxwell Building, Mayfield Road, Edinburgh EH9 3JZ, UK
a.krause@epcc.ed.ac.uk

Abstract. Because of the nature of Grid to be heterogeneous and distributed environment, the database systems which should works on Grid must support this architecture. OGSA-DAI is an example of such extensible service-based framework that allow data resources to be incorporated into Grid fabrics. On the other side, many algorithms (for example, for data mining) are not built in Java, they aren't open source projects and can't be easily incorporated in OGSA-DAI workflows. For that reason, we propose OGSA-DAI extension with new computational resources and activities that allow executing of external jobs, and returning data into OGSA-DAI workflow. In this paper, we introduce heterogeneous and distributed databases, then we discuss our proposed model, and finally, we report our initial implementation.

Keywords: Grid computing, heterogeneous distributed databases, OGSA-DAI.

1 Introduction

The computational needs of today's science and engineering are constantly increasing. The Grid provides both computational and storage resources to the scientific applications [1]. E-Science is computationally intensive science that is carried out in highly distributed network environments, or science that uses immense data sets [2]. Digital data play major role in e-Science applications, from small projects [3] to The Large Hadron Collider [4]. Grid computing is one of the most exploited platforms for e-Science and provide computational and data resources. Significant effort is made in developing of Grid infrastructure in South-Eastern Europe region [5].

Because of mainly file-based organization of Grid data, there is a need how to use data which are originally organized as shared and structured collections, stored in

S. Markovski and M. Gusev (Eds.): *ICT Innovations 2012*, AISC 207, pp. 151–160.
DOI: 10.1007/978-3-642-37169-1_15

relational databases, or in some cases in structured documents or in assemblies of binary files. These requirements of Grid applications require developers of Grid systems to provide practical implementations of "relational Grid databases" (or data access and integration [6]). One of the widely used middleware for heterogeneous and distributed database is OGSA-DAI [7] build on WS-DAI standards for data access and integration.

While other Grid middleware's use data in file based manner, OGSA-DAI workflows consists of activities that use relational databases. One of the open issues in OGSA-DAI is how to execute external jobs. By external jobs we assume programs that are already compiled and can be executed in current operating system without user interaction. These programs can be useful for lightweight data processing, like data pre-processing in data mining algorithms, creating data structures for fast data search or different algorithms which have need from data stored in heterogeneous distributed relational databases. External jobs take input, usually via input file, and return output in output file. Input and output files usually are text files in CSV, XML or JSON format. Some applications defined their type of text files. Good to mention example of such external job is widely used Weka [8][9] application for data mining, which have algorithms for preprocessing, clustering, classification etc. Weka takes input in textual ARFF file format and produce output in text file (also there is support for XML and other file types).

The goal of this paper is to present architecture for execution of external jobs in OGSA-DAI workflows and to present the current status of our implementation. With this upgrade, in OGSA-DAI can be added more complex algorithms for data analyze, for example, calculations for OLAP cube, which are not supported in databases by default.

The rest of the paper is organized as follow. In Section 2 we describe standards for data access and integration which are key concepts for practical implementations of relational database middleware for Grid. We described OGSA-DAI, a practical release of WS-DAI specifications in Section 3. In Section 4 we present extension of OGSA-DAI that we propose, and in Section 5, the current status of practical implementation. We conclude paper and give recommendations for future work in Section 6.

2 Data Access and Integration

Grid is made of different type of resources with different characteristics, and as whole must act as one system. For example, on the Grid we can find two clusters with different type of processors or installed software, even on the same cluster we can find machines with different resources. This property is true for databases. There are several types of databases which can be used in applications: relational, XML, even CSV files or just files. But the Grid users (applications, clients) must see data as one integrated part, no matter where they are located or which their physical form is.

Another key moment in organizing databases on Grid is the service-oriented architecture of Grid and Grid services, defined by Open Grid Service Architecture (OGSA) [10]. OGSA has a one of the central roles in Grid computing, because it simplifies the development of secure, robust systems and enable the creation of interoperable, portable, and reusable components. The main idea of transient Grid services is to extend well-defined Web service standards (WSDL, XML schemas, SOAP, etc.), to support basic Grid behaviors. These standards are proposed by Open Grid Forum [11] and this organization have key role in development of new Grid technologies.

OGF issued the Web Service Data Access and Integration (WS-DAI) family of specifications [12]. This specifications defines web service interfaces to data resources, such as relational or XML databases. In particular, the WS-DAI specifications are part of a wide ranging activity to develop the Open Grid Services Architecture (OGSA) within the OGF. The specification WS-DAI is on the top level and it defines the core data access services which are web service interfaces to data resources.

WS-DAI specifications have five main components: interfaces, data services, properties, messages and data resources. Interfaces are used to interact with data services. Data services are web services that implements interfaces to access distributed data resources. Properties describe data services, messages are sent to data services and data resources are system that can act as a source of data. Data resources are already existent database management systems, for example, MySQL, and this system has no real connection with WS-DAI specifications.

On Fig. 1 is given the proposed messages flow between client and data stored on Grid, by Open Grid Forum. This scheme enables clients to access or modify the data and their organization through Grid data service.

In order to implement these specifications on real system, on Fig. 2 is given proposed architecture of Grid database system [13]. During the fact that these traditional DBMSs exist for many years and are well developed and tested, in this case can be used some of the existing. On top of this system Grid Data Services must be implemented, as part of the middleware that is responsible for communications among heterogeneous distributed database.

Grid Data Service Middleware is one kind of transparent wrapper for complex Grid database. Transparency refers to: location, name of database, distribution, replication, ownership and cost, heterogeneity and schema. Clients exchange data with Grid database, even if they don't know where that data are stored or in which format. Important for client is that in Grid database must be implemented all (or common) functionalities of regular, mainly relational, database management systems and in natural way. That ensures that clients (users, software developers) can easily use Grid database and easily adapt their applications with new kind of database.

Ideally, the database vendors would themselves agree on, and support, such a standard. This would remove the need for third parties to provide wrappers for each different DBMS, and would also open the way for internal performance optimizations, e.g. in generating XML result sets directly from the internal representation [14].

As a conclusion, a heterogeneous distributed database system (or simply Grid database) should provide service-based access to existing relational database management systems that are located on Grid nodes. This database will be some form of virtual database and clients will access to data on transparent way no mater of physical organization of data [13].

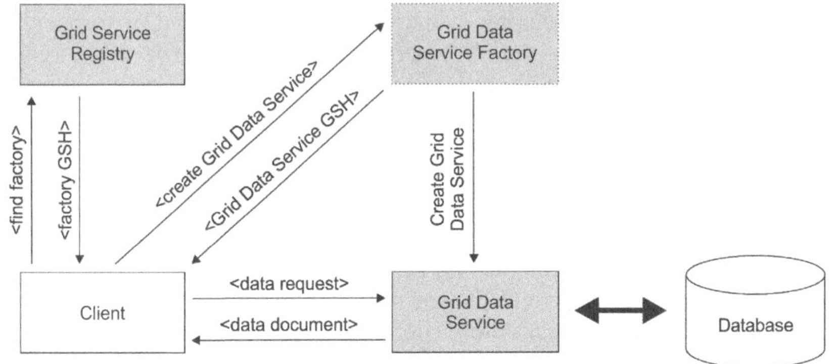

Fig. 1. Accessing and modifying data stored on Grid using a Grid Data Service

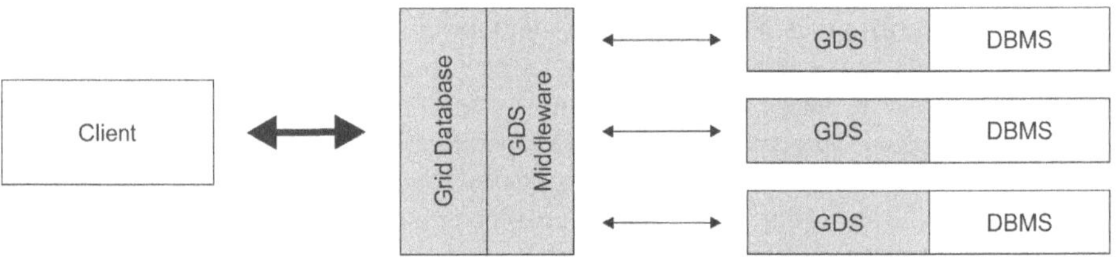

Fig. 2. Architecture of Grid Database System

3 OGSA-DAI Heterogeneous Distributed Databases

OGSA-DAI stands for Open Grid Services Architecture - Data Access and Integration. It is a Java open source project and it is a one of the most comprehensive heterogeneous database system, which can be used in Grid.

One of the primary goals of data access and integration is to treat these data, whatever their origin, within a uniform framework. OGSA-DAI make the DBMSs, which can be heterogeneous and distributed on Grid, transparent to the users [15]. It hides the underlying drivers that are needed to communicate with the database management and as name tells, OGSA-DAI provide service-oriented interface to the clients. This process naturally will add some overhead on top of the communication itself but it is not that significant [16].

It supports several kinds of databases: relational (Oracle, DB2, SQL server, MySQL, Postgress), XML databases like xindice and eXist and files. There is an opportunity for programmers to develop new drivers for DBMS which are not supported in original version. Acting in the top of local DMBS that are located on different sites, OGSA-DAI provides uniform Grid database.

Today, OGSA-DAI has a several thousand users and a lot of scientific projects that are using it as a Grid database (i.e. ADMIRE1, BIRN2, GEO Grid [17], MESSAGE [18], BEinGRID5, LaQuAT6, Database Grid7, mantisGRID [19] etc.). List of most important projects can be found in [20].

Architecture of OGSA-DAI is in compliance with needs described in Section 2. In OGSA-DAI, *data resource* represents externally managed database system and data service provide interface to access to this data resources. A *data request execution resource* (DRER) executes OGSA-DAI workflows (and termed requests) provided by clients. A DRER can execute a number of such requests concurrently and also queue a number more. An OGSA-DAI server must have at least one DRER. Clients execute *workflows* using OGSA-DAI as follows: a client submits their workflow (or request) to a *data request execution service* (DRES). This is a web service which provides access to a DRER. A workflow consists of a number of units called *activities*. An activity is an individual unit of work in the workflow and performs a well-defined data-related task such as running an SQL query, performing a data transformation or delivering data. Activities are connected. The outputs of activities connect to the inputs of other activities. Data flows from activities to other activities and this is in one direction only. Different activities may output data in different formats and may expect their input data in different formats [21]. Main activities are organized in three types: Statement Activity, Translation Activity and Delivery Activity. On Fig. 3 a) is shown example of OGSA-DAI workflow with three activities. The interaction between client and data through DRES, is shown on Fig. 3 b).

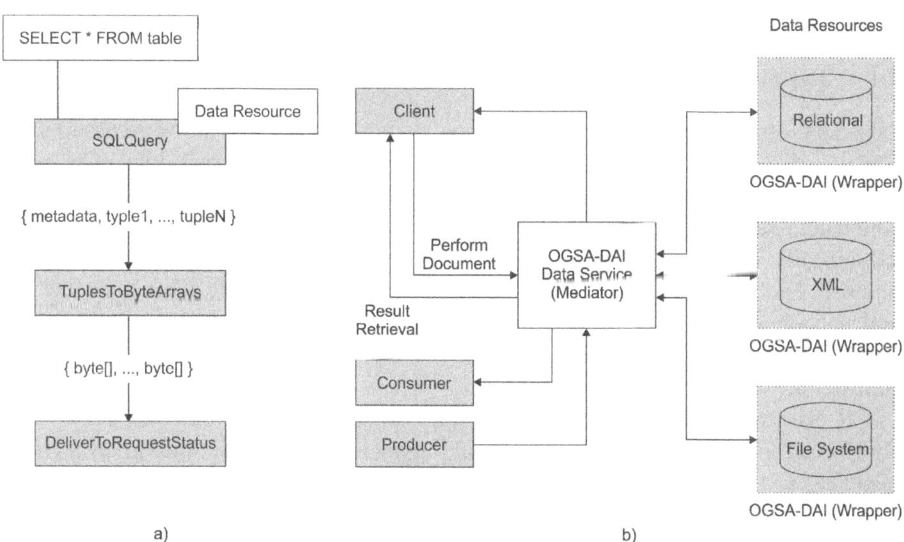

Fig. 3. a) Example of OGSA-DAI query-transform-deliver workflow. b) Interaction between client and data.

4 Model for Executing External Jobs in OGSA-DAI Workflows

At the beginning we analyze several systems for scheduling of jobs in distributed environments [22][23]. As we explain in previous section, key components, or extensibility points, of OGSA-DAI are resources and workflow activities. The idea of our proposed model is to define *computational resource* (CR) and *execution activity* (EA), as shown on Fig. 4.

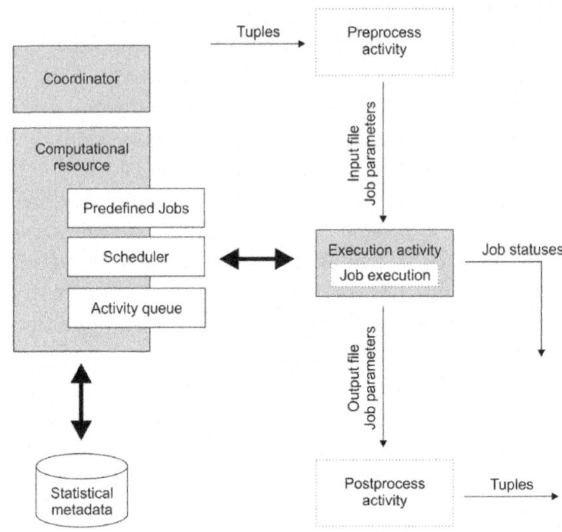

Fig. 4. Architecture of proposed model for executing external jobs

4.1 Defining the Model

Like data resources which are connected to existing database and allow access to data, CR are places (computers) on which can be executed external jobs. CR have necessary configuration for executing jobs in the operating system. It has parameters which are read from configuration file (maxNoJobs, OperatingSystem, etc.) or created dynamically in addiction of current system status (if processor usage is high or number of jobs is equal to number of maximum allowed jobs, CR doesn't execute job and put job in *activity queue*). CR has list of *predefined jobs* with parameters and *statistics* of previous executions. If job is not previously defined, parameters and statistics are not present.

CR has *scheduler* which is responsible for starting the external jobs. This component read statistics, CR static and dynamic properties, and gives system resources to EA. Scheduler have activity queue and keep all activities that should be executed. For each DBMS (MySQL, Oracle, XML), is defined specific wrapper. In this case, each type of CR must have different implementation (for example, CR for executing batch jobs in Windows has different implementation from CR in Linux).

EA is activity which is connected to the CR and is wrapper to job/application that should be executed in workflow. Activity can be part of regular OGSA-DAI workflow but must be specified with one computational resource. OGSA-DAI in general don't have workflow coordinator which can choose appropriate resource for activity and user must specify computational resource. In our model, we propose *coordinator* which can choose on which resource to execute EA. For example, if activity is executed in several seconds (data about execution are read from statistics), it is executed on same computer, but if activity last for several hours it is executed on other CR.

On Fig. 4 EA is surrounded by two activities, one for preprocess data and one for postprocess. These activities are specific for different type of applications. First activity convert and export data in file format, while last activity read data from output file of external job and convert it into stream of data. EA have other pipe that report job statuses and rest of the workflow know the current system status and make decisions from this information - workflow can have maximum waiting time and if job is not finished, workflow can continue without that data (or job can be started again).

4.2 Scenarios of Practical Usage of the Model

In this section we present two scenarios of executing external jobs, which can commonly occurred in real situations. These scenarios are defined on cost bases of data transfer and computational capabilities of OGSA-DAI servers.

Communication cost can be very expensive when data are transferred between servers. If application is data-intensive; have need of distributed queries; or algorithms are iterative and exchange data between them very often; then a lot of messages must be exchanged. While this is a general rule to be followed in the design of distributed systems in which there is the potential for large amounts of information to be moved over networks, the problem is accentuated in service oriented architectures, like Grid and OGSA-DAI concretely, due to the fact that exchanging messages between web services is more expensive. Currently, the main expense is due to the fact that data held within a service (e.g. in a database) is converted to and from a character-based representation (of XML or JSON) so that it can be packaged in a SOAP message and sent between services. This imposes a CPU cost for converting to and from the specific format and also results in a large increase in size between an efficient representation within a service (e.g. within a database) and the tagged, character-based representation, incurring increased transfer costs as more bytes are shipped [14].

The idea of first scenario is to try to avoid moving data as much as possible through moving the computation to the data, rather than vice versa. This is the main way to minimize the number of messages; the amount of data exchanged and this approach avoid communication costs. Here, computations are done on the database servers and data transfer is avoided as much as possible. This scenario can be apply when algorithms use data from one server; when algorithms don't have need from high computational power - like preprocessing of data; etc.

Our second scenario of using OGSA-DAI with external jobs is to separate computation tasks from database computers. This can be done from several reasons: database servers don't have enough computational power, algorithms work with small amount of data from overall database but impose high computational power - therefore database servers can be bottleneck; data are highly distributed and must be transferred to one location for computation to be done; etc.

For example, data can be distributed on several locations in different organizations, but their computers don't have enough computational resources. Although data transfer can be expensive in such a situation, the computation can be done on other computers, or clusters, adopted for that purpose.

Another possible situation is when computation is near the data (but not on same server), the data transfer is realized with great speed. In this scenario computation can be done on dedicate cluster nodes in the same cluster.

5 Current Status and Practical Implementation

With cooperation with EPCC institute at University of Edinburgh, we start to implement proposed architecture and now is work in progress. OGSA-DAI is open source project build in Java and have good documentation for developers. We extend latest OGSA-DAI 4.2 release with:

- Computational resources
- Activities connected with CR

Currently CR support only static properties for execution of activities. These properties are read from configuration file (in OGSA-DAI manner).

Until now we concentrate on implementation of EA that runs batch jobs or standalone applications on CR computer. These jobs have one input and one output file. EA is extending MatchedIterativeActivity and implements ResourceActivity interface. Matched iterative activities take input into blocks, in our case jobs, that should be executed, and process each block individually. Inputs of activity are consisting of job parameters and input file. Scheduler is in early phase of development and check maximum number of jobs that can run in parallel. If number of active jobs is lower than maximum number, the job is started in separate thread. For defining job we use Callable interface, instead of Runnable, because Callable allow more control on threads - canceling thread if error occur and can return result. Threads are submitted to ExecutorService which automatically take care of maximum number of active threads (jobs). Output of activity is jobs states, which have job description and job current status (submitted, started, executing, and finished). Job state for each job is produced periodically on several seconds. With this output, rest of the workflow (postprocess activity) knows the current state of the jobs.

Our test case is consisting of crawling (downloading) articles from news web sites. External job is made in Python, have one textual input file and one XML output file. With our workflow we want to collect news articles from several web sites. Preprocess activity take tuples as input, export them into text file and create jobs for executing. These jobs are followed to the EA. With several executions of workflow we conclude that the maximum number of jobs parameter depends on CPU usage of external program, and in our case we set it to 20. This is the main reason why we propose list of predefined jobs in CR, and keeping different parameters for each job. Another moment is overall CPU usage, because several workflows can be run in parallel in OGSA-DAI. For that reason we propose dynamic properties of CR, and if CPU usage is above 90%, EA will be put into waiting queue.

6 Conclusion and Future Work

We can highlight three conclusions. First, Grid applications have an increasing need of database systems. Combining Grid and database technologies is an essential approach to meet the requirements of large-scale Gird applications. Our second conclusion is that today there are many efforts to enable organizing and querying data in Grid environment. OGSA-DAI is a good example. And, finally, we show that there is possibility to implement OGSA-DAI extension which can allow execution of external jobs.

Our proposed model allows different types of algorithms for data analyzing, built in different languages or like separate applications, to be incorporated in heterogeneous distributed databases. More complex cases cover execution of external jobs on other platforms, like desktop grid, Condor clusters, cloud services etc.

Acknowledgements. This work is done in collaboration with Ally Hume, Michael Jackson and Mario Antonioletti from EPCC at the University of Edinburgh.

References

1. Foster, I., Kesselman, C.: The Grid: Blueprint for a New Computing Infrastructure. Elsevier (2004)
2. Taylor, J.: Defining e-Science, `http://www.nesc.ac.uk/nesc/define.html`
3. Kakasevski, G., et al.: Grid enabled system for gathering and retrieving images from WEB. In: ETAI 2007 (2007)
4. The Large Hadron Collider Homepage, `http://lhc.web.cern.ch/lhc/`
5. Balaz, A., Prnjat, O., Vudragovic, D., Slavnic, V., Liabotis, I., Atanassov, E., Jakimovski, B., Savic, M.: Development of Grid E-Infrastructure in South-Eastern Europe. J. Grid Comput. 9, 135–154 (2011)
6. Hong, N.P.C., et al.: Grid Database Service Specification. GGF Informational Document (February 2003)
7. Antonioletti, M., Atkinson, M., Baxter, R., Borley, A., Hong, N., Collins, B., Hardman, N., Hume, A., Knox, A., Jackson, M., et al.: The design and implementation of Grid database services in OGSA-DAI. Concurrency and Computation: Practice and Experience 17(2-4), 357–376 (2005)
8. Holmes, G., Donkin, A., Witten, I.H.: Weka: A machine learning workbench. In: Proc. Second Australia and New Zealand Conference on Intelligent Information Systems, Brisbane, Australia (1994)
9. The University of Waikato; WEKA - Open source datamining software (2012), `http://www.cs.waikato.ac.nz/ml/weka/`
10. Foster, I., Kesselman, C., Nick, J.M., Tuecke, S.: The physiology of the grid: An open grid services architecture for distributed systems integration. Open Grid Service Infrastructure WG, Global Grid Forum 22, 1–5 (2002)
11. Open Grid Forum - community of users, developers, and vendors for standardization of Grid computing (2012), `http://www.ogf.org/`

12. Atkinson, M.P., Dialani, V., Guy, L., Narang, I., Paton, N.W., Pearson, D., Storey, T., Watson, P.: Grid Database Access and Integration: Requirements and Functionalities. DAIS-WG, Global Grid Forum Informational Document, GFD.13 (March 2003)
13. Lee, A., Magowan, J., Dantressangle, P., Bannwart, F.: Bridging the Integration Gap, Part 1: Federating Grid Data. IBM Developer Works (August 2005)
14. Watson, P.: Databases in Grid Applications: Locality and Distribution. In: Jackson, M., Nelson, D., Stirk, S. (eds.) BNCOD 2005. LNCS, vol. 3567, pp. 1–16. Springer, Heidelberg (2005)
15. Jackson, M., et al.: OGSA-DAI 3.0–the whats and the whys. In: Proc. UK e-Science All Hands Q6 Meeting, pp. 158–165 (2007)
16. Abhinav, K.: Evaluation of Grid Middleware OGSA-DAI. School of Computing Science. Vellore Institute of Technology, Deemed University (May 2006)
17. Tanimura, Y., Yamamoto, N., Tanaka, Y., Iwao, K., Kojima, I., Nakamura, R., Tsuchida, S., Sekiguchi, S.: Evaluation of Large-Scale Storage Systems for Satellite Data in GEO GRID. In: The International Archives of the Photogrammetry, Remote Sensing and Spatial Information Sciences, Beijing, vol. XXXVII(Pt. B4), pp. 1567–1574 (2008)
18. MESSAGE (Mobile Environmental Sensing System Across Grid Environments) Project (2010), http://bioinf.ncl.ac.uk/message/
19. Garcia Ruiz, M., Garcia Chaves, A., Ruiz Ibañez, C., Gutierrez Mazo, J.M., Ramirez Giraldo, J.C., Pelaez Echavarria, A., Valencia Diaz, E., Pelaez Restrepo, G., Montoya Munera, E.N., Garcia Loaiza, B., Gomez Gonzalez, S.: mantisGRID: A Grid Platform for DICOM Medical Images Management in Colombia and Latin America. J. Digit. Imaging (Febuary 2010)
20. OGSA-DAI open source project publications (2012), http://sourceforge.net/apps/trac/ogsa-dai/wiki/Publications
21. OGSA-DAI user documentation (2012), http://sourceforge.net/apps/trac/ogsa-dai/wiki/UserDocumentation
22. Singh, G., Kesselman, C., Deelman, E.: Optimizing Grid-Based Workflow Execution. J. Grid Comput. 3(3-4), 201–219 (2005)
23. Luckow, A., Lacinski, L., Jha, S.: SAGA BigJob: An Extensible and Interoperable Pilot-Job Abstraction for Distributed Applications and Systems. In: CCGRID 2010, pp. 135–144 (2010)

Quasigroup Representation of Some Feistel and Generalized Feistel Ciphers

Aleksandra Mileva[1] and Smile Markovski[2]

[1] University "Goce Delčev", Faculty of Computer Science, "Krste Misirkov" bb, 2000, Štip, Republic of Macedonia
aleksandra.mileva@ugd.edu.mk

[2] University "Ss Cyril and Methodius", Faculty of Computer Science and Engineering, Rudjer Boškovikj 16, P.O. Box 393, 1000, Skopje, Republic of Macedonia
smile.markovski@finki.ukim.mk

Abstract. There are several block ciphers designed by using Feistel networks or their generalization, and some of them allow to be represented by using quasigroup transformations, for suitably defined quasigroups. We are interested in those Feistel ciphers and Generalized Feistel ciphers whose round functions in their Feistel networks are bijections. In that case we can define the wanted quasigroups by using suitable orthomorphisms, derived from the corresponding Feistel networks. Quasigroup representations of the block ciphers MISTY1, Camellia, Four-Cell^{+} and SMS4 are given as examples.

Keywords: Feistel network, Feistel cipher, orthomorphism, quasigroup, MISTY1, Camellia, Four-Cell^{+}, SMS4.

1 Introduction

A block cipher is a type of symmetric-key cryptographic system that consists of two algorithms: one for encryption and the other for decryption. Encryption algorithm takes two inputs: a fixed-length block of plaintext data and a secret key, and produces a block of ciphertext data of the same length. Decryption algorithm is performed by applying the reverse transformation to the ciphertext block using the same secret key.

For his Lucifer block cipher, H. Feistel [11] invented a special transformation that takes any function f (known as round function) and produces a permutation. The one half of the input swaps with the result obtained from XOR-ing the output of the function f applied to this half, and the other half of the input. This became a round of so called Feistel structure for construction of block ciphers, known as Feistel network or Feistel cipher.

Feistel networks and their modifications have been used in many block cipher designs since then: DES [14], MISTY1 [17], Camellia [1], Khufu and Khafre [18], LOKI [3], ICE [13], CAST-128 [2], Blowfish [22], etc. Different generalizations of Feistel networks are also known that split the input blocks into $n > 2$ sub-blocks

S. Markovski and M. Gusev (Eds.): *ICT Innovations 2012*, AISC 207, pp. 161–171.
DOI: 10.1007/978-3-642-37169-1_16

(cells), such as the *type-1*, *type-2* and *type-3* Extended Feistel networks from Zheng et al [23], the Generalized Feistel-Non Linear Feedback Shift Register (GF-NLFSR) from Choy et al [4], etc. They find application in several block ciphers, such as Four-Cell$^+$ [4], CAST-256 [2], SMS4 [9], etc, sometimes known as Generalized Feistel ciphers. Primary advantage of Feistel and Generalized Feistel ciphers to most of the other block ciphers is using the same algorithm for both encryption and decryption, with reverse order of subkeys in decryption.

A quasigroup is a groupoid $(Q, *)$ with the property that each one of the equations $a * x = b$ and $y * a = b$ have a unique solution x, respectively y. When Q is a finite set, the main body of the Cayley table of the quasigroup $(Q, *)$ represents a Latin square.

There are research activities for representing some cryptographic primitives, or their building blocks by quasigroups and quasigroup string transformations. The idea for connection between block cipher's modes of operation and quasigroups can be found in the paper [7], which describes some modes of operation as quasigroup operations. Paper [19] describes an algorithm for generating an optimal 4×4 S-boxes (a building block of some lightweight block ciphers) by non-linear quasigroups of order 4 and quasigroup string transformations. This is a promising methodology for more optimized hardware implementations of S-boxes. We go one step further. In this paper we examine several block ciphers that use Feistel networks or their generalizations, and we rewrite their rounds by quasigroup operations, using ideas from [20]. Special attention is given to the block ciphers Misty1, Camellia, Four-Cell, Four-Cell$^+$ and SMS4. For all of them, one feature is the same - they use bijections as round functions in their Feistel networks. We represent them by quasigroup string transformations, by applying different quasigroups of orders 2^b, where b is the bit length of the block.

We emphasize here that there are several Feistel ciphers that **do not use** bijections as round functions, like DES, Triple DES, DESX, ICE, Blowfish, CAST-128, Knufu, Khafre, LOKI, etc.

2 Feistel Networks as Orthomorphisms

Sade [21] proposed the following construction of quasigroups, known as *diagonal method.* Consider the group $(\mathbb{Z}_n, +)$ and let θ be a permutation of the set $\mathbb{Z}_n$, such that $\phi(x) = x - \theta(x)$ is also a permutation. Define an operation $\circ$ on $\mathbb{Z}_n$ by:

$$x \circ y = \theta(x - y) + y \tag{1}$$

where $x, y \in \mathbb{Z}_n$. Then $(\mathbb{Z}_n, \circ)$ is a quasigroup (we say that $(\mathbb{Z}_n, \circ)$ is derived by θ).

Definition 1. *[8], [10] A* **complete mapping** *of a group $(G, +)$ is a permutation $\phi : G \to G$ such that the mapping $\theta : G \to G$ defined by $\theta(x) = x + \phi(x)$ ($\theta = I + \phi$, where I is the identity mapping) is again a permutation of G. The mapping θ is the* **orthomorphism** *associated to the complete mapping ϕ. A group G is* **admissible** *if there is a complete mapping $\phi : G \to G$.*

A generalization of the diagonal method by using complete mappings and orthomorphisms is given by the following theorem.

Theorem 1. *Let ϕ be a complete mapping of the admissible group $(G,+)$ and let θ be an orthomorphism associated to ϕ. Define operations $\circ$ and $*$ on G by*

$$x \circ y = \phi(y-x) + y = \theta(y-x) + x, \tag{2}$$

$$x * y = \theta(x-y) + y = \phi(x-y) + x, \tag{3}$$

where $x, y \in G$. Then $(G,\circ)$ and $(G,*)$ are quasigroups, opposite to each other, i.e., $x \circ y = y * x$ for every $x, y \in G$.

In [20] we have defined parameterized versions of the Feistel network, the *type-1* Extended Feistel network, and the Generalized Feistel-Non Linear Feedback Shift Register (GF-NLFSR), and we have proved that if a bijection f is used for their creation, then they are orthomorphisms of abelian groups. Now we are going to modify slightly these definitions in order to be suitable for our next purposes. All of the proofs given in [20] are almost immediately applicable for the modified versions as well. As an illustration, we will give only the proof of Theorem 2.

Definition 2. *Let $(G,+)$ be an abelian group, let $f_C : G \to G$ be a mapping, where C is an arbitrary constant and let $A, B \in G$. The* **Parameterized Feistel Networks (PFN)** *$F^d_{A,B,C}$ and $F^l_{A,B,C} : G^2 \to G^2$ created by f_C are defined for every $l, r \in G$ by*

$$F^d_{A,B,C}(l,r) = (r + A, l + B + f_C(r)) \text{ and } F^l_{A,B,C}(l,r) = (r + A + f_C(l), l + B).$$

The PFNs $F^d_{A,B,C}$ and $F^l_{A,B,C}$ are bijections with inverses

$$(F^d_{A,B,C})^{-1}(l,r) = (r - B - f_C(l-A), l - A),$$

$$(F^l_{A,B,C})^{-1}(l,r) = (r - B, l - A - f_C(r-B)).$$

Theorem 2. (Theorem 3.3 from [20]) *Let $(G,+)$ be an Abelian group and let $f_C : G \to G$ be a bijection for some constant C and $A, B \in G$. Then the Parameterized Feistel Networks $F^d_{A,B,C}$ and $F^l_{A,B,C} : G^2 \to G^2$ created by f_C are orthomorphisms of the group $(G^2,+)$.*

Proof. Let $\Phi = F^d_{A,B,C} - I$, i.e.,

$$\Phi(l,r) = F^d_{A,B,C}(l,r) - (l,r) = (r + A - l, l + B + f_C(r) - r)$$

for every $l, r \in G$. Define the function $\Omega : G^2 \to G^2$ by

$$\Omega(l,r) = (f_C^{-1}(l + r - A - B) - l + A, f_C^{-1}(l + r - A - B)).$$

We have $\Omega \circ \Phi = \Phi \circ \Omega = I$, i.e., Φ and $\Omega = \Phi^{-1}$ are bijections.

Similarly we can proof that $F^l_{A,B,C}$ is an orthomorphism of the group $(G^2,+)$.

Definition 3. *Let an abelian group $(G,+)$, a mapping $f_C : G \to G$ for some constant C, constants $A_1, A_2, \ldots, A_n \in G$ and an integer $n > 1$ be given. The* type-1 **Parameterized Extended Feistel Network (PEFN)** $F_{A_1,A_2,\ldots,A_n,C} : G^n \to G^n$ *created by f_C is defined for every $(x_1, x_2, \ldots, x_n) \in G^n$ by*

$$F_{A_1,A_2,\ldots,A_n,C}(x_1, x_2, \ldots, x_n) =$$

$$= (x_2 + f_C(x_1) + A_1, x_3 + A_2, \ldots, x_n + A_{n-1}, x_1 + A_n).$$

Theorem 3. (Theorem 3.4 from [20]) *Let $(G,+)$ be an abelian group, $A_1, A_2, \ldots, A_n \in G$ and $n > 1$. If $F_{A_1,A_2,\ldots,A_n,C} : G^n \to G^n$ is a* type-1 *PEFN created by a bijection $f_C : G \to G$, then $F_{A_1,A_2,\ldots,A_n,C}$ is an orthomorphism of the group $(G^n, +)$.*

Definition 4. *Let an abelian group $(G,+)$, a mapping $f_C : G \to G$ where C is an arbitrary constant, constants $A_1, A_2, \ldots, A_n \in G$ and an integer $n > 1$ be given. The* **PGF-NLFSR (Parameterized Generalized Feistel-Non Linear Feedback Shift Register)** $F_{A_1,A_2,\ldots,A_n,C} : G^n \to G^n$ *created by f_C is defined for every $(x_1, x_2, \ldots, x_n) \in G^n$ by*

$$F_{A_1,A_2,\ldots,A_n,C}(x_1, x_2, \ldots, x_n) =$$

$$= (x_2 + A_1, x_3 + A_2, \ldots, x_n + A_{n-1}, x_2 + \ldots + x_n + A_n + f_C(x_1)).$$

Theorem 4. (Theorem 3.5 from [20]) *For any abelian group $(\mathbb{Z}_2^m, \oplus)$ and for any even integer n, the PGF-NLFSR $F_{A_1,A_2,\ldots,A_n,C} : (\mathbb{Z}_2^m)^n \to (\mathbb{Z}_2^m)^n$ created by a bijection $f_C : \mathbb{Z}_2^m \to \mathbb{Z}_2^m$ is an orthomorphism of the group $((\mathbb{Z}_2^m)^n, \oplus)$.*

Here we define a new generalization of the Feistel network as *type-4* PEFN, which 4-cell version was first presented in the SMS4 block cipher.

Definition 5. *Let an abelian group $(G,+)$, a mapping $f_C : G \to G$, where C is an arbitrary constant, constants $A_1, A_2, \ldots, A_n \in G$ and an integer $n > 1$ be given. The* type-4 **Parameterized Extended Feistel Network (PEFN)** $F_{A_1,A_2,\ldots,A_n,C} : G^n \to G^n$ *created by f_C is defined for every $(x_1, x_2, \ldots, x_n) \in G^n$ by*

$$F_{A_1,A_2,\ldots,A_n,C}(x_1, x_2, \ldots, x_n) =$$

$$= (x_2 + A_1, x_3 + A_2, \ldots, x_n + A_{n-1}, x_1 + A_n + f_C(x_2 + \ldots + x_n)).$$

If f_C is a bijection, then *type-4* PEFN $F_{A_1,A_2,\ldots,A_n,C}$ is a bijection with inverse

$$F^{-1}_{A_1,A_2,\ldots,A_n,C}(y_1, y_2, \ldots, y_n) = \\ = (y_n - A_n - f_C(y_1 + y_2 + \ldots + y_{n-1} - A_1 - A_2 - \ldots - A_{n-1}), y_1 - A_1, \\ y_2 - A_2, \ldots, y_{n-1} - A_{n-1}).$$

Theorem 5. *For any abelian group $(\mathbb{Z}_2^m, \oplus)$ and any even integer n, the* type-4 *PEFN $F_{A_1,A_2,\ldots,A_n,C} : (\mathbb{Z}_2^m)^n \to (\mathbb{Z}_2^m)^n$ created by a bijection $f_C : \mathbb{Z}_2^m \to \mathbb{Z}_2^m$ is an orthomorphism of the group $((\mathbb{Z}_2^m)^n, \oplus)$.*

Proof. Let $\Phi = F_{A_1,A_2,\ldots,A_n,C} \oplus I$ and $x_i \in (\mathbb{Z}_2^m)^n$. Then
$\Phi(x_1, x_2, \ldots, x_n) = F_{A_1,A_2,\ldots,A_n,C}(x_1, x_2, \ldots, x_n) \oplus (x_1, x_2, \ldots, x_n)$
$= (x_2 \oplus A_1 \oplus x_1, x_3 \oplus A_2 \oplus x_2, \ldots, x_n \oplus A_{n-1} \oplus x_{n-1}, x_1 \oplus x_n \oplus A_n \oplus f_C(x_2 \oplus \ldots$
$\ldots \oplus x_n)) = (y_1, y_2, \ldots, y_{n-1}, y_n)$.

First we observe that

$$\begin{aligned}
&y_1 = x_2 \oplus x_1 \oplus A_1,\\
&y_1 \oplus y_2 = x_3 \oplus x_1 \oplus A_1 \oplus A_2,\\
&\ldots\ldots\ldots\ldots\\
&y_1 \oplus y_2 \oplus \ldots \oplus y_{n-2} = x_{n-1} \oplus x_1 \oplus A_1 \oplus A_2 \oplus \ldots \oplus A_{n-2},\\
&y_1 \oplus y_2 \oplus \ldots \oplus y_{n-1} = x_n \oplus x_1 \oplus A_1 \oplus A_2 \oplus \ldots \oplus A_{n-1},\\
&y_1 \oplus y_2 \oplus \ldots \oplus y_n = f_C(x_2 + \ldots + x_n) \oplus A_1 \oplus A_2 \oplus \ldots \oplus A_n.
\end{aligned}$$

The sum of right-hand sides of the first $n-2$ equations, when $n =$ is even, is $x_2 \oplus x_3 \oplus \ldots \oplus x_{n-1} \oplus A_2 \oplus A_4 \oplus \ldots \oplus A_{n-2}$.

Define the function $\Omega : (\mathbb{Z}_2^m)^n \to (\mathbb{Z}_2^m)^n$ by

$$\Omega(y_1, \ldots, y_n) = (z \oplus y_1 \oplus \ldots \oplus y_{n-1} \oplus A_1 \oplus \ldots \oplus A_{n-1}, z \oplus y_2 \oplus \ldots \oplus y_{n-1} \oplus$$

$$A_2 \oplus \ldots \oplus A_{n-1}, \ldots, z \oplus y_{n-2} \oplus y_{n-1} \oplus A_{n-2} \oplus A_{n-1}, z \oplus y_{n-1} \oplus A_{n-1}, z),$$

where
$z = f_C^{-1}(y_1 \oplus y_2 \oplus \ldots \oplus y_n \oplus A_1 \oplus A_2 \oplus \ldots \oplus A_n) \oplus y_1 \oplus (y_1 \oplus y_2) \oplus \ldots \oplus (y_1 \oplus y_2 \oplus \ldots \oplus y_{n-2}) \oplus A_2 \oplus A_4 \oplus \ldots \oplus A_{n-2}$.

We have $\Omega \circ \Phi = \Phi \circ \Omega = I$, i.e., Φ and $\Omega = \Phi^{-1}$ are bijections.

3 Quasigroup Representation of Some Feistel and Generalized Feistel Ciphers

Given a quasigroup $(Q, *)$, an adjoint binary operation $\backslash$ (known as left division) on the set Q can be defined by $x\backslash y = z \iff x * z = y$. Then $(Q, \backslash)$ is also a quasigroup and the following equalities are identities over the algebra $(Q, *, \backslash)$:

$$x\backslash(x * y) = y, \qquad x * (x\backslash y) = y. \tag{4}$$

Consider the finite set Q as an alphabet with word set $Q^+ = \{x_1x_2\ldots x_t \mid x_i \in Q, t \geq 1\}$. For fixed letter $l \in Q$ (called a leader) the transformations $e_l, d_l : Q^+ \to Q^+$ are defined in [16], as follows.

$$e_l(x_1 \ldots x_t) = (z_1 \ldots z_t) \iff z_j = z_{j-1} * x_j, \; 1 \leq j \leq t,$$

$$d_l(z_1 \ldots z_t) = (x_1 \ldots x_t) \iff x_j = z_{j-1}\backslash z_j, \; 1 \leq j \leq t,$$

where $z_0 = l$.

Here, we define a generalized version of these transformations. Let $*_1, *_2, \ldots, *_t$ be t (not necessarily distinct) quasigroup operations on a finite set Q, $l \in Q$ and let $\backslash_i$ be the left division adjoint operation corresponding to $*_i$. Then the generalized transformations $e_{l,*_1,*_2,\ldots,*_t}$ and $d_{l,\backslash_1,\backslash_2,\ldots,\backslash_t}$ are defined as follows.

$$e_{l,*_1,*_2,\ldots,*_t}(x_1 \ldots x_t) = (z_1 \ldots z_t) \iff z_j = z_{j-1} *_j x_j, \; 1 \leq j \leq t, \tag{5}$$

$$d_{l,\backslash_1,\backslash_2,\dots,\backslash_t}(z_1 \dots z_t) = (x_1 \dots x_t) \iff x_j = z_{j-1} \backslash_{t-j+1} z_j,\ 1 \le j \le t \quad (6)$$

where $z_0 = l$.

It follows from the identities (4) that $e_{l,*_1,\dots,*_t}(d_{l,\backslash_t,\backslash_{t-1},\dots,\backslash_1}(z_1 \dots z_t)) = z_1 \dots z_t$ and $d_{l,\backslash_1,\backslash_2,\dots,\backslash_t}(e_{l,*_t,*_{t-1},\dots,*_1}(x_1 \dots x_t)) = x_1 \dots x_t$. So, we have the following proposition.

Proposition 1. *The generalized transformations $e_{l,*_1,*_2,\dots,*_t}$ and $d_{l,\backslash_1,\backslash_2,\dots,\backslash_t}$ are permutations on Q^+.*

We have noticed the following. In some Feistel ciphers there are r rounds of r different PFNs $F^d_{A_i,B_i,C_i}$ or $F^l_{A_i,B_i,C_i}$, applied iteratively on a starting block X_0. The output of the i-th round can be written as $X_i = F^d_{A_i,B_i,C_i}(X_{i-1})$ (or $X_i = F^l_{A_i,B_i,C_i}(X_{i-1})$). If the used round functions f_{C_i} are bijections, then the PFNs $F^d_{A_i,B_i,C_i}$ and $F^l_{A_i,B_i,C_i}$ are orthomorphisms on some group $(G^2, +)$ (with zero $\mathbf{0}$), and we can define r different quasigroups $(G^2, *_i)$ as $X *_i Y = F^d_{A_i,B_i,C_i}(X - Y) + Y$ (or $X *_i Y = F^l_{A_i,B_i,C_i}(X - Y) + Y$). So, we can write the output of the i-th round as

$$X_i = X_{i-1} *_i \mathbf{0}.$$

The output X_r of the final r-th round can be written as

$$X_r = X_{r-1} *_r \mathbf{0} = (X_{r-2} *_{r-1} \mathbf{0}) *_r \mathbf{0} = ((\dots(X_0 *_1 \mathbf{0})\dots) *_{r-1} \mathbf{0}) *_r \mathbf{0}.$$

In fact, this is an application of a generalized $e_{l,*_1,*_2,\dots,*_t}$ transformation on string of r zeros $\mathbf{0}$ with leader $l = X_0$ and r different quasigroups $(G^2, *_i)$:

$$X_r = e_{X_0,*_1,*_2,\dots,*_r}(\underbrace{\mathbf{0}, \mathbf{0}, \dots, \mathbf{0}}_{r}).$$

The same notice holds for Generalized Feistel ciphers that as rounds use different *type-1* PEFNs, *type-4* PEFNs, or PGF-NLFSRs, with bijections as round functions.

In this paper, quasigroup representations of MISTY1, Camellia, Four-Cell, Four-Cell$^+$ and SMS4 are given. Other block ciphers, that can be represented by quasigroups are MISTY2 (2-cell PGF-NLFSRs), KASUMI (PFNs), etc.

3.1 Quasigroup Representation of Misty1

The MISTY1 is 64-bit block cipher, designed by Matsui [17] and others from Mitsubishi Electric Corporation, with 128-bit user key and a variable number (multiply of 4) of rounds (recommended number of rounds is 8). It is one of the recommended block ciphers by Japanese CRYPTREC (Cryptography Research and Evaluation Committees) [6], NESSIE (New European Schemes for Signatures, Integrity and Encryption) project in 2003 [15], and as an ISO [12] international standard in 2005 and 2010.

MISTY1 has 5 FL layers divided by two Feistel rounds with FO round function. FL function is a key-dependent and linear, and it is applied to both 32-bit halves of the input word. FO is a key-dependent non-linear function which is a bijection, for fixed subkeys. For our discussion, we are not interested how FO is generated. MISTY1 uses 10 32-bit subkeys KL_i, 8 64-bit subkeys KO_i and 8 48-bit subkeys KI_i, generated by the secret key K.

Let the plaintext be denoted by $M = (l_0, r_0) = X_0 \in (\{0,1\}^{32})^2$. The MISTY1 algorithm can be represented as follows:

1. For $i = 1, 3, 5, 7$ do
   $(l'_i, r'_i) = (FL(l_{i-1}, KL_i), FL(r_{i-1}, KL_{i+1}))$,
   $(l_i, r_i) = (r'_i \oplus FO_{(KO_i, KI_i)}(l'_i), l'_i)$,
   $(l_{i+1}, r_{i+1}) = (r_i \oplus FO_{(KO_{i+1}, KI_{i+1})}(l_i), l_i)$.
2. The ciphertext is $C = (FL(r_8, KL_{10}), FL(l_8, KL_9))$.

In fact, in step 1, 8 PFN $F^l_{0,0,(KO_j,KI_j)}$, $j \in \{1, 2, \ldots, 8\}$, are used, and because their round functions $FO_{(KO_i,KI_i)}$ are bijections (for fixed constant (KO_i, KI_i)), $F^l_{0,0,(KO_j,KI_j)}$ are orthomorphisms of the group $((\mathbb{Z}_2^{32})^2, \oplus)$. We can define quasigroup operations as

$$X *_j Y = F^l_{0,0,(KO_j,KI_j)}(X \oplus Y) \oplus Y,$$

where $X, Y \in (\mathbb{Z}_2^{32})^2$. The quasigroups are of order 2^{64}.

Now we can rewrite MISTY1 with quasigroups as follows:

1. For $i = 1, 3, 5, 7$ do
   $X'_i = (l'_i, r'_i) = (FL(l_{i-1}, KL_i), FL(r_{i-1}, KL_{i+1}))$
   $X_{i+1} = (l_{i+1}, r_{i+1}) = e_{X'_i, *_i, *_{i+1}}(\mathbf{0}, \mathbf{0})$
2. The ciphertext is $C = (FL(r_8, KL_{10}), FL(l_8, KL_9))$.

3.2 Quasigroup Representation of Camellia

Camellia is 128-bit block cipher, jointly developed by NTT and Mitsubishi Electric Corporation in 2000 [1]. It was selected as one of recommended algorithms by CRYPTREC in 2002 [6], NESSIE project in 2003 [15], and as an ISO [12] international standard in 2005 and 2010.

Camellia uses variable key length of 128, 192 and 256 bits, known as Camellia-128, Camellia-192 and Camellia-256. Camellia-128 has 18-rounds, and Camellia-192/256 has 24 rounds. It is a Feistel cipher, with FL/FL^{-1}–function layer after every 6 rounds, and additional input/output whitenings. By the key shedule algorithm, the following subkeys are generated from the secret key K: $kw_{t(64)}$ $(t = 1, 2, 3, 4)$, $k_{i(64)}$ $(i = 1, 2, \ldots, p)$ and $kl_{v(64)}$ $(v = 1, 2, \ldots, r)$, where $p = 18, r = 4$ for Camellia-128 and $p = 24, r = 6$ for Camellia-192/256.

Let the plaintext be denoted by $M = (m_1, m_2) \in (\{0,1\}^{64})^2$ and let

$$X_0 = (x_0, x_1) = (m_1 \oplus kw_{1(64)}, m_2 \oplus kw_{2(64)}).$$

Let $(x_{i-1}, x_i) \in (\{0,1\}^{64})^2$ denote the input of the i-th round $(i = 1, 2, \ldots, p)$, then the output of the i-th round, except for $i = 6, 12$ (and 18 for Camellia-192/256) is

$$X_i = (x_i, x_{i+1}) = (x_i \oplus f(x_{i-1}, k_{i(64)}), x_{i-1}),$$

where $f(x,k) = f_k(x) = P(S(x \oplus k))$ $(x,k \in \{0,1\}^{64})$, S is a non-linear bijection and P is a linear bijection. So, f_k is a bijection for fixed k. For $i = 6, 12$ (and 18 for Camellia-192/256), the output is

$$(x'_i, x'_{i+1}) = (x_i \oplus f(x_{i-1}, k_{i(64)}), x_{i-1})$$

$$(x_i, x_{i+1}) = (FL(x'_i, kl_{2i/6-1}), FL^{-1}(x'_{i+1}), kl_{2i/6}).$$

The ciphertext is $C = (c_1, c_2) = (x_p \oplus kw_{3(64)}, x_{p+1} \oplus kw_{4(64)})$.

We can rewrite the i-th round of Camellia by using quasigroups $((\mathbb{Z}_2^{64})^2, *_i)$ of order 2^{128}. Let $F^l_{0,0,k_{i(64)}}$ be p PFNs created by the bijections $f_{k_{i(64)}}$, respectfully, (where 0 is zero in $(\mathbb{Z}_2^{64}, \oplus)$) for rounds $i = 1, 2, \ldots, p$. $F^l_{0,0,k_{i(64)}}$ are orthomorphisms of the group $((\mathbb{Z}_2^{64})^2, \oplus)$. We define quasigroup operations as

$$X *_i Y = F^l_{0,0,k_{i(64)}}(X \oplus Y) \oplus Y,$$

where $X, Y \in (\mathbb{Z}_2^{64})^2$.

Now, the output of the i-th round, except for $i = 6, 12$ (and 18 for Camellia-192/256) represented by quasigroups is

$$X_i = (x_i, x_{i+1}) = X_{i-1} *_i \mathbf{0}.$$

For $i = 6, 12$ (and 18 for Camellia-192/256), the output is

$$X'_i = (x'_i, x'_{i+1}) = X_{i-1} *_i \mathbf{0}$$

$$X_i = (x_i, x_{i+1}) = (FL(x'_i, kl_{2i/6-1}), FL^{-1}(x'_{i+1}), kl_{2i/6}).$$

Camellia-128 can be represented by quasigroup transformations as

$$X_0 = M \oplus (kw_{1(64)} || kw_{2(64)})$$

$$X'_6 = (x'_6, x'_7) = e_{X_0, *_1, *_2, \ldots, *_6}(\underbrace{\mathbf{0}, \mathbf{0}, \ldots, \mathbf{0}}_{6})$$

$$X_6 = (FL(x'_6, kl_1), FL^{-1}(x'_7), kl_2)$$

$$X'_{12} = (x'_{12}, x'_{13}) = e_{X_6, *_7, *_8, \ldots, *_{12}}(\underbrace{\mathbf{0}, \mathbf{0}, \ldots, \mathbf{0}}_{6})$$

$$X_{12} = (FL(x'_{12}, kl_3), FL^{-1}(x'_{13}), kl_4)$$

$$X_{18} = e_{X_{12}, *_{13}, *_{14}, \ldots, *_{18}}(\underbrace{\mathbf{0}, \mathbf{0}, \ldots, \mathbf{0}}_{6})$$

$$C = (X_{18} \oplus (kw_{3(64)} || kw_{4(64)})).$$

Similarly holds for Camellia-192/256.

3.3 Quasigroup Representation of Four-Cell and Four-Cell$^+$

The block cipher Four-Cell is first presented in [4] and its revisited version Four-Cell$^+$ is presented in [5]. Both block ciphers use r rounds which are 4-cell PGF-NLFSR structure and have 128-bit block and key size. Four-Cell has $r = 25$ rounds and Four-Cell$^+$ has $r = 30$ rounds, with whitening at the end.

Let the plaintext be denoted by $X_0 = (x_0, x_1, x_2, x_3) \in (\{0,1\}^{32})^4$, and let $(x_{i-1}, x_i, x_{i+1}, x_{i+2}) \in (\{0,1\}^{32})^4$ denote the input of the i-th round. Then the output of the i-th round is

$$X_i = (x_i, x_{i+1}, x_{i+2}, x_{i+3}) = (x_i, x_{i+1}, x_{i+2}, x_i \oplus x_{i+1} \oplus x_{i+2} \oplus f_{sk_i}(x_{i-1})).$$

It uses two types of nonlinear functions for round i, defined as follows:

$f_{sk_i}(x) = MDS(S(x \oplus sk_i))$, for rounds $i = 1, 2, \dots, p$ and $i = 21, 22, \dots, r$ ($p = 5$ for Four-Cell and $p = 10$ Four-Cell$^+$), and

$f_{sk_i}(x) = S(MDS(S(x \oplus sk_{i0})) \oplus sk_{i1})$, for rounds $i = p, p+1, \dots, 20$, where $sk_i = (sk_{i0}, sk_{i1})$. Every f_{sk_i} is a bijection for fixed sk_i, and how S and MDS are defined is not important for this discussion.

Let $F_{0,0,0,0,sk_i} : (\{0,1\}^{32})^4 \to (\{0,1\}^{32})^4$, $i = 1, 2, \dots, r$, be r 4-cell PGF-NLFSRs, created by the bijections f_{sk_i}, respectfully (0 is zero in $((\{0,1\}^{32})^4, \oplus)$). The quasigroup operations are defined by

$$X *_i Y = F_{0,0,0,0,sk_i}(X \oplus Y) \oplus Y,$$

where $X, Y \in (\{0,1\}^{32})^4$.

If $\mathbf{0} = (0,0,0,0)$, the output of the i-th round is

$$X_i = (x_i, x_{i+1}, x_{i+2}, x_{i+3}) = X_{i-1} *_i \mathbf{0}.$$

The output of the r-th round can be written as generalized $e_{l,*_1,*_2,\dots,*_r}$ transformation of string that consists of r zeros $\mathbf{0}$ and $l = X_0$, with p different quasigroups of order 2^{128}:

$$X_r = e_{X_0,*_1,*_2,\dots,*_r}(\underbrace{\mathbf{0}, \mathbf{0}, \dots, \mathbf{0}}_{r}).$$

3.4 Quasigroup Representation of SMS4

The SMS4 is a Chinese 128-bit block cipher standard, mandated for use in protecting wireless networks, with 128-bit user key. The encryption algorithm written in Chinese was released in January 2006 and its English translation was released in 2008 [9].

This Generalized Feistel cipher uses 32 rounds which are *type-4* PEFNs. Let the plaintext be denoted by $X_0 = (x_0, x_1, x_2, x_3) \in (\{0,1\}^{32})^4$, and let the input of the i-th round be denoted by $X_{i-1} = (x_{i-1}, x_i, x_{i+1}, x_{i+2}) \in (\{0,1\}^{32})^4$. Then the output of the i-th round is

$$X_i = (x_i, x_{i+1}, x_{i+2}, x_{i+3}) = (x_i, x_{i+1}, x_{i+2}, x_{i-1} \oplus f_{sk_i}(x_i \oplus x_{i+1} \oplus x_{i+2})).$$

The ciphertext is $C = (x_{35}, x_{34}, x_{33}, x_{32})$.

The i-th round function $f_{sk_i} : \mathbb{Z}_2^{32} \to \mathbb{Z}_2^{32}$ is defined as $f_{sk_i}(x) = L(S(x \oplus sk_i))$, where L is a linear permutation, S is a non-linear permutation which applies the same 8×8 S-box 4 times in parallel, and sk_i is a subkey generated by the secret key K. So f_{sk_i} is a bijection, and the *type-4* PEFN $F_{0,0,0,0,sk_i}$ is an orthomorphism in the group $((\{0,1\}^{32})^4, \oplus)$. The corresponding quasigroup $((\{0,1\}^{32})^4, *_i)$ is defined as

$$X *_i Y = F_{0,0,0,0,sk_i}(X \oplus Y) \oplus Y,$$

where $X, Y \in (\{0,1\}^{32})^4$.

If $\mathbf{0} = (0,0,0,0)$, the output of the last 32-th round can be written as generalized $e_{l,*_1,*_2,\ldots,*_{32}}$ transformation of string that consists of 32 zeros $\mathbf{0}$ and $l = X_0$, with 32 different quasigroups of order 2^{128}:

$$X_{32} = e_{X_0,*_1,*_2,\ldots,*_{32}}(\underbrace{\mathbf{0},\mathbf{0},\ldots,\mathbf{0}}_{32}).$$

4 Future Work and Open Problems

Representing some block ciphers as quasigroup string transformations can be used to analyze the block ciphers from a totally new perspective. Instead of analyzing the strength of the block ciphers via different number of rounds, the described block ciphers can be instantiated with smaller quasigroups and then those miniature block ciphers that have the same structure as the original ones can be analyzed. Additionally, the possibility to construct distinguishers for block ciphers with quasigroup representation need to be explored. In investigating of the connections between cryptographic primitives and quasigroups, one can try to find quasigroup representations of MDS matrices and optimal 8×8 S-boxes. Finding hidden quasigroups in other block ciphers is also a challenge.

References

1. Aoki, K., Ichikawa, T., Kanda, M., Matsui, M., Moriai, S., Nakajima, J., Tokita, T.: Camellia: A 128-bit block cipher suitable for multiple platforms - design and analysis. In: Stinson, D.R., Tavares, S. (eds.) SAC 2000. LNCS, vol. 2012, pp. 39–56. Springer, Heidelberg (2001)
2. Adams, C.M., Tavares, S.E.: Designing S-boxes for Ciphers Resistant to Differential Cryptanalysis. In: 3rd Symposium on State and Progress of Research in Cryptography, Rome, Italy, pp. 181–190 (1993)
3. Brown, L., Kwan, M., Pieprzyk, J., Seberry, J.: Improving Resistance to Differential Cryptanalysis and the Redesign of LOKI. In: Imai, H., Rivest, R.L., Matsumoto, T. (eds.) ASIACRYPT 1991. LNCS, vol. 739, pp. 36–50. Springer, Heidelberg (1993)
4. Choy, J., Chew, G., Khoo, K., Yap, H.: Cryptographic Properties and Application of a Generalized Unbalanced Feistel Network Structure. In: Boyd, C., González Nieto, J. (eds.) ACISP 2009. LNCS, vol. 5594, pp. 73–89. Springer, Heidelberg (2009)

5. Choy, J., Chew, G., Khoo, K., Yap, H.: Cryptographic Properties and Applications of a Generalized Unbalnced Feistel Network Structure. Cryptography and Communications 3(3), 141–164 (2011) (revised version)
6. CRYPTREC report 2002, Archive (2003), `http://www.ipa.go.jp/security/enc/CRYPTREC/fy15/doc/c02e_report2.pdf`
7. Gligoroski, D., Andova, S., Knapskog, S.J.: On the Importance of the Key Separation Principle for Different Modes of Operation. In: Chen, L., Mu, Y., Susilo, W. (eds.) ISPEC 2008. LNCS, vol. 4991, pp. 404–418. Springer, Heidelberg (2008)
8. Denes, J., Keedwell, A.D.: Latin squares: New developments in the theory and applications. Elsevier science publishers (1991)
9. Diffie, W., Ledin, G. (trans.): SMS4 encryption algorithm for wireless networks. Cryptology ePrint Archive, Report 2008/329 (2008)
10. Evans, A.B.: Orthomorphism Graphs of Groups. Journal of Geometry 35(1-2), 66–74 (1989)
11. Feistel, H.: Cryptography and computer privacy. Scientific American 228(5), 15–23 (1973)
12. International Standard - ISO/IEC 18033-3, Information technology - Security techniques - Encryption algorithms - Part 3: Block ciphers (2005)
13. Kwan, M.: The Design of the ICE Encryption Algorithm. In: Biham, E. (ed.) FSE 1997. LNCS, vol. 1267, pp. 69–82. Springer, Heidelberg (1997)
14. NBS FIPS PUB 46: Data Encryption Standard. National Bureau of Standards, U.S. Department of Commerce (1977)
15. NESSIE-New European Schemes for Signatures, Integrity, and Encryption, final report of European project IST-1999-12324, Version 0.15, Archive (April 19, 2004), `https://www.cosic.esat.kuleuven.be/nessie/Bookv015.pdf`
16. Markovski, S., Gligoroski, D., Andova, S.: Using quasigroups for one-one secure encoding. In: VIII Conf. Logic and Computer Science, LIRA 1997, Novi Sad, Serbia, pp. 157–162 (1997)
17. Matsui, M.: New block encryption algorithm MISTY. In: Biham, E. (ed.) FSE 1997. LNCS, vol. 1267, pp. 54–68. Springer, Heidelberg (1997)
18. Merkle, R.C.: Fast Software Encryption Functions. In: Menezes, A., Vanstone, S.A. (eds.) CRYPTO 1990. LNCS, vol. 537, pp. 476–500. Springer, Heidelberg (1991)
19. Mihajloska, H., Gligoroski, D.: Construction of Optimal 4-bit S-boxes by Quasigroups of Order 4. In: SECURWARE 2012, Rome, Italy (2012)
20. Mileva, A., Markovski, S.: Shapeless quasigroups derived by Feistel orthomorphisms. Glasnik Matematicki (accepted for printing)
21. Sade, A.: Quasigroups automorphes par le groupe cyclique. Canadian Journal of Mathematics 9, 321–335 (1957)
22. Schneier, B.: Description of a New Variable-Length Key, 64-Bit Block Cipher (Blowfish). In: Fast Software Encryption, Cambridge Security Workshop Proceedings, pp. 191–204. Springer (1994)
23. Zheng, Y., Matsumoto, T., Imai, H.: On the construction of block ciphers provably secure and not relying on any unproved hypotheses. In: Brassard, G. (ed.) CRYPTO 1989. LNCS, vol. 435, pp. 461–480. Springer, Heidelberg (1990)

Telemedical System in the Blood Transfusion Service: Usage Analysis

Marko Meža, Jurij Tasič, and Urban Burnik

Faculty of Electrical Engineering,
Tržaška 25, 1000 Ljubljana, Slovenia
{marko.meza,jurij.tasic,urban.burnik}@fe.uni-lj.si
http://ldos.fe.uni-lj.si

Abstract. Partners FE, KROG and ZTM (alphabetical order) as stated in acknowledgemetns, have developed, manufactured and installed a telemedical system into the blood transfusion service of Slovenia. The system was installed in eleven hospitals, offering blood transfusion services and two blood transfusion centers. The system was in use for nearly seven years. After period of operation, system state snapshot was performed and analyzed. The analysis was focused on per hospital usage preferences through time. Distribution of patients ABO RhD blood typing was also analyzed. In the paper the telemedical system is presented. The method of data collection and data analysis methods are described. The final section presents results accompanied with discussion where economical impact of the telemedical system in comparison to the manual operation is briefly presented.

Keywords: telemedicine, blood transfusion, teleconsulting, telemedical system, blood typing, pre-transfusion testing, economical impact of telemedicine, usage analysis

1 Introduction

The blood transfusion service is a distributed service. Blood products are delivered from the central blood bank to the final consumer – local hospitals. Before a blood transfusion is given, obligatory pre-transfusion tests are performed at the hospital laboratories for confirmation of the donor/patient compatibility. They are normally performed by specially trained personnel. Normally, these agglutination-based compatibility tests have straightforward results, enabling easy diagnostic decisions of the local staff. In approx 1-5% of tests, various serological difficulties and ambiguities occur [1]. In these cases, transfusion is delayed and an expert opinion is needed for the resolution of test results. In such cases, patient's blood samples are sent to the reference transfusion laboratory to a qualified immunohaematology expert by a courier. The expert performs additional tests and interprets them. The consultation between the expert and local technician is then performed by telephone or fax. Notably, this method is error prone procedure with a poor safety, quality and traceability performance, resulting in a time consuming and expensive operation.

S. Markovski and M. Gusev (Eds.): *ICT Innovations 2012*, AISC 207, pp. 173–182.
DOI: 10.1007/978-3-642-37169-1_17

In case of Slovenia, transfusion service is provided by two central national transfusion centres with reference laboratories (with Blood transfusion Center of Slovenia - ZTM being main one) and 9 hospital transfusion departments with their transfusion laboratories [5][3]. Transfusion centers are expert centers coordinating all transfusion actions on national level. They are responsible for blood collection, testing, production of blood components and performing pre-transfusion tests. Hospital transfusion departments with their transfusion laboratories are responsible for blood collecting and pre-transfusion testing [2]. On the average, transfusion service of Slovenia handles 400 transfusion cases per day [1]. Since it is impossible to predict when transfusion services are required, it is necessary to provide transfusion experts 24/7 in all hospitals depending on transfusion service. Therefore an adequate number of experts are required. It is difficult to satisfy these demands due to lack of the qualified experts and lack of funds. Furthermore, in smaller hospitals only a few transfusions per day occur. In these hospitals transfusion expertise is provided without dedicated personnel by expanding a set of assignments of other on duty doctors. These doctors attended additional transfusion medicine training course. They are able to solve straightforward cases occurring during their shift, when dedicated transfusion specialist is not present. But these doctors solve relatively small amount of transfusion cases and an obvious problem is lack of practical experience of these multipurpose doctors.

The core problem, addressed by our telemedical system is the remote readout and interpretation of the agglutination tests, performed prior each blood transfusion (pre-transfusion tests), following the EU Directive 2002/98/EC that introduced quality management and haemovigilance into the practice of the current blood supply [7]. The ability to perform remote readout and interpretation of pre-transfusion tests solved the problem of the delayed pre-transfusion procedure in cases of ambiguous cases which required attendance of reference laboratories. Another problem addressed by the system is provision of remote immunohaemathology expertise to laboratories with insufficient personnel resources. Furthermore, the law requires issue of a signed document as a result of the pre-transfusion test. Since the telemedical system does not transmit signed paper documents, our system had to provide legally sound documents. To provide documents, legally equivalent to their signed paper counterparts, a digital signature infrastructure was introduced into the system.

We analysed seven years of telemedical system usage in Slovenian transfusion practice. The paper presents the telemedical system [6], methods of system usage and results of usage analysis.

2 Telemedical System

Our system enables remote interpretation of pre-transfusion tests performed on gel-cards. Considering advantages of both store-and-forward [12] and real-time approach [8][9] a hybrid telemedical system was developed using store-and-forward as base mode of operation with possibility to switch to real-time mode when necessary. In real-time mode audio/video communication between its users is offered. A special device Gelscope was developed to capture images of gel-cards [4].

Our system follows high level of security, user identification and data protection requirements. The result of the system usage is prompt exchange of immunohaematology expertise in form of legally valid electronic document containing the pre-transfusion test readout, interpretation and diagnose.

2.1 System Usage Workflow

The system usage workflow can be explained on the following use case scenario. Let us assume that in a hospital transfusion laboratory there is no immunohaematology expert present and a need for pre-transfusion testing emerges. Since no expert is present, a remote pre-transfusion test interpretation is required. Users involved into the process are divided into two roles. Users belonging to the role consulting user are laboratory technicians. These are users requesting the remote interpretation. Users belonging to the consultant role are immunohaematology experts. They are legible to issue legally valid pre-transfusion test interpretations. Technicians as well as consultants both use the terminal with the teleconsulting application. Screenshot of the teleconsulting application can be observed on Figure 3. All users must log into the application using their unique credentials. Based on the credentials the system determines user's role and rights to access different data. Roles and rights for each user are defined on the teleconsulting server. On Figure 1, a laboratory setup of the technician side is photographed [6].

At first the laboratory technician prepares the required samples according to the protocol of the pre-transfusion testing [13]. Result of sample preparation is a series of gel-cards with administered blood samples after centrifugation with agglutinates fixed in the gel of gel-cards. After gel-cards are prepared, the laboratory technician sets up a teleconsulting session by setting up a question. Teleconsulting session consists of a question and if answered of the answer to that question [6].

Setting up the teleconsulting question is divided into several steps. The first step is entering a blood sample number into the teleconsulting application by means of scanning sample number from the sample vial, using the barcode reader. After capturing the sample number, patient's data is automatically obtained by the teleconsulting application and appended to the teleconsulting question. In the next step, the technician uses the Gelscope device to capture images of all gel-cards used in the test. For each gel-card, card type is selected and inserted into the teleconsulting application from provided dropdown menu. Users can add up to 10 gel-cards to each question. In the next step of the question setup, the case specific question is typed into the teleconsulting application. Then the available immunohaematology expert - the consultant is selected from the list of available consultants. After all question data has been entered, the technician finishes question data entry. The teleconsulting application then relays the entered data to the teleconsulting server [6].

All teleconsulting applications continuously pool the server for any new questions/answers. When a new question to the currently logged consultant emerges, the teleconsulting application immediately notifies the consultant about the incoming interpretation request. Notification is accomplished by displaying an alert window on the terminal and by sending an SMS to the consultant's mobile phone. Using the telemedical application, consultant reviews the question. He/she interprets the test results for each of the gel-cards and enters them into the telemedical application. After all gel-cards are interpreted, consultant determines and types the answer to the test case specific question set by the technician. After all data has been entered, the consultant can decide to digitally sign the answer. In case of digital signature the answer becomes signed document, which is legally equal to the paper document required for the blood transfusion procedure. After data has been entered into the teleconsulting application it relays it to the teleconsulting server [6].

Since all active teleconsulting applications pool the server for new questions/answers, a new answer to his/her question is noticed by the technician's teleconsulting application. When new question is noticed, the telemedical application immediately notifies the technician about the incoming answer by displaying an alert window on the terminal screen. It also sends an SMS message to his/her mobile phone. Technician then reviews the consultant's answer along with the question. He/she can then verify the digital signature of the document and proceeds with the transfusion procedure as if the consultant was actually present in the laboratory and has signed the paper document. In cases, when immediate response is needed and verbal discussion is required between the technician and the consultant, users can establish a videoconference call using the telemedical application [6].

2.2 System Architecture

The basic telemedical system architecture is client-server architecture. The system consists of several components, illustrated on Figure 2. The first component of the system is the telemedical terminal, comprising of a personal computer, running the MS Windows XP SP3, running the telemedical application. Screen shot of the telemedical application can be observed on Figure 3. Telemedical terminal component is used at the hospital laboratories (consulting side) and at the transfusion centre (consultant side). Along with standard PC peripherals, the telemedical terminal is equipped with special propose hardware for capture of gel-card images - Gelscope device, shown on Figure 1, web camera, headphones with microphone, bar code reader, colour laser printer and VPN router. Existing hospital information system also exists in all hospital laboratories and is also shown in Figure 2. Since consultant side does not need the equipment for question set up, terminals are lacking the Gelscope device and the connection to the existing hospital information system [6].

The second component is the server side of the system with main server running Debian Linux operating system, running the telemedical server application, certificate authority server application and database. The server also has GSM

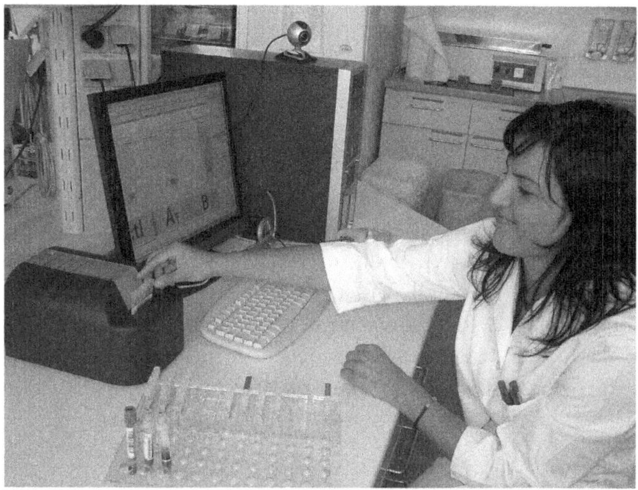

Fig. 1. Laboratory setup of the telemedical system. Main application window can be observed on the terminal. Gelscope device is visible right next to the terminal monitor. Videoconference call is not established.

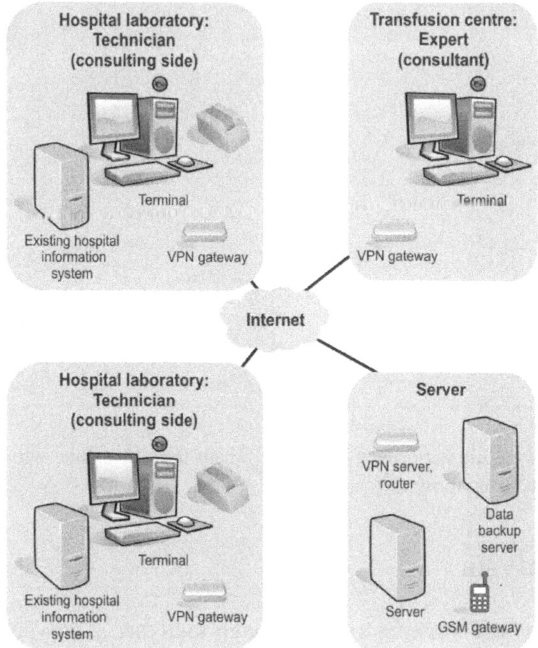

Fig. 2. Teleconsulting system structure. Three terminal setups, server setup and connecting infrastructure are illustrated.

module attached to it and runs SMS notification application. The server side also includes VPN server - router. All nodes of the system are securely connected through the secure virtual private network (VPN) tunnels via the public Internet established by the VPN gateways and the VPN server [6].

The telemedical system is integrated with the existing hospital information system through a special integration module to obtain patient basic data (name surname, birth date, and blood type), previous patient anamnesis and patient transfusion history. Basic operation mode of the telemedical system is a store-and-forward mode. If required, users can switch to the real-time mode by establishing a videoconference call. Videoconference calls are established using the videoconference module which is a part of the telemedical application. A digital signature infrastructure was also developed and integrated into our telemedical system. Therefore through system issued readouts, interpretations and diagnoses have the same legal value as their signed paper counterparts [6].

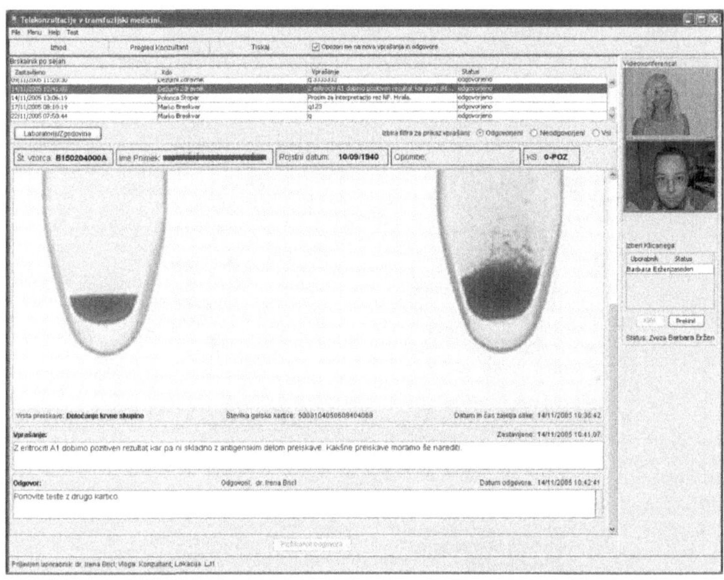

Fig. 3. Telemedical application user interface. Main user interface window with a videoconference in progress is shown.

3 Data Analysis

This section describes working system usage analysis. Analysis was performed by first capturing a snapshot of the system's database, preprocessing the data from the database and analyzing the data using IBM SPSS Statistics, version 19.

3.1 Methods

In order to perform data analysis and to obtain the data we formally requested the access to the data from the Blood Transfusion Center of Slovenia - ZTM. Since the data contains personal data about patients, the access was granted under strict conditions. Results were reviewed by the Blood Transfusion Center of Slovenia prior publication.

The data was captured from working system using Heidi SQL tool and exported in form of csv (coma separated values) file. Interesting tables from database were selected. Since we analyzed patient data and system usage data through time, tables containing patient data and session creation data were selected for analysis. For each of the selected tables csv files were imported into the SPSS tool for further analysis.

Patient Data Analysis. Patients blood type and year of birth distribution were analyzed.

Blood type data is clearly indicated in the system's database and is one of main pieces of information regarding blood transfusion process. It represents classification of blood based on the presence or absence of inherited antigenic substances on the surface of red blood cells [14]. Our data was classified using combination of ABO blood group system and RhD blood group system. ABO group system is the most important blood–group system in human–blood transfusion, which provides possible classification of blood as O, AB, A and B. RhD system is the second most significant blood-group system in human-blood transfusion where the most significant Rh antigen is the D antigen, because it is the most likely to provoke an immune system response of the five main Rh antigens [14]. Blood type classification Using RhD is marked either positive either negative regarding the presence of D antigen bodies. Our data had RhD classification superimposed on ABO classification, thus we had O-NEG, O-POZ, AB-NEG, AB-POZ, A-NEG, A-POZ, B-NEG and B-POZ possible classifications.

Blood type analysis was performed by computing the histogram of blood types in our database.

To analyse patients year of birth, new field containing only year of birth was computed from patient's birth date. Then histogram was computed and drawn.

System Usage Data Analysis. System usage spread across time across different institutions was analyzed by counting teleconstulting cases handled by the system. For each case in the system, creation time, session origination institution and session destination institution are recorded in the session data table.

At first overall telemedical system usage was analyzed by computing case frequency across time. In order to do that cases were at first clustered by on monthly basis using their creation time stamp. Then frequencies of cases for each month were calculated. Using already on monthly basis clustered data, frequency of system usage among different institutions was also analyzed. The analysis was performed the same way as in overall system analysis with addition of clustering by session originating institution.

At last overall system usage per institution location was analyzed by counting number of sessions originating from institutions.

3.2 Results

Using our system 35650 cases of teleconsultation were performed in timespan from September 2005 to middle of June 2012, based on data obtained from the Blood Transfusion Center of Slovenia - ZTM. Originating institution can be observed in table 1.

Table 1. Per hospital system usage

Hospital	BR	CE	IZ	JE	MS	MB	NM	NG	PT	SG	TR
Number	2698	57	2551	4682	7205	65	2183	77	4980	5208	3098
Ratio	8,22%	0,17%	7,78%	14,27%	21,96%	0,20%	6,65%	0,23%	15,18%	15,88%	9,44%

3.3 Patient Data

In our system recorded patients birth data is indicated on figure 4a.

Patient blood type distribution can be observed in figure 4b. As can be observed from figure 4b, the most common blood type of with our system handled patients is A-POZ followed by O-POZ. On the other side, the rarest blood type is AB-NEG blood type.

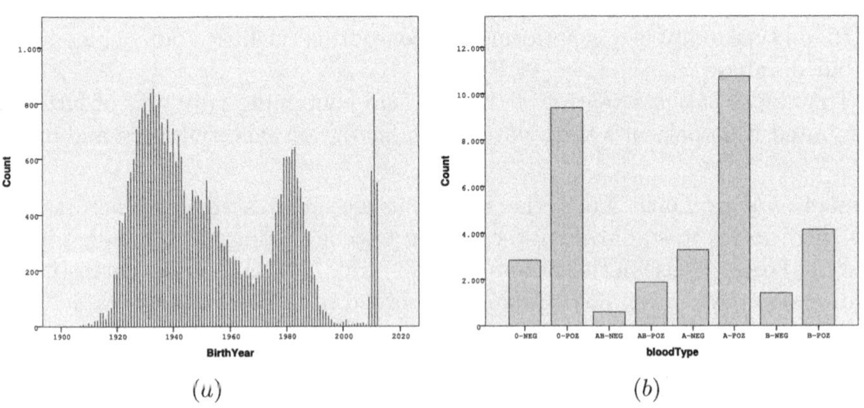

(a) (b)

Fig. 4. (a) Patient year of birth distribution and (b) Patient blood type distribution

3.4 System Usage Analysis

Per hospital system usage analysis can be seen on figure 5a. Data represent every single workstation location. In some hospitals, several workstations are present. Per location aggregated data can be observed in table 1. Overall telemedical system usage can be observed on figure 5b.

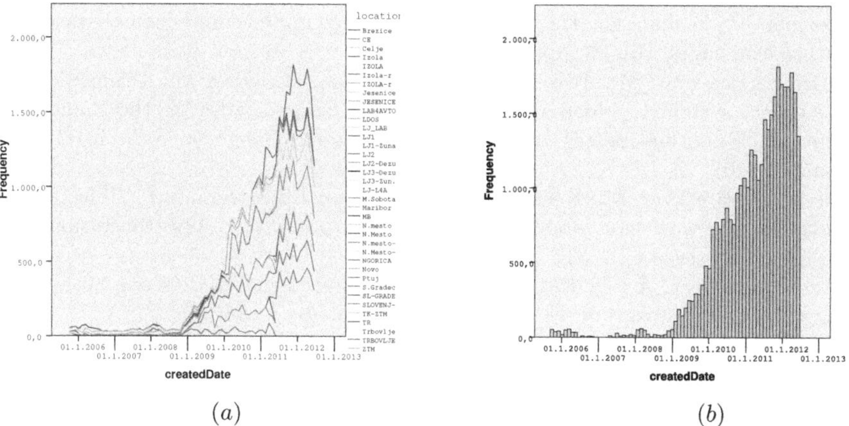

Fig. 5. (a) Per hospital system usage dynamics. (b) Overal telemedical system usage.

4 Discussion

Seven years of system usage proven that it is a helpful tool in the blood transfusion practice of Slovenia. The system was well accepted, and is being regularly used which can be observed from steady increase in number of cases/month. Using our system, a significant savings of were introduced to the blood transfusion service of Slovenia. In study [15] authors analysed economical impact of the telemedicine for four transfusion wards and found out that savings of using telemedical system were 609.000 EUR per year, for year 2009. The cost analysis was not final, since not all transfusion wards were included in study.

By observing results from our study presented in figure Overall telemedical system usage can be observed on figure 5b we can observe, that volume of performed telemedical sessions in year 2009 when the study [15] was performed was relatively small compared to the current volume of performed telemedical sessions. Therefore we can claim, the telemedical system is mature and helpful tool in the blood transfusion service of Slovenia, appropriate for use in other national transfusion services organized similar than Slovene (according to the EU Directive 2002/98/EC).

Acknowledgments. The system was financed and developed within cooperation of the Faculty of Electrical Engineering, University of Ljubljana, KROG-mit d.o.o. company and the Blood Transfusion Centre of Slovenia. Special thanks go to mag. Marko Breskvar, prim. Irena Bricl and prof. dr. Primož Rožman. This work is supported in part within the research group "Algorithms and optimization methods in telecommunications".

References

1. Rozman, P., Domanovic, D.: Transfusion medicine in Slovenia - current status and future challenges, vol. 33, pp. 420–426 (2006)
2. Bricl, I., Breskvar, M., Tasi, J., Mea, M., Roman, P., Jeras, M.: Telemedicine as a support system to blood transfusion service reorganization in the Republic of Slovenia. Published poster on XXXIst International Congress of the ISBT, Germany (2010)
3. Meza, M., Breskvar, M., Kosir, A., Bricl, I., Tasic, J.F., Rozman, P.: Telemedicine in the blood transfusion laboratory - remote interpretation of pre-transfusion tests. J. Telemed Telecare 13, 357–362 (2007)
4. Meza, M., Kosir, A.: Gelcard illumination enhancement in Gelscope 80 by LED distribution optimization. Elektrotehniki Vestnik 76, 79–84 (2009)
5. Meza, M., Pogacnik, M., Tkalcic, M., Jere, A., Breskvar, M., Rozman, P., Bricl, I., Tasic, J.F., Leban, M.: Description of pilot implementation of telemedicine system in blood transfusion practice. In: Proceedings of the COST. Thessaloniki: Informatics and Telematics Institute, Centre for Research and Technology, Thessaloniki, pp. 61–65 (2004)
6. Meza, M.: Development and introduction of the telemedical system into the blood transfusion practice. In: Advances in Telemedicine: Applications in Various Medical Disciplines and Geographical Regions, pp. 179–200. Intech, Rijeka (2011)
7. Faber, J.C.: Work of the European Haemovigilance Network (EHN). Transfus. Clin. Biol. 11, 2–10 (2004)
8. Bonnardot, L., Rainis, R.: Store-and-forward telemedicine for doctors working in remote areas. J. Telemed. Telecare 15, 1–6 (2009)
9. Wootton, R., Blignault, I., Cignoli, J.: A national survey of telehealth activity in Australian hospitals. J. Telemed. Telecare 16, 176–180 (2010)
10. Wootton, R.: Realtime telemedicine. J. Telemed. Telecare 12, 328–336 (2006)
11. Bahaadinbeigy, K., Yogesan, K., Wootton, R.: A survey of the state of telemedicine in Western Australia. J. Telemed. Telecare 16, 176–180 (2010)
12. Harnett, B.: Telemedicine systems and telecommunications. J. Telemed. Telecare 12, 4–15 (2006)
13. Langston, M.M., Procter, J.L., Cipolone, K., Stroncek, D.: Evaluation of the gel system for ABO grouping and D typing. Transfusion 39, 300–305 (1999)
14. Anthea, M., Hopkins, J., McLaughlin, C.W., Johnson, S., Quon Warner, M., LaHart, D., Wright, J.: Human Biology and Health. Prentice Hall, Englewood Cliffs (1993)
15. Breskvar, M., Veluscek, I., Bric, I., Peterlin, S.: The economical impact of introducing telemedicine system into the Slovenian blood transfusion service. Informatica Medica Slovenica 15, 11–12 (2010)

On the Strong and Weak Keys in MQQ-SIG

Håkon Jacobsen, Simona Samardjiska, and Danilo Gligoroski

Department of Telematics,
NTNU, Trondheim, Norway
{hakoja,simonas,danilog}@item.ntnu.no

Abstract. In this paper we describe a methodology for identifying strong and weak keys in the recently introduced multivariate public-key signature scheme MQQ-SIG. We have conducted a large number of experiments based on Gröbner basis attacks, in order to classify the various parameters that determine the keys in MQQ-SIG. Our findings show that there are big differences in the importance of these parameters. The methodology consists of a classification of different parameters in the scheme, together with introduction of concrete criteria on which keys to avoid and which to use. Finally, we propose an enhanced key generation algorithm for MQQ-SIG that generates stronger keys and will be more efficient than the original key generation method.

Keywords: Multivariate Cryptography, Multivariate Quadratic Quasigroups, MQQ-SIG, Quasigroup String Transformations, Public-Key Cryptography.

1 Introduction

The notion of *weak keys* is an important concept in cryptography. Let $\mathcal{C}$ be a cryptographic scheme and k a key in its key space $\mathcal{K}$. We say that k is a weak key for $\mathcal{C}$, if it has some properties that cause the cipher to behave in some undesirable way; normally by making cryptanalysis easier. The opposite notion of weak keys, is *strong keys*. While it would be preferable to only use strong keys, the designers of a cipher are usually more concerned with avoiding weak keys in the system.

We can extend the notion of weak and strong keys to also include parameters that are not technically cryptographical keys, but are important for the behavior of the cryptosystem nonetheless. An example is in elliptic curve cryptography, where there class of supersingular curves yields groups in which the related decisional Diffie-Hellman problem is easy to solve [7]. These curves represent a class of public parameters that elliptic curve cryptosystems should avoid.

1.1 Our Contribution

Recently, a new multivariate public-key signature scheme MQQ-SIG was suggested in [8], using the concept of quasigroup string transformations. In this

S. Markovski and M. Gusev (Eds.): *ICT Innovations 2012*, AISC 207, pp. 183–193.
DOI: 10.1007/978-3-642-37169-1_18

work we describe a methodology for identifying strong and weak keys in that scheme.

We have conducted a large number of experiments, based on Gröbner basis attacks, in order to classify the various parameters that determine the keys in MQQ-SIG. So far, little is known about what each parameter actually contributes to the security of the system. Our findings show that there are big differences in the importance of these parameters. The methodology consists of a classification of different parameters in the scheme, together with an introduction of concrete criteria on which keys to avoid, and which to use. We collect all our findings in a concrete proposal for an enhancement of the MQQ-SIG key generation algorithm, which both generates stronger keys and are more efficient than the original method.

2 A Brief Description of MQQ-SIG

MQQ-SIG is, like other $\mathcal{MQ}$-schemes, based on two bijective linear/affine transformations and one central multivariate quadratic map:

$$\mathcal{P} = \mathcal{S} \circ \mathcal{P}' \circ \mathcal{S}' \colon \{0,1\}^n \to \{0,1\}^n. \tag{1}$$

Here, $\mathcal{S}'$ is a bijective affine transformation defined as $\mathcal{S}'(\mathbf{x}) = \mathcal{S}(\mathbf{x}) + \mathbf{v}$, where $\mathcal{S}$ is given by a bijective linear transformation $\mathcal{S}(\mathbf{x}) = S \cdot \mathbf{x}$, while the central quadratic map $\mathcal{P}'$ is the application of the quasigroup operation. The specific procedure on how to apply the quasigroup operation is an intrinsic part in the design of MQQ-SIG and not actually relevant for its key generation. Hence, it will not be considered further here (all the details can be found in [8]) . Additionally, $\mathcal{S}$ and $\mathcal{S}'$ are completely determined by the matrix S, so in reality there are only two main factors to consider in MQQ-SIG; the matrix S and the quasigroup itself. First we look at the quasigroup construction. The quasigroups in MQQ-SIG are created by the following expression:

$$\mathbf{x} * \mathbf{y} = B \cdot U(\mathbf{x}) \cdot A_2 \cdot \mathbf{y} + B \cdot A_1 \cdot \mathbf{x} + \mathbf{c}, \tag{2}$$

where $\mathbf{x} = (x_1, \ldots, x_d)$, $\mathbf{y} = (y_1, \ldots, y_d)$, and A_1, A_2 and B are invertible $d \times d$ matrices over $\mathbb{F}_2$, and the vector $\mathbf{c}$ is an element in $\mathbb{F}_2^d$. The matrix $U(\mathbf{x})$ is an upper triangular matrix with all diagonal elements equal to 1, and where the elements above the main diagonal are linear expressions in the variables $x_1, \ldots, x_d$. A quasigroup constructed in this way, is called a *Multivariate Quadratic Quasigroup (MQQ)*.

The matrices A_1, A_2 and B, together with the vector $\mathbf{c}$, can all be generated by a uniformly random process. The matrix $U(\mathbf{x})$ is computed as a matrix of column vectors as follows:

$$U(\mathbf{x}) = I_d + \left(U_1 \cdot A_1 \cdot \mathbf{x} \cdots U_d \cdot A_1 \cdot \mathbf{x} \right), \tag{3}$$

where the U_i's are strictly upper triangular matrices having all elements in the rows $\{i, \ldots, d\}$ zero. The elements in the rows $\{1, \ldots, i-1\}$ (and which lays above the main diagonal) may be either 0 or 1, generated uniformly at random. In particular, U_1 is the all zero matrix. We see immediately that $U(\mathbf{x})$ is a very central parameter in the quasigroup construction. More specifically, $U(\mathbf{x})$ defines an isotopy class on which A_1, A_2 and B all give linearly isotopic quasigroups within the same equivalence class. The concept of isotopy classes lead naturally to some questions regarding their security properties.

1. How will the isotopies created by A_1, A_2 and B influence the behavior of the system? We look at A_1, A_2 and B separately in Section 3.4 and Section 3.5, in order to answer these questions.
2. Building upon the previous observation, it is also prudent to ask if there exist isotopy classes which have inherently better or worse properties than others. This question will be studied in Section 3.3.

Lastly, we look at the matrix S. The standard way to hide the inner structure of the central map in most $\mathcal{MQ}$ schemes, is to use two random affine/linear transformations $\mathcal{S}$ and $\mathcal{T}$. In MQQ-SIG, these transformations are determined by the specially constructed matrix S. How will this choice affect the security of the system? Is it better to follow the standard approach? We investigate these questions in Section 3.6.

3 Experimental Procedure

3.1 Hardware and Software

For the experiments we used a 64-bit Ubuntu Linux server, version 10.04.4 / 2.6.32, with a 64 core Intel Xeon 2.27 GHz processor and 1 TB of RAM. To run the attacks we used the computer algebra system Magma [1], (version 2.17-3). Magma implements several algorithms for computing Gröbner bases, including the original algorithm by Buchberger [2] and the new (and faster) F_4-algorithm due to Faugère [6]. F_4 is chosen in the special case of $\mathbb{F}_2$.

3.2 Experiment Algorithm

The public key in MQQ-SIG consists of $\frac{n}{2}$ randomly generated multivariate quadratic equations with $n \in \{160, 192, 224, 256\}$. Our experiment procedure consisted of running the F_4 algorithm on many different instances of MQQ-SIG, trying to obtain the Gröbner basis for each instance's public key. The details are given in Algorithm 3.1. We note, however, that in order to make the experiments feasible, we had to scale down on the number of equations. In particular we created 100 MQQ-SIG instances of sizes $n = 32, 40, 48$ and 56. In the rest of the paper we will refer to these 100 unmodified instances as the "original instances".

Notice that the system $\widetilde{\mathcal{P}}$, given to the Gröbner basis algorithm in Step. 6 is not an instance of a real MQQ-SIG public key, where the system is actually underdetermined ($n = 2m$). To solve an underdetermined system of multivariate quadratic equations, Courtois et. al [4] suggested to “fix” some of the n variables (e.g. by choosing their values at random) to obtain a system with $n \leq m$. The resulting system can then be solved using an algorithm which is efficient at solving (over)-determined systems, for instance the XL algorithm [5]. In Algorithm 3.1 we simulate this approach by evaluating r of the variables in the public key at random values (Step. 4).

Algorithm 3.1. Gröbner basis attack used for all experiments

Input: r; the number of equations to remove from the public key.
Output: A Gröbner basis for the public key.

1. Generate the public key $\mathcal{P}$ according to the specification of the experiment.
2. Remove the first r equations of $\mathcal{P}$.
3. Generate a random Boolean vector $\mathbf{x} = (x_1, x_2, \dots, x_r) \in \{0,1\}^r$.
4. Evaluate each of the remaining $n - r$ polynomials $P_i(x_1, \dots, x_n)$, at the random vector $\mathbf{x}$, for $i = r+1, \dots, n$.
5. Obtain a system $\widetilde{\mathcal{P}} = \{P_i(x_{r+1}, \dots, x_n) | i = r+1, \dots, n\}$ of $n-r$ equations in $n-r$ variables.
6. Compute a Gröbner basis for the system $\widetilde{\mathcal{P}}$, using Magma.

3.3 The Matrix $U(\mathbf{x})$

The importance of analyzing the matrix $U(\mathbf{x})$ is apparent since it is the sole factor in determining which linear isotopy class the quasigroup will belong to. Hence, in MQQ-SIG, $U(\mathbf{x})$ defines a specific isotopy class, while A_1, A_2 and B have a role as the isotopy triples. They send the quasigroup created by $U(\mathbf{x})$ to different quasigroups within the same equivalence class. In MQQ-SIG these classes are linear, that is, defined by linear permutations, but we remark that isotopies do not have to be linear in general.

Since $U(\mathbf{x})$ is the single source for different isotopy classes, it is interesting to investigate the characteristics of the classes it might create. Recall that $U(\mathbf{x})$ is an upper triangular matrix containing linear expressions in $\mathbf{x}$. If a row does not contain any terms in $\mathbf{x}$, we call it a *linear row*. It is easy to see that without the isotopy (A_1, A_2, B), every MQQ will have at least one linear row.

It is an undesirable property to have an MQQ that contains only linear terms in some of its coordinates, since it reduces the resulting system complexity. Unfortunately, simply specifying that the quasigroup should contain no linear coordinates is not enough to guarantee a strong system. In [8] the authors point out that the quadratic polynomials of an MQQ of type $\text{Quad}_{d-k}\text{Lin}_k$, might cancel each other when combined linearly, yielding a system of less then $d-k$ quadratic polynomials.

Later Chen et al. [3] ascertained that this was in fact the case, and suggested an alternative way of classifying the MQQ's; introducing the notion of a strict MQQ type.

Definition 1. *A quasigroup $(Q, *)$ of order 2^d is called an* MQQ of strict type, *denoted $Quad^s_{d-k}Lin^s_k$, if there are at least $d-k$ quadratic polynomials whose linear combinations do not result in a linear form, where $0 \leq k < d$.*

The definition of $U(\mathbf{x})$ in MQQ-SIG is based on a procedure for creating quasigroups of strict types, taken from [3]. The number of linear rows in $U(\mathbf{x})$ correspond to the number k in the definition of strict MQQ's. In particular, the quasigroups in MQQ-SIG are of the type $\mathrm{Quad}_7^s\mathrm{Lin}_1^s$ for one linear row, $\mathrm{Quad}_6^s\mathrm{Lin}_2^s$ for two linear rows, and so on. The specification does not allow for the creation of MQQ's of type $\mathrm{Quad}_8\mathrm{Lin}_0$. How to construct quasigroups of this type is an open problem.

To evaluate the effect of having linear rows in $U(\mathbf{x})$, we took 40 of the original instances with $n = 48$, and created new $U(\mathbf{x})$ matrices for them, having between one to six linear rows. Then we ran Algorithm 3.1 on them.

Table 1. Average run time of 40 different instances of 48 variables, on different numbers of linear rows in $U(\mathbf{x})$

Linear rows	$\bar{t}$	σ
1	177.94	25.65
2	149.37	29.16
3	55.95	14.56
4	3.44	0.13
5	0.12	0.00
6	0.01	0.00

The results are shown in Table 1 and clearly demonstrates that more linear rows in $U(\mathbf{x})$ affect the behavior of the system in a negative way. It is rather conclusive that the $U(\mathbf{x})$'s with several linear rows should be avoided, while those with only one linear row should be preferred. Fortunately, the vast majority of generated $U(\mathbf{x})$ matrices contains only one linear row, so adding this check to the key generation algorithm does not add any significant cost.

Recommendation 1. *The isotopy classes used in MQQ-SIG should be those that are determined by $U(\mathbf{x})$ matrices having only one linear row.*

3.4 The Matrices A_1 and A_2

It is interesting to see how a linear isotopy affects the properties of the MQQ's within the equivalence class of linearly isotopic quasigroups. It can be shown that the degrees of MQQ's are invariant under linear isotopies. If the complexity of Gröbner basis attacks also remain the same it would alleviate the need for the matrices A_1, A_2 and B. On the other hand, if the strength of the system varies with different isotopies, we should investigate their features to understand why they might lead to weak keys.

To gauge the contribution of either A_1, A_2 or B, we picked out three instances of 48 variables from the original experiments. We chose the instances so as to include one with very good performance, one with bad performance and one being close to the average of all the 100 instances. For ease of reference, we will refer to them as the "Good", the "Bad" and the "Average" instance, respectively.

For each instance we generated 20 new values for A_1 and used them to create new public keys where all values, except A_1, was fixed. Similarly, for A_2 and B. We then ran Algorithm 3.1 on all these modified instances. The results are shown in Figure 1.

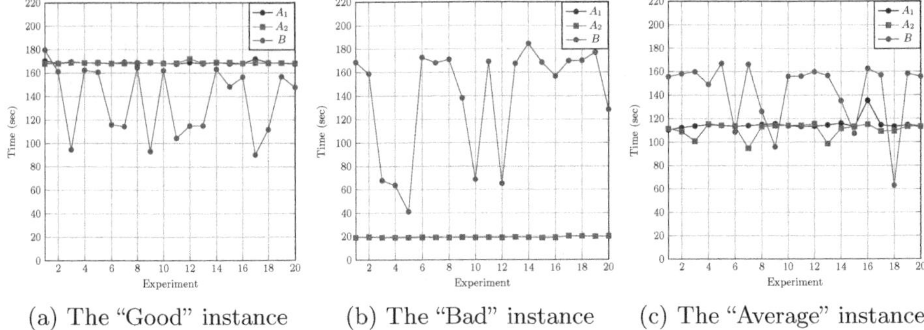

(a) The "Good" instance (b) The "Bad" instance (c) The "Average" instance

Fig. 1. Run times of three instances of 48 variables using 20 different values for the matrices A_1, A_2 and B

As apparent from these graphs, changing either A_1 or A_2, does not to alter the behavior of the given isotopy class. For all three cases the performance is highly consistent across the 20 different values of A_1 and A_2. More importantly, if given a weak quasigroup by $U(\mathbf{x})$ and B, changing A_1 or A_2 will not make it any stronger. This motivates the following conjecture.

Conjecture 1. Fix $U(\mathbf{x})$, B and $\mathbf{c}$ in equation (2). Then the equivalence class of quasigroups generated by the isotopy (A_1, A_2, B), for randomly generated A_1 and A_2, is invariant with respect to their security against Gröbner basis attacks.

The matrix B, on the other hand, is seen to heavily affect the run times of the Gröbner basis computations. This indicates that there are some decisive attributes of B which influence the performance of the system. Hence, subdividing the isotopy class based on different B's, seems to be the most relevant when classifying the quasigroups (up to linear isotopies). We will study B in more detail in the next section. Based on Conjecture 1, we give the following recommendation:

Recommendation 2. *The matrices A_1 and A_2 can be excluded from the definition of a quasigroup in MQQ-SIG (equation (2)), without affecting the security of the system against Gröbner basis attacks.*

3.5 The Matrix B

We saw in the previous section that the matrix B is very influential for the behavior of a quasigroup. The significance of B can be seen from the expression

$$B \cdot (U(\mathbf{x}) \cdot A_2 \cdot \mathbf{y}), \tag{4}$$

which is taken from the equation defining the MQQ's in MQQ-SIG (cf. equation (2)). Because $U(\mathbf{x}) \cdot A_2 \cdot \mathbf{y}$ is a vector of quadratic terms, B will be a decisive factor in how those terms will be combined linearly in the resulting MQQ. This relates it to the ranks of the $\mathbf{B}_{f_i}$ matrices (equation (9) in [8]), of which entries correspond to the quadratic terms in each polynomial f_i in the MQQ vector.

The designers of MQQ-SIG placed two restrictions on the ranks of these $\mathbf{B}_{f_i}$ matrices (see equation (8a) and (8b) in [8] for details) in order to counter certain cryptanalytic attacks based on MinRank [9]. However, the ranks are actually relevant for the resistance against Gröbner basis attacks. Thus, we wanted to investigate whether the original criteria could be weakened, while maintaining the same security.

We tested three cases of MQQ's: (1) MQQ's where at least one of the $\mathbf{B}_{f_i}$ matrices had rank zero (which means that f_i has no quadratic terms); (2) MQQ's where all $\mathbf{B}_{f_i}$'s satisfied the specified criteria in [8]; (3) MQQ's where no $\mathbf{B}_{f_i}$ matrix had rank zero, but did not have to satisfy the criteria in (2). The quasigroups were derived from the original set of MQQ-SIG instances, by randomly creating new B matrices until the criteria were met. The average run times are summarized in Table 2.

Table 2. Average run times for 100 MQQ-SIG instances of n variables, conditioned on the matrices $\mathbf{B}_{f_i}$

	No conditions (original system)		At least one rank zero		No rank zero		All satisfy equations (8a) and (8b) in [8]	
n	$\bar{t}$	σ	$\bar{t}$	σ	$\bar{t}$	σ	$\bar{t}$	σ
40	2.88	0.75	1.90	0.93	2.86	0.72	2.82	0.81
48	138.75	45.66	74.94	54.93	142.39	42.62	138.38	46.09
56	2666.15	709.56	1434.31	894.83	2593.06	711.56	2721.13	637.70

The MQQ's that satisfy equations (8a), (8b) and (9) in [8], do not appear to perform significantly better than the "no-zero-rank"-experiments. This suggests that the main advantage of those criteria, when it comes to resistance against Gröbner basis attacks, is to filter out the quasigroups with linear polynomials.

Since the "no-zero-rank" is a much weaker requirement that of equations (8a) and (8b) from [8], more quasigroups will satisfy it. This can lead to faster generation times. To quantify this, we also recorded the number of different B matrices we had to try before finding a suitable MQQ in the above experiments. Our findings showed, that on average, we had to generate 400 different B matrices before the criteria in equations (8a) and (8b) from [8], were met. The "no-zero-rank" criterion, on the other hand, was satisfied after only one or two tries. Hence, we end with the following recommendation.

Recommendation 3. *The criteria in equations (8a) and (8b) in [8] can be replaced with the following:*

$$\forall i \in \{1, \ldots, d\},\ \mathit{Rank}(\mathbf{B}_{f_i}) > 0. \tag{5}$$

That is, all f_i's in the MQQ-vector must contain quadratic terms.

3.6 The Matrix S

The last parameter of MQQ-SIG we consider is the matrix S. Recall that S defines the linear and affine transformations $\mathcal{S}$ and $\mathcal{S}'$ respectively. Their purpose is to hide the internal structure of the central map $\mathcal{P}'$. If S is randomly created, it requires n^2 bits to store, which for the proposed sized of n in MQQ-SIG (160, 192, 224 and 256), would add between 3 to 8 KB to the private key.

The designers of MQQ-SIG tried to mitigate this problem by creating S in very special manner, using circulant matrices. We omit the details here (see [8]), but remark that this design makes it possible to simply store two permutations on n symbols. The reason for using two circulant matrices instead of just one, was to avoid the regular structure exhibited in circulant matrices. Even so, the current construction is still highly structured. There is no guarantee that none of its internal structure might leak through in a Gröbner basis attack. Thus, we decided to evaluate the behavior of S compared to a truly random matrix.

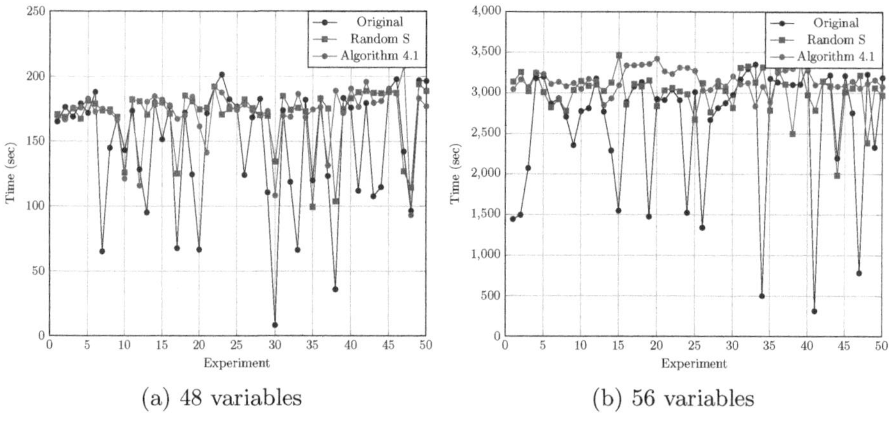

(a) 48 variables (b) 56 variables

Fig. 2. Comparison of the run times of 50 of the original instances, with the same instances where either S is set to a random matrix, or adjusted in order to comply with Algorithm 4.1.

In Figure 2 we compare the run times of the first 50 original instances ($n = 48$ and 56), with the same instances, but using randomly created S matrices instead. These results indicate that the specially constructed S is not able to hide the internal algebraic structure of the system quite as well as a randomly created S.

We have not been able to determine exactly which properties of the S matrix in MQQ-SIG that differentiates it from a random matrix, but there is a definite difference. In the end, the strong and consistent performance of using a random S, and the uncertainty of why the special construction behaves the way it does, we err on the side of caution.

Recommendation 4. *The matrix S, and the vector $\mathbf{v}$, should be created uniformly at random.*

4 Enhancing the MQQ-SIG Key Generation Algorithm

In the previous section we offered four concrete recommendations on how the various parameters of MQQ-SIG should be created in order to avoid weak keys from being generated. However, they are based on measurements were we had adjusted a single system parameter individually. Here we will put all these recommendations together and propose an enhanced key generation algorithm.

4.1 The New Key Generation Algorithm

First we summarize the recommendations we made in the previous chapter.

1. When creating $U(\mathbf{x})$ it should be checked that it only contains one linear row (Recommendation 1).
2. A_1 and A_2 can be dropped (Recommendation 2).
3. Only MQQ's of type $\text{Quad}_8\text{Lin}_0$ should be used (Recommendation 3).
4. The matrix S should be created as a random (invertible) matrix, not using the circulant matrix design (Recommendation 4).

The modified key generation algorithm, taking these recommendations into account, is given in Algorithm 4.1.

Algorithm 4.1. Enhanced MQQ-SIG key generation

Input: The number of variables n, where $n \in \{160, 192, 224, 256\}$.
Output: A public key $\mathcal{P} = \{P_i(x_1, \dots, x_n) | i = 1 + \frac{n}{2}, \dots, n\}$. A private key, consisting of an MQQ of order 2^8, the matrix S and a random vector $\mathbf{v}$ in $\mathbb{F}_2^n$.

The central map $\mathcal{P}'$
The MQQ is defined just as in equation 2, but with $A_1 = A_2 = I_d$. Additionally, $U(\mathbf{x})$ and B are subject to the following restrictions:

- $U(\mathbf{x})$ must have no linear rows other than the last;
- B must be chosen so that the resulting MQQ is of type $\text{Quad}_8\text{Lin}_0$.

The linear transformations $\mathcal{S}$ and $\mathcal{S}'$
The matrix S is constructed as a random invertible $n \times n$ matrix over $\mathbb{F}_2$, and $\mathbf{v}$ is constructed as a random element in $\mathbb{F}_2^n$.

In Figure 2 we compare the run times of 50 of the original instances ($n = 48$ and 56), with the same instances modified according to the procedure given in Algorithm 4.1. These results show that the MQQ-SIG instances that were modified in order to comply with Algorithm 4.1, perform noticeably better than the original instances. In particular, for the case $n = 56$, the "weakest" keys generated are relatively much closer to the average, than for $n = 40$. If this behavior continues for higher number of variables, then a real MQQ-SIG system, i.e. with $n = 160, 192, 224$ or 256, would probably be very unlikely to generate weak keys using Algorithm 4.1. In the end, these results are strong evidence in support of modifying the MQQ-SIG key generation algorithm according to the suggestions in this paper.

5 Conclusions

Our methodology consisted of a classification of different parameters in the scheme, together with an introduction of concrete criteria on which keys to use (and which to avoid), based on a large number of numerical experiments. Our findings show that there are big differences in the importance of the various parameters of MQQ-SIG. Additionally, we identified an unnecessary requirement in the original specification, requiring the quasigroups to fulfill a certain condition. Removing this restriction can potentially speed up the key generation process by a large factor. Finally, we verified experimentally that our enhanced key generation algorithm yielded better performing instances.

References

1. Bosma, W., Cannon, J.J., Fieker, C., Steel, A.: Handbook of Magma functions. Computational Algebra Group, School of Mathematics and Statistics, University of Sydney, 2.17-3 edn. (2010)
2. Buchberger, B.: Ein Algorithmus zum Auffinden der Basiselemente des Restklassenringes nach einem nulldimensionalen Polynomideal. Ph.D. thesis, University of Innsbruck (1965)
3. Chen, Y., Knapskog, S.J., Gligoroski, D.: Multivariate quadratic quasigroups (MQQ): Construction, bounds and complexity. In: Inscrypt - 6th International Conference, Shanghai, China. Science Press of China (October 2010)
4. Courtois, N., Goubin, L., Meier, W., Tacier, J.-D.: Solving underdefined systems of multivariate quadratic equations. In: Naccache, D., Paillier, P. (eds.) PKC 2002. LNCS, vol. 2274, pp. 211–227. Springer, Heidelberg (2002)
5. Courtois, N., Klimov, A., Patarin, J., Shamir, A.: Efficient algorithms for solving overdefined systems of multivariate polynomial equations. In: Preneel, B. (ed.) EUROCRYPT 2000. LNCS, vol. 1807, pp. 392–407. Springer, Heidelberg (2000)
6. Faugère, J.C.: A new efficient algorithm for computing Gröbner bases (F4). Journal of Pure and Applied Algebra 139(1-3), 61–88 (1999)
7. Galbraith, S.D.: Supersingular curves in cryptography. In: Boyd, C. (ed.) ASIACRYPT 2001. LNCS, vol. 2248, pp. 495–513. Springer, Heidelberg (2001)

8. Gligoroski, D., Ødegård, R.S., Jensen, R.E., Perret, L., Faugère, J.-C., Knapskog, S.J., Markovski, S.: MQQ-SIG, an ultra-fast and provably CMA resistant digital signature scheme. In: Chen, L., Yung, M., Zhu, L. (eds.) INTRUST 2011. LNCS, vol. 7222, pp. 184–203. Springer, Heidelberg (2012)
9. Kipnis, A., Shamir, A.: Cryptanalysis of the HFE public key cryptosystem by re-linearization. In: Wiener, M. (ed.) CRYPTO 1999. LNCS, vol. 1666, pp. 19–30. Springer, Heidelberg (1999)

An Approach to Both Standardized and Platform Independent Augmented Reality Using Web Technologies

Marko Ilievski and Vladimir Trajkovik

Faculty of Electrical Engineering and Information Technologies, Skopje, Macedonia
marko@vint.com.mk, trvlado@finki.ukim.mk

Abstract. Augmented reality has become very popular in the mobile computing world. Many platforms have emerged, that offer a wide variety of tools for creating and presenting content to the users, however all of them use proprietary technologies that lack standardization. On the other hand the latest developments in the suite of web standards offer all the capabilities required for building augmented reality applications. This paper offers a different approach of implementing augmented reality applications using only web technologies which enables standardization and platform independence.

Keywords: augmented reality, augmented reality development, mobile augmented reality, mobile computing.

1 Introduction

The augmented reality has established itself as one of the prominent concepts in the field of mobile applications. This is due to the context oriented approach of defining what the user wants to see and how it should be displayed. Determining the context can be categorized as location-based and marker-based. The location-based uses positioning data from the device's GPS and other sensors such as the digital compass to find the location and orientation of the user. On the other hand the marker-based uses image recognition algorithms in order to determine where the virtual information should be rendered.

These concepts were the basis for today's popular AR browser applications such as Layar [1] and Wikitude [2]. The increased popularity in these applications and the development efforts transformed them into platforms with wide variety of tools and APIs so that third party developers and content publishers can add their own content into the augmented world [3]. Being developed by different companies however, leads to a proprietary technologies and data formats for transferring and presenting the content. The content publishers who want to support augmented reality are left to choose which platform to take and consequently which group of users they will target. The standardization of augmented reality would provide interoperability between the platforms and the content publishers, and would ease the development of client applications.

S. Markovski and M. Gusev (Eds.): *ICT Innovations 2012*, AISC 207, pp. 195–203.
DOI: 10.1007/978-3-642-37169-1_19

This paper takes the suite of web standards as a basis for developing AR applications purely on web technologies for both location-based and marker-based tracking. The following section defines the challenges behind building AR applications and how they can be solved using the APIs offered in the latest drafts of the web standards. Two prototype applications are presented in section 3 which are used to prove the feasibility and evaluate the usability of the proposed solution.

2 Requirements and Reference Model

There are many standards which can be used for developing standardized AR applications and services. Some of them are already used [4] in platforms such as Layar and Wikitude, but they still rely heavily on proprietary technologies. The benefit of using existing standards is avoiding costly errors by using the experience from those who already implemented and optimized many of today's popular standards, such as HTML, JSON, WebGL, X3D etc.

Another strong motivation for using existing standards is time-to-market. Re-purposing existing content and applications is critical. The use of existing standards or profiles of these standards is driven by the need to avoid making mistakes. Also the use of currently deployed and proven (and emerging) technologies should solve/address urgent issues for AR publishers, developers and users. [5]

The AR browsers can be compared with web browsers in a sense that they are both HTTP user agents which consume content from the Web. However, unlike AR browsers, web browsers follow strict set of standards for describing the content and therefore different browsers can retrieve and render the same content in a consistent manner. Also, there is already a lot of research and development effort put in the web standards. Having this in mind a different view of how AR is used in today's applications can be defined. Instead of building separate browser applications for accessing and presenting AR content, the existing web browsers can be used for that purpose. Effort needs to be made in creating open source projects and thus libraries that will handle the dirty work behind the scenes, so that the developers can focus on the value that their application and content will bring. This introduces a new way of designing and interacting with web applications. What is more important, popular web platforms such as Wikipedia [6] can provide their content through the use of augmented reality.

The main challenge behind building web based mobile applications is the inability to interact with the mobile device's built-in hardware components such as camera, compass, gyroscope and similar. The W3C Device APIs Working Group [7] is working on these requirements and there are already working drafts for the main device APIs. What is even more important is the early support in the mobile web browsers for these draft standards. This enables early development of proof of concepts for standardized web based AR applications.

In order to define our approach of using existing standards for standardized AR, there has to be a clear understanding of the different aspects that require standardization and what is the best approach of achieving this. The doctoral thesis of Asa MacWilliams [8] provides a reference model, illustrated in Figure 1, which gives a nice overview of the building blocks used in a typical AR system.

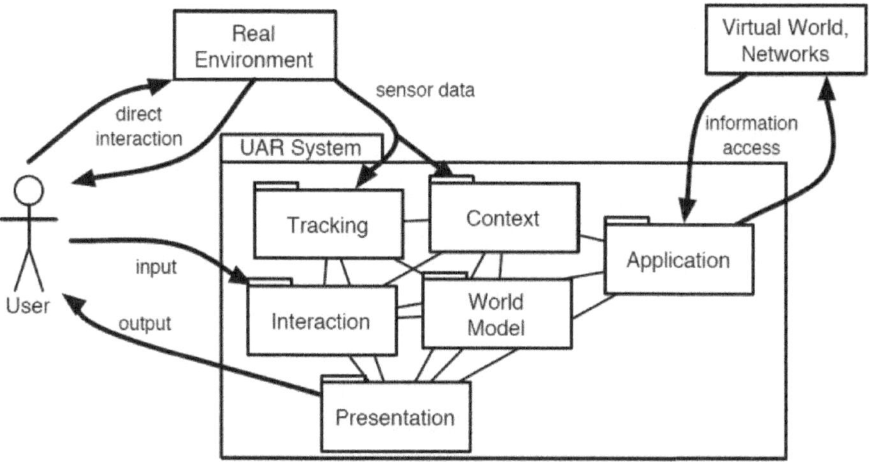

Fig. 1. Reference AR model (Source: A Decentralized Adaptive Architecture for Ubiquitous Augmented Reality Systems, page 14)

In the following sections we provide a detailed explanation of the individual components presented in the reference model and which standards can be applied in order to implement those components.

2.1 Tracking

The tracking component handles the changes in the user's location and orientation or responds to markers, so that virtual information can be rendered on the reality view in a way that makes digital objects seem part of the natural environment. The tracking in today's AR browsers is usually based on location sensors such as GPS, compass and accelerometer. Some AR browsers implement image recognition in order to support marker-based tracking [3].

2.1.1 Location-Based Tracking

The location based tracking depends heavily on sensor data. To render the POIs on screen properly, the rendering algorithm has to know the user's location, orientation and viewing angle. This information is provided by different APIs and the challenge here is to synchronize the data that is passed to the rendering algorithm.

The Geolocation API [9] shows the location of the users and keeps track of them as they move around. The API is device-agnostic, i.e. it does not depend on how the browser determines location, as long as the clients request and receive location data in a standard way. The mechanism for getting the user's location can be via GPS, Wi-Fi, or simply asking the user to enter the location manually. The API is asynchronous using callback methods, because the location lookups can take some time.

Orientation data can be obtained using the DeviceOrientation Event API [10]. This API introduces new DOM events that provide information about the physical orientation and motion of the device.

2.1.2 Marker-Based Tracking

This type of tracking works by recognizing special markers which the user sees through the device's camera. The challenge is to design algorithm which can recognize the markers and render appropriate information on those markers. There is already work done for marker-based tracking using web technologies. A JavaScript port of the FLARToolKit exists called JSARToolkit which operates on canvas images and video element content [11].

The library works by passing a canvas for analysis, which returns a list of markers found in the image and the corresponding transformation matrices. To render an object on top of the marker, the transformation matrix can be passed to a rendering library so that the object is transformed using the matrix.

To achieve real time tracking every frame of a video input needs to be drawn on a canvas and then the canvas can be passed to the JSARToolKit. On modern JavaScript engines JSARToolKit works fast even on 640x480 video frames. However, the larger the video frame, the longer it takes to process. The mobile web browsers although run the latest JavaScript engines the hardware capabilities on the majority of devices can affect the performance of the JSARToolkit. Section 3.1 provides more detailed explanation and test results for the JSARToolkit used in smartphones.

2.2 Context

First of all point needs to be taken that tracking is a special case of context and it is a key concept in AR applications. For that reason it is presented as a separate component in the architecture. The context component however, should provide the application with additional information regarding the user's context, such as name, avatar, friends, or near real time statuses (in office, busy, sitting down, at gym etc.)

2.3 Application

The application contains the main logic and coordinates the other subsystems.

2.4 World Model

This component provides access and storage for the points of interest in the world and possibly other metadata and 3D objects. The endpoints which provide points of interest depending on the platform in use are usually called "channels", "layers" or "worlds" [3].

2.5 Presentation

The presentation component handles the output to the screen, i.e. superimposing 2D or 3D virtual content on the reality view stream.

A common solution to rendering virtual information in the real world is splitting the UI in two layers. The bottom layer is live feed from the device's camera and the layer above is where the digital content is being rendered.

2.5.1 Camera Layer

The WebRTC [12] standard allows web browsers to conduct real-time communication. This enables applications such as voice calls, video chat, P2P file sharing and so on. The getUserMedia API [13] of this standard is used for accessing the device's camera. Some of the major browsers already offer support for this API which makes it suitable for implementing the camera layer of the presentation component.

2.5.2 Content Representation and Discovery

The augmented reality standards being in their very early stages of development also affect the interoperability of the different AR platforms. This mainly refers to the content that cannot be shared between the AR browsers and there are no discovery mechanisms for such content. At the moment the publishing of AR content is done through multiple interfaces and data formats for each platform that is to be supported.

The efforts for standardizing POIs were initiated by the W3C in June 2010. A workshop "Augmented Reality on the Web" was held in Barcelona [14], which attracted many attendees and papers. From this workshop a Working Group was set up to work in the area of Point of Interest (POI) representation. [15]

Although there is no standardized representation of POIs, a common protocol for their retrieval and presentation is adopted by most AR browsers. The protocol can be defined as following, initial request is sent to retrieve minimal information about the POIs in a specified radius of the user depending on his location, after which the content is displayed in a form of short descriptions, images or objects. The full data is loaded only when the user selects a particular POI from the screen. This helps to reduce the application's use of the mobile network.

This protocol only relates to location-based AR tracking, due to the nature of marker-based tracking, i.e. a single marker corresponds to a single POI. As mentioned earlier there are significant differences in the representation of POIs. This reflects the varied capabilities and approaches used in the AR browsers. However, this representation also depends on the domain of the AR application that is being built. Being able to provide standardized AR content that is platform independent is a crucial aspect when building AR applications. A modern web browser supports different types of content. It can be HTML/CSS, 2D or 3D graphics rendered on canvas using JavaScript, SVG vector graphics, etc. There is also collaboration between the Web3D Consortium and the W3C in order to create standard for native support of X3D scenes in HTML [16]. This covers the content and styling of individual POIs. However, their representation is yet to be defined by the working group. This is necessary so that proper querying and discovery by the search engines can be possible. The First Public Working Draft [17] of the POI WG offers a structure for describing POIs which can be used as a starting point for developing proof of concepts.

The goal of this paper is not to go into the details of implementing standardized POI representation and retrieval. Therefore the location-based AR prototype discussed in section 3.2 uses the Wikipedia's Geolocation Services [18] in order to cover the process of retrieving POIs.

3 Prototypes

This section describes the prototype applications that were developed to prove the feasibility and evaluate the usability of the proposed solution described in the previous section. Two separate applications were developed, one for location-based and the other for marker-based AR. The applications were tested on Opera Mobile 12 for Android, because it currently provides the best support for the required standards [19]. The chipset of the device used in the testing is Qualcomm MSM8255 Snapdragon which includes a 1GHz CPU and a separate GPU. The testing is based on measuring frames per second (FPS) with the help of stats.js[20]. FPS is a crucial measure for AR systems since it determines the responsiveness of the application.

3.1 Marker-Based AR

The marker-based AR prototype was developed using the JSARToolkit and Three.js [21] as a rendering library. The analyses are done on a video frame size of 320x240. When the marker is detected a simple cube is rendered on top of it. As illustrated in the picture below, the FPS count is between 0 and 3 which means that the image

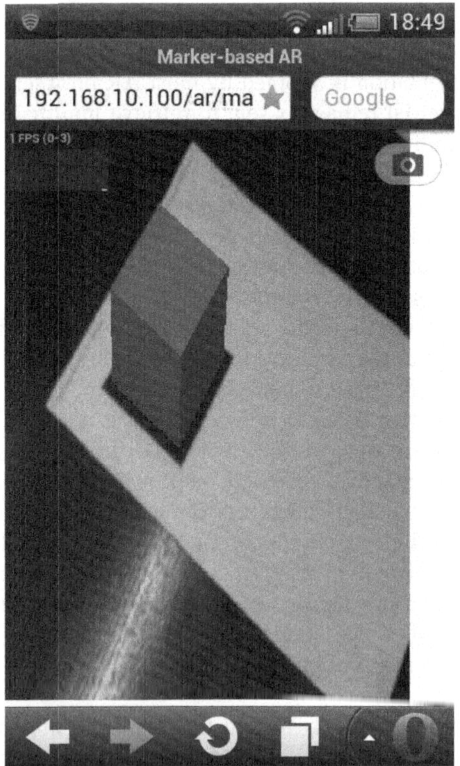

Fig. 2. Marker-based AR prototype

processing and rendering of 3D object in real time drastically affects the performance of such application and consequently its usability.

As already discussed image recognition algorithms tend to be computationally intensive. A possible improvement would be implementing them as native functionality of the web browser which can be far more efficient than in an interpreted language such as JavaScript. Another option for improving performance is running the recognition algorithms on a separate thread using Web Workers [22] and then synchronizing the results.

3.2 Location-Based AR

The main focus in this prototype was to test the critical building blocks of location-based AR such as obtaining proper location and orientation data, and rendering the POIs on a camera feed in real time. For simplification as a source of POIs the Wikipedia Geolocation API was used. The API is queried by means of jQuery JSONP GET requests. After the necessary data is received a continuous loop is being executed which attempts to position the POIs correctly in real time depending on the orientation data. The performance test results are shown in Figure 3.

Fig. 3. Location-based AR prototype

With an average FPS count of 10, the location-based AR prototype provides a far better user experience. Having in mind that this is an initial prototype, with improvements it can become stable for commercial use. Critical component for performance in location-based AR is the rendering algorithms. Again, as in the case of the marker-based prototype, the computationally intensive algorithms should be executed on a different thread using Web Workers. Additionally, more work needs to be done in optimizing the rendering algorithms. On the other hand better support of the used web standards by the web browser vendors would be beneficial in terms of performance.

4 Conclusion and Future Work

In this paper we have presented the benefits of building standardized and platform independent AR application based on web technologies. We have discussed the current status of web standards that can be used for building such applications. By using the discussed standards two prototype applications were developed and evaluated to prove the paper's goal that the web standards have reached a point where AR can be implemented using only web technologies. The initial test results show that the performance of the applications is far from satisfactory, although there are many options for future improvements.

Future work would include optimizing the application by using multiple threads and improvement of the algorithms which are critical for performance. What can also be expected as natural improvement is the support for the necessary standards in the web browsers as well as the improved hardware capabilities in the mobile devices.

References

1. Layar AR Browser, http://www.layar.com/
2. Wikitude AR Browser, http://www.wikitude.com/
3. Butchart, B.: Augmented Reality for Smartphones. A Guide for Developers and Content Publishers. JISC Observatory (2011)
4. de Smit, J.: Towards Building Augmented Reality Web Applications. In: W3C Workshop: Augmented Reality on the Web (2010)
5. Perey, C., Engelke, T., Reed, C.: Current Status of Standards for Augmented Reality. In: Recent Trends of Mobile Collaborative Augmented Reality Systems, pp. 21–38. Springer, New York (2011)
6. Wikipedia, http://www.wikipedia.org/
7. W3C Device APIs Working Group, http://www.w3.org/2009/dap/
8. MacWilliams, A.: A Decentralized Adaptive Architecture for Ubiquitous Augmented Reality Systems. Technischen Universität München (2005)
9. Geolocation API Specification, http://www.w3.org/TR/geolocation-API/
10. DeviceOrientation Event Specification, http://dev.w3.org/geo/api/spec-source-orientation
11. JSARToolkit, a JavaScript port of FLARToolKit, https://github.com/kig/JSARToolKit/

12. WebRTC: Real-time Communication Between Browsers,
http://www.w3.org/TR/webrtc/
13. Media Capture and Streams,
http://dev.w3.org/2011/webrtc/editor/getusermedia.html
14. Augmented Reality on the Web workshop, Barcelona (2010),
http://www.w3.org/2010/06/w3car/report.html
15. W3C Point of interest Working Group (POI WG), http://www.w3.org/2010/POI/
16. Augmented Reality Roadmap for X3D,
http://www.web3d.org/x3d/wiki/index.php/Augmented_Reality_Roadmap_for_X3D
17. Points of Interest Core Recommendation, http://www.w3.org/TR/poi-core/
18. Geolocation API for Wikipedia, http://wikilocation.org/
19. HTML5 Compatibility Tables, http://mobilehtml5.org/
20. A JavaScript Performance Monitor-stats.js,
https://github.com/mrdoob/stats.js/
21. A JavaScript 3D library - three .js, https://github.com/mrdoob/three.js
22. Web Workers, http://www.w3.org/TR/workers/

Dynamics of Global Information System Implementations: A Cultural Perspective

Marielle C.A. van Egmond, Dilip Patel, and Shushma Patel

London South Bank University, London, United Kingdom
{vanegmom,dilip,shushma}@lsbu.ac.uk

Abstract. This paper presents the results of an exploration of theoretical views on the role of culture related to dynamics and processes within global information system implementations, as well as the preliminary results of our case study. Several previous studies show that at the intersection of information system implementation and culture, one must address a construct that may exist at one or more organisational levels simultaneously. We look at global information system implementation processes and dynamics from a qualitative perspective whilst observing a situation in its own context in a case study to gain further insights into key cultural elements as variables which are external to information system technology and seemingly emergent elements of global and multi-sited information systems.

Keywords: global information systems, information system implementation, dynamics, transformation cloud, culture, key cultural elements, organisational culture.

1 Introduction

A recent survey done by Gartner identified that 20% to 35% of all information system implementations fail and 80% exceed time and budget estimates. An average of 65% of IT implementation projects is delivered but does not meet initial expectations or requirements.

Information technology projects in organisations are generally focussed on the technological development and implementation of information systems. As companies increasingly engage in international business activities, examination of global information system implementation projects becomes more significant [16]. A vast number of project success models has been proposed however, information technology project failure and deficiency rates continue to be high and it is generally acknowledged that information technology investments are expensive and of high risk.

Over the past decennia we see a focus on research into process of information system implementation in particular in relation to organisational aspects and emergent elements of global and multi-sited information system implementations [4]. These dynamics are seen to be largely dependent on a number of variables external to the technology itself [12].

S. Markovski and M. Gusev (Eds.): *ICT Innovations 2012*, AISC 207, pp. 205–214.
DOI: 10.1007/978-3-642-37169-1_20

This paper presents the results of an exploration of theoretical views on the role of culture related to dynamics and processes within global information system implementations. We adopt the definition of a global information system "as an organizational system that consists of technical, organizational and semiotic elements which are all re-organized and expanded to serve an organizational purpose" [9]. We aim to gain further insights on the dynamics which comprise cultural elements and consequently on key cultural elements as variables which are external to information system technology and seemingly emergent elements of global and multi-sited information system implementation.

2 Culture and Key Cultural Elements

2.1 Defining Culture

Culture is a term that has many various definitions. In this paper we build on previous research [2], and adhere to the following definition of culture: Culture is a set of basic assumptions that exist, whether learned, adopted and/or emerging, at several levels [11] of human existence in a social system, representing the shared beliefs, norms and/ or values of human behaviour and behavioural patterns, identifying what is important to a particular group and defining rules, knowledge, artefacts and creations.

2.2 Organisational Culture and Software Implementations

The concept of organisational culture has emerged a few decennia ago. Most contemporary authors agree on the following characteristics of the organizational/corporate culture construct: holistic, historically determined, related to anthropological concepts, socially constructed, soft, and difficult to change. According to Hofstede [6] the difference between national and organizational cultures is that the differences between national cultures are mainly found in the values of the different cultures, whereas differences between corporate cultures are mainly found in the practices between different companies. Hofstede introduced his pioneering five dimensions of culture in relation to organisations and doing business in different nations: power-distance, individualism, masculinity, uncertainty avoidance and long-term orientation. These were further developed into a set of national level cultural differences measurable in terms of scores. Stewart [19] states that the 'desired' organization's culture and the 'actual' organization's culture often are worlds apart. He claims that it is important to understand how these variations affect information system implementation processes. Walsham [23] takes the concept of culture a step further by relating information system deployment to cross cultural working. McGrath [14] reports on the relationship between information systems implementation and cultural change within an organisation and what this means for those touched by it.

Tan and Myers [15] mention existing criticisms on the work of Hofstede such as that it is rather crude and simplistic, since it is focussed on national cultural values that influence work-related actions and attitudes. Tan [21] as well as Martinsons and

Davison [13] recommend that further research is needed to understand the impact of culture on information system development and deployment. They also highlight that global information system implementation is complex and subject to different variables, such as business goals, information requirements and IT and user requirements. These are not static but dynamic; they are evolving and changing constantly. Vice versa, information system implementation also triggers a need for change. Both are influenced by human interaction and perception, which are often manifested in culture. Recently Leidner and Kayworth [12] identified culture as a key management variable for successful deployment and management of information systems in organizations.

2.3 Cultural Elements

There are numerous definitions and descriptions of elements that enable further operationalization of the concept of culture. We expect that the results of our case study will characterise operant and emerging key cultural elements in global information system implementations. Yet as a point of departure and based on preceding anthropological research [2] we have taken the approach to define six basic cultural elements as universal and conceptual characteristics of humanity and her social systems that demarcate cultural differences. These are caused by interpretation of and adaptation to varying environments, changing livelihood systems and a diversity of historical stances. We defined the following preliminary description of basic cultural elements: environment (i.e. life equipment system and technology), history and legacy, religion/ beliefs/ goals/ mission, social organisation/ kinship/ networks/ power/ politics and hierarchy; laws and regulations and formal and informal information, knowledge and language.

3 Information System Implementation Dynamics

3.1 Punctuated Equilibrium

In the process of effectuating our investigation on culture and its subsequent key cultural elements in relation to global information system implementations, we have contemplated to the concept of an information system implementation causing a punctuated equilibrium within an organisation [9].

3.2 Organisational Elements and Dynamics in Global Information System Implementations

Within the settings and dynamics of any existing organisation, information system implementation processes start in an AS IS situation (homeostasis X) aiming to develop and emerge towards a TO BE situation (homeostasis Y). Hence the start of an implementation causes a disruption of an organisational equilibrium and induces a phase of transformation. Lyytinen and Newman [1] view information systems development and implementation as a socio-technical change process in which technologies, human actors, organizational relationships and tasks change, driven by configurations in work systems, building systems and the environment, and their misalignments and gaps. It is much more difficult for major transformational change to occur, or be implemented, because it typically involves a profound reformulation of an organisation's mission, structure and management, and fundamental changes in the basic social, political, and cultural aspects of an organisation [16].

Relatively little attention has been paid to global or multi-site organisations, global information systems and global information system implementation. When we address global information systems in this paper, we refer to organisations, which are multinational and have information systems with components and functions that span those national borders as defined by Nickerson, Ho, and Eng [16]. Global information system implementations in multinational organisations generally display added complexity created by the legal, political, economic, cultural, and technical differences at different sites. In order to study global information system implementations and the impact of these external variables, we designed a concept to approach an organisation in transformational change due to a global information system implementation.

3.3 Information System Implementation Transformation Cloud

In this paper we proceed on three constructs. (1) The Ishikawa diagram [7], that depicts six organisational elements affecting an overall organisational problem, Equipment, Process, People, Materials, Environment and Management, and that reveals key relationships between the elements as well as the possible causes, providing additional insight into process behaviour. (2) Sammon and Adam [17] integrate software implementation project elements and congruent organisational prerequisites to define five areas that are critical to data warehousing implementation and organisational readiness: system factors, data factors, skills factors, organisational factors and project management factors. (3) The concept of THIO-model [3][11][18] that originates in global technology transfer studies and describes four inextricably interrelated components of technology production and development processes Technoware, Humanware, Infoware and Orga-ware (see table 1.)

Our conceptual framework to approach an organisation in transformational change due to a global information system implementation has emerged from conjoining these three constructs. We define a global information system implementation as a process that comprises of dynamic macro organisational elements technoware, infoware and humanware, which are –due to an punctuated equilibrium- subject to transformation within the information system implementation transformation cloud and are encapsulated within the organisational setting (orga-ware) (fig. 1). We expect this breakdown to actualise culture and key cultural elements as external variables in global information system implementations in our case study.

Table 1. Elements of changing organisational settings in global information system implementation

Ishikawa (1991) *Fishbone*	Sammon and Adam (2004) *Critical project categories*	Smith and Sharif (2007) *THIO*	Comprising elements
Environment	Organisational factors	Orga-ware	location, time, temperature, and culture in which processes operate, organizational settings management practices
Materials Equipment	Systems factors	Technoware	Technologies, tools and infrastructure, equipment, computers, machinery, transportation and physical infrastructure.
Management (Measurements)	Data factors	infoware	Data, data related processes, procedures, techniques, methods, and specifications management practices
Process (Methods)	Project management factors	Infoware	Processes, procedures, techniques, methods, theories and specifications
People	Skill factors	Humanware	Knowledge, skills, experience, values and cultural elements

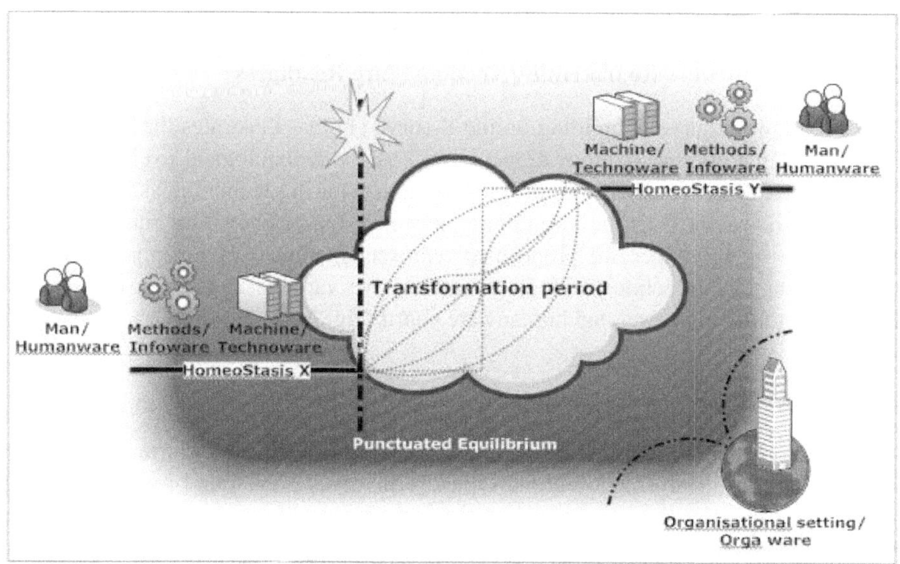

Fig. 1. Punctuated Equilibrium in an Organisational Setting

In our concept of breaking down three distinctive macro dynamic elements technoware, infoware and humanware within a global organisational setting, the fourth macro dynamic element, orga-ware, each is represented as a curve each being subject to a process of development and change within the information system implementation transformation cloud, that spans an aggregation of episodes and sub processes. Based on our research results so far, we present an interpretative depiction of each element's readiness [19] throughout the transformation cloud against time: the techno-curve representing technoware, the processes-curve representing infoware and the human-curve representing humanware.

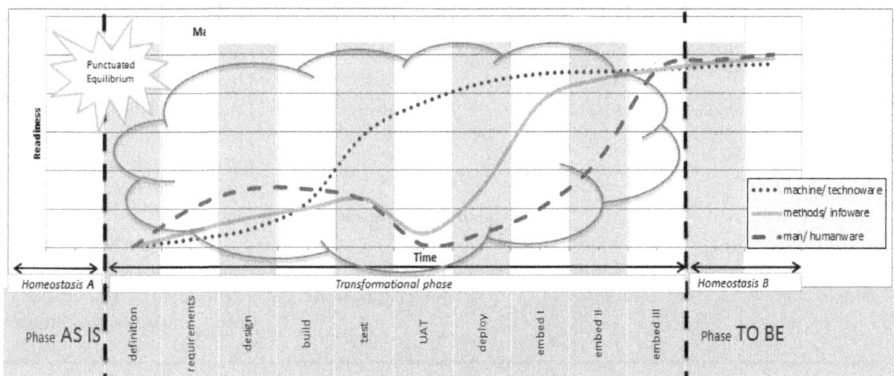

Fig. 2. Macro-dynamics in an information system implementation transformation cloud

3.4 Techno-Curve Representing Techno-Ware Readiness

The Technoware curve is building on the S-curve [1] that illustrates the introduction, growth and maturation of innovations as well as the technological cycles that most industries experience: emerging, developing and maturing technologies. Usually the S-curve is represented as the variation of performance or costs as a function of the time/effort. In this paper we propose to represent readiness of information system development in the techno-curve as an interpretive variable in relation to time and equivalent to the infoware- and humanware charts (fig. 2).

3.5 Process-Curve Representing Infoware Readiness

We postulate that organisational change occurring within the information system implementation transformation cloud typically involves a profound reformulation of an organisation's mission, structure and management, and fundamental changes in the basic social, political, and cultural aspects of an organisation [10][20][21]. Hence, the concept of transformation within this phase covers both operational processes and socio-cultural dimensions of organisations.

When using the process-curve, we refer to infoware-readiness like corporate operational processes and practices within an organisation that are impacted by information system implementation.

Not many studies have resulted in a chart that is depicting the operational process readiness related to time and project phases. The preliminary results of our case study show that the process-curve represents the readiness of the macro element infoware within the information system implementation transition cloud, and runs a different course compared to the techno-curve. It appears that preparing operational processes is not undertaken during the initial phases of the information system implementation project lifecycle (fig. 2). Little operational involvement is observable during the standard implementation lifecycle phases, such as definition and requirements capturing, design, build and technical testing. When operational business engagement is required –generally in the testing phase- the system appears to be not fit for purpose because the corporate operational processes have not been aligned and redesigned to the new system; operational process readiness takes a deep dive at this moment. At this point readying operational processes will be prioritised, which causes an emerging boost of the graph in relation to time and effort. The process-curve in figure 2 shows our interpretation of the infoware readiness based on the data we have gathered so far, depicted as an interpretive variable of readiness in relation to time equivalent to the techno-curve and human-curve. This interpretation will be explored further in our continuing research.

3.6 Human-Curve Representing Humanware Readiness

We have pointed out that humanware is the third organisational element that should be observed as it is subject to a process of change in the transformation cloud. This concurs with earlier work presenting socio-cultural dimensions in relation to information system implementation transformations [10][20][21].

Humanware represents the knowledge, skills, experience, values, habits and normative elements of the human factor within an organisational setting and is of vital importance within information system implementation transformation processes. In relation to global information system implementation processes we refer to humanware readiness in the human-curve.

The human-curve (fig. 2) is based on Elizabeth Kubler-Ross' [8] extended grief cycle, which is widely used and amply developed within change management disciplines. The depiction of our interpretation of the human-curve in this paper is based on our findings so far. It shows a certain peak of inflated expectations, which in the testing phase drops dramatically and seems to be uplifted again slowly. A potential relation between humanware and info-ware has been reported [2][18]. This seems also the case in both curves in our case. We are expecting that our continuing research will further clarify these observations.

4 Dynamics of Global Information System Implementations; a Cultural Perspective

As we combine these three graphs (fig. 2), a distinct difference between the courses of the three macro organisational elements emerges. In the next phase of our research, we aim to develop a thorough understanding of these graphs, individually and as a whole, from the point of view that a set of value-based key cultural elements is related to the course of each, to gain further insights into the nature and dynamics of global information system implementations, and further investigate the similarities and differences between the info-ware and humanware graphs.

Table 2. Macro-Dynamic processes vs. Basic Value-Based Cultural Elements

Basic value-based cultural elements / **Macro-dynamic processes in Transformation cloud**	Environment life *equipment system, technology*	History and legacy	Religion/ beliefs/ goals/ mission	Social organisation/ kinship/ networks/ power/ politics and hierarchy	Laws and regulations	Formal and informal information, knowledge and language.
Orga-ware *Location, time, temperature, and culture in which processes operate, organizational settings*	V	V	V	V	V	V
Technoware *Technologies, tools and infrastructure, equipment, computers, machinery, transportation and physical infrastructure.*	V	V		V		V
Infoware *Data, processes, procedures, techniques, methods, theories, management and specifications*		V	V	V	V	V
Humanware *Knowledge, skills, experience, values and cultural elements*		V	V	V		V

The development of our conceptual framework as presented in this paper conceptualises global information system implementations as a process of a punctuated equilibrium of an organisational setting and introduces the transformation cloud, which is a dynamic phase spanning an aggregation of episodes, processes and emerging elements. We seek to further materialise Lyytinen's [1] concept of socio-technical change to distil key cultural elements that have an effect on or are emerging from the global information system implementation. To support and substantiate our observations in and findings of our case study in the next phase of our research project, we combine the six categories of value-based cultural elements and macro-dynamic process elements within the transformation cloud.

5 Conclusion and Future Directions

Our assumptions and hypothesis will be further investigated in a longitudinal case study in a multinational organisation, which is implementing a global information system. We will use the case(s) to investigate and further define the processes within the transformation cloud, to map the key cultural elements in global information systems implementations and to gain understanding of how key culture elements actually can be materialized as independent variable in global information system implementations

Ultimately, the aim of our research is to increase the understanding and knowledge of global information system implementations and related change management and communications activities, as well as the development of a practical and usable tool for senior stakeholders and project managers for successful information system implementation.

References

1. Dattee, B.: Challenging the S-curve: patterns of technological substitution, conference paper. In: DRUID Summer Conference 2007, Copenhagen (2007)
2. van Egmond, M.C.A.: Home of the spirits; A Case Study on the Impact of Culture on the Acceptation and Implementation of Bamboo as Construction Material and its New or Innovated Technologies in West-Java, Indonesia. MSc thesis, University of Utrecht (2004)
3. van Egmond-de Wilde de Ligny, E.L.C.: Technology Mapping for Technology Management. IOS Press, Incorporated (1999)
4. van Fenema, P.C., Koppius, O.R., van Baalen, P.J.: Implementing packaged enterprise software in multi-site firms: intensification of organizing and learning. European Journal of Information Systems 16, 584–598 (2007)
5. Gardner, D.G.: Operational readiness - is your system more ready than your environment. In: Proceedings of Project Management Institute 32nd Annual Seminars and Symposium, Nashville (2001)
6. Hofstede, G.: Allemaalandersdenkenden; omgaan met cultuurverschillen, Contact, Amsterdam (1991)
7. Ishikawa, K.: Introduction to Quality Control. In: Loftus, J.H. (trans.) Tokyo: 3A Corporation (1990)

8. Kübler Ross, E.: On Death and Dying: What the Dying Have to Teach Doctors, Nurses, Clergy, and Their Own Families. Tavistock, London (1973)
9. Lyytinen, K., Newman, M.: Punctuated equilibrium, process models and information system development and change: towards a socio-technical process analysis. Case Western Reserve University, USA, Sprouts: Working Papers on Information Systems 6(1) (2006), `http//sprouts.aisnet.org/6-1`
10. Lyytinen, K., Newman, M.: Explaining information systems change: a punctuated socio-technical change model. European Journal of Information Systems (EJIS) 17, 589–613 (2008)
11. Leidner, D.: A Review of Culture in Information Systems Research: Towards a Theory of IT-Culture Conflict. MIS Quarterly, 357–399 (June 2006)
12. Leidner, D., Kayworth, T.: Global Information systems: the implications of Culture for IS management (2008)
13. Martinsons, M.G., Davison, R.M.: Culture's consequences for IT application and business process change: a research agenda. International Journal of Internet and Enterprise Management 5(2) (2007)
14. McGrath, Organisational culture and information systems implementation: a critical perspective, London School of Economics and Political Science, Information Systems Department (2003)
15. Myers, M.D., Tan, F.B.: Beyond models of national culture in information system research. Journal of Global Information Management (January/March 2002)
16. Nickerson, R.C., Ho, L.C., Eng, J.: An exploratory study of strategic alignment and global information system implementation success in fortune 500 companies. In: Proceedings AMCIS 2003, `http://online.sfsu.edu/~rnick/dauphine/amcis03a.pdf` (retrieved)
17. Sammon, D., Adam, F.(n.d.): Towards a model for evaluating organisational readiness for ERP and Data warehousing projects. Working paper, University Collage Cork, Cork Ireland (2004)
18. Smith, R., Sharif, N.: Understanding and acquiring technology assets for global competition. Technovation 27(11), 643–649 (2007)
19. Stewart, G.: Organisational Readiness for ERP Implementation. In: AMCIS 2000 Proceedings. Paper 291 (2000), `http://aisel.aisnet.org/amcis2000/291`
20. Sundarasaradula, D., Hasan, H.: Model synthesis and discussion, ch. 11 (2005), `http://epress.anu.edu.au/info_systems/mobile_devices/ch11s08.html#d0e5079` (retrieved March 15, 2012)
21. Tan, F.B.: Advanced topics in global information management, vol. 2, p. 321 (2003)
22. Trompenaars, F.: Riding the waves of culture; understanding cultural diversity in business (1993)
23. Walsham, G.: Cross-Cultural Software production and use: a structurational Analysis. MIS Quarterly, 329–380 (December 2002)

Compute and Memory Intensive Web Service Performance in the Cloud

Sasko Ristov, Goran Velkoski, Marjan Gusev, and Kiril Kjiroski

Ss. Cyril and Methodious University,
Faculty of Information Sciences and Computer Engineering,
Rugjer Boshkovikj 16, 1000 Skoipje, Macedonia
{sashko.ristov,marjan.gushev,kiril.kjiroski}@finki.ukim.mk,
velkoski.goran@gmail.com

Abstract. Migration of web services from company's on-site premises to cloud provides ability to exploit flexible, scalable and dynamic resources payable per usage and therefore it lowers the overall IT costs. However, additional layer that virtualization adds in the cloud decreases the performance of the web services. Our goal is to test the performance of compute and memory intensive web services on both on-premises and cloud environments. We perform a series of experiments to analyze the web services performance and compare what is the level of degradation if the web services are migrating from on-premises to cloud using the same hardware resources. The results show that there is a performance degradation on cloud for each test performed varying the server load by changing the message size and the number of concurrent messages. The cloud decreases the performance to 71.10% of on-premise for memory demand and to 73.86% for both memory demand and compute intensive web services. The cloud achieves smaller performance degradation for greater message sizes using the memory demand web service, and also for greater message sizes and smaller number of concurrent messages for both memory demand and compute intensive web services.

Keywords: Cloud Computing, JAVA, Apache.

1 Introduction

Web services are the most commonly used technology as a standardized mechanism to describe, locate and communicate with web applications. They are used for collaboration between loosely bound components. Effective and ubiquitous B2B systems are being built using web services [4]. Additionally, independence of the underlying development technology enhances web services usage due to the mitigation to development process time and effort [6]. SOAP and REST are two main approaches for interfaces between web site and web services. A high-level comparison of these approaches is realized in [3]. RESTful web services are more convenient to be hosted on mobile devices than SOAP [10].

Research results about web service performance can be found in many papers in different domains. Web services can be simulated and tested for various performance metrics before they are deployed on Internet servers, which give results

S. Markovski and M. Gusev (Eds.): *ICT Innovations 2012*, AISC 207, pp. 215–224.
DOI: 10.1007/978-3-642-37169-1_21

close to the real environment [17]. The authors in [16] propose a deserialization mechanism to reuse matching regions from the previously deserialized application objects from previous messages, and performs deserialization only for a new region that would not be processed before. Web service performance in wireless environments and implementing WS-Security are analyzed in [15]. Web server performance parameters response time and throughput are analyzed via web services with two main input factors message size and number of messages in [13]. Here we extend this research to compare the web service performance with the same input factors in the cloud.

Web servers are usually underutilized since IT managers plan the strategy for hardware resources in advance for the period of several years. Servers are overutilized in peaks which can enormously increase web service response time or even make the services unavailable. Companies can benefit if they migrate their services in the cloud since it offers flexible, scalable and dynamic resources.

Although a public cloud can be a good solution for small and medium enterprises, it provides several open issues: Software Licensing; Security, Privacy and Trust; Cloud Lock-In worries and Interoperability; Application Scalability Across Multiple Clouds; Dynamic Pricing of Cloud Services; Dynamic Negotiation and SLA Management; Regulatory and Legal Issues [2]. It is not an optimal solution for many-tasks scientific computing [7]. The cloud and virtual environments are also worse than on-premise environment for cache intensive algorithms when the data exceeds the cache size [12]. EC2 is slower than a typical mid-range Linux cluster and a modern HPC system for high performance computing (HPC) applications due to interconnection on the EC2 cloud platform which limits performance and causes significant variability [8]. However, the cloud provides better performance in distributed memory per core [5].

The goal of research in this paper is the performance analysis of web services hosted on cloud. We perform series of experiments for compute and memory web services on the same hardware infrastructure hosted on-premise and in the cloud. The rest of the paper is organized as follows. Section 2 describes the realized experiments, infrastructure and platform environments. In Section 3 we present the results of the experiments and analyze how the message size and the number of concurrent messages impact the web service performance in the cloud and on-premise. The conclusion and future work are specified in Section 4.

2 The Methodology

This section describes testing methodology including identification of environment, infrastructure and platform, test plan and design implementation details. Several steps were performed to create efficient and effective tests and results.

2.1 Test Environment Identification

We realize the experiments on traditional client-server architecture on the same hardware infrastructure but different platform. Two same web servers are used as hardware infrastructure with Intel(R) Xeon(R) CPU X5647 @ 2.93GHz with 4 cores and 8GB RAM. The other server with the same hardware infrastructure is used as a client. Linux Ubuntu 64 bit Server 11.04 is installed on the machines on both the server and the client side. Apache Tomcat 6.0 is used as web server where RPC style web services are being deployed. SOAPUI [14] is used to create various server load tests. Client and server are in the same LAN segment to exclude the network impact shown in [9].

Two different platforms are deployed. *On-premise* platform environment consists of traditional Linux operating system installed as host. *Cloud environment* is developed using OpenStack Compute project deployed in dual node [11]. We use one Controller Node and one Compute Node. KVM virtualization standard is used for instancing virtual machine. The cloud consists of the same hardware and operating system as previously described.

2.2 Performance Criteria Identification

We measure response time for various experiments with different number and sizes of concurrent requests for both platforms. Client is on the same VLAN as the web server, with network response time smaller than 1 ms, and none of the packets are lost during the test. This means that we can assume that the response time measured with SOAPUI is the same as the server response time.

2.3 Test Data

The basic goal is to measure the performance drawbacks caused by migration of web services in the cloud.

Test data consists of Concat and Sort web services. The *Concat web service* accepts two string parameters and returns a string which is concatenation of the input. This is a memory demand web service that depends on the input parameter size M with complexity $O(M)$. The *Sort web service* also accepts two string parameters and returns a string that is concatenation of the two input strings which is then alphabetically sorted using sort function in [1]. This is also a memory demand service that depends on the input parameter size M. In addition it is a computational intensive web service with complexity $O(M \cdot log_2 M)$.

Experiments are repeated for parameter sizes M that change values from $256B$, $768B$, $1280B$, $1792B$, $2304B$ to $2816B$. The generated SOAP messages have the following sizes $786B$, $1810B$, $2834B$, $3858B$, $4882B$ and $5908B$ correspondingly. The server is loaded with various number of messages (requests) N in order to retain server normal workload mode, that is, 500, 1000, 1500 and 2000 requests per second for each message size.

2.4 Test Plan

The first part of the experiment consists of series of test cases that examine the impact of increasing the message size to the server response time. The second part of the experiment consists of series of test cases that examine the impact of increasing the number of concurrent messages to the server response time. All test cases are performed on: 1) web services hosted on-premise; and 2) web services hosted in the cloud.

Each test case runs for 60 seconds, N messages are sent with M bytes each, with variance 0.5. The accent is put on server response time in regular mode, and neither burst nor overload mode.

We expect that response time will be increased while increasing the number of messages and their size. We would like to determine which parameter impacts the server performance most? Is it the number of concurrent messages or message sizes and has the platform any influence if it is on-premise or in the cloud?

Monitors are checked before each test. All server performance parameters are examined if their status is returned to nominal state after execution of each test. If not, the server is restarted and returned into it's nominal state. Network latency is measured to ensure proper response time results during the tests.

3 The Results and Analysis

This section describes the results of testing the performance impact of cloud virtualization layer. We also analyze the results to understand the performance impact of different message sizes and number of concurrent messages on both web services described in 2.3.

3.1 Web Service Performance Hosted On-Premise

The performance of web services is measured while hosted on-premise with different payload: 1) different message size for constant number of concurrent messages and 2) different number of concurrent messages for a constant message size.

Figure 1 depicts the response time of Concat web service hosted on-premise. We can conclude that both input factors are important for Concat web service performance, i.e. response time increases when message size or number of concurrent messages increase.

Response time of Sort web service hosted on-premise is presented in Figure 2. We can conclude that only input factor message size is important for Sort web service performance. That is, response time increases only if message size increases regardless of number of concurrent messages.

3.2 Web Service Performance Hosted in the Cloud

We measure the performance of web services hosted in the cloud with different payload: 1) different message size for constant number of concurrent messages and 2) different number of concurrent messages for constant message size.

The results for response time of Concat web service hosted in the cloud is shown in Figure 3. Both input factors are important for Concat web service performance, i.e. response time increases for greater message sizes or number

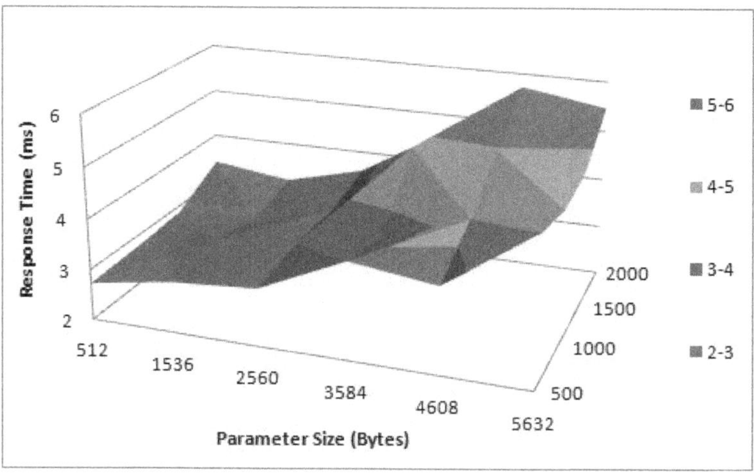

Fig. 1. Concat web service response time while hosted on-premise

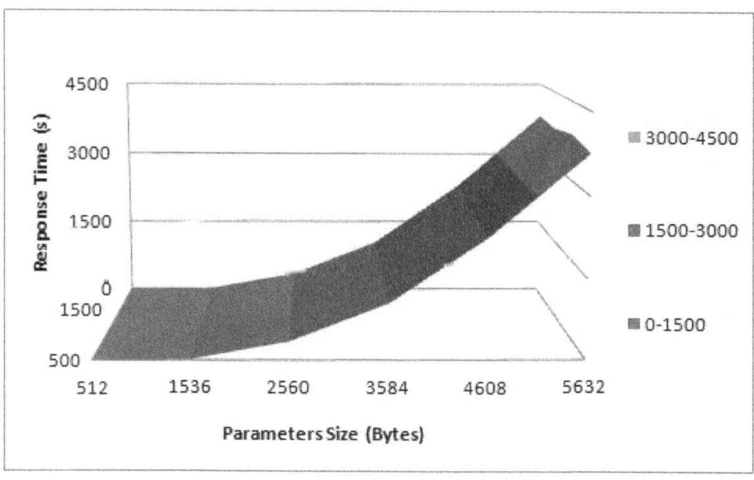

Fig. 2. Sort web service response time while hosted on-premise

of concurrent messages. However, there are performance drawbacks due to additional virtualization layer and cloud software, and small response time in ms comparable to network latency which will be the subject in our further research.

Figure 4 presents the response time of Sort web service hosted in the cloud. Only the message size impacts its performance (the response time increases as message size increases regardless of the number of concurrent messages).

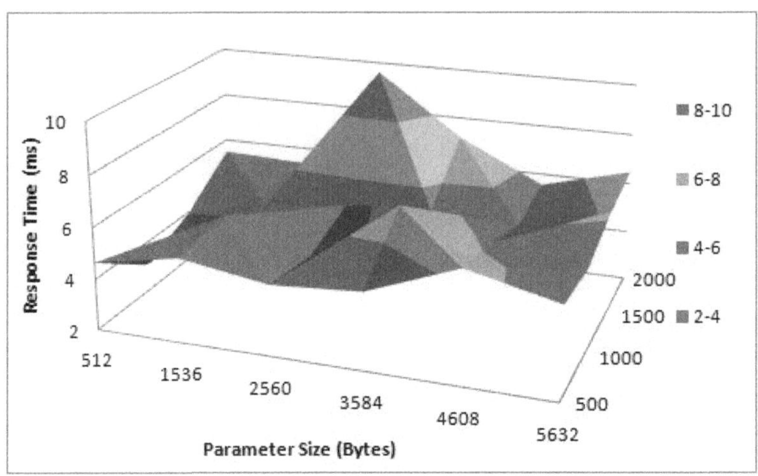

Fig. 3. Concat web service response time while hosted in the cloud

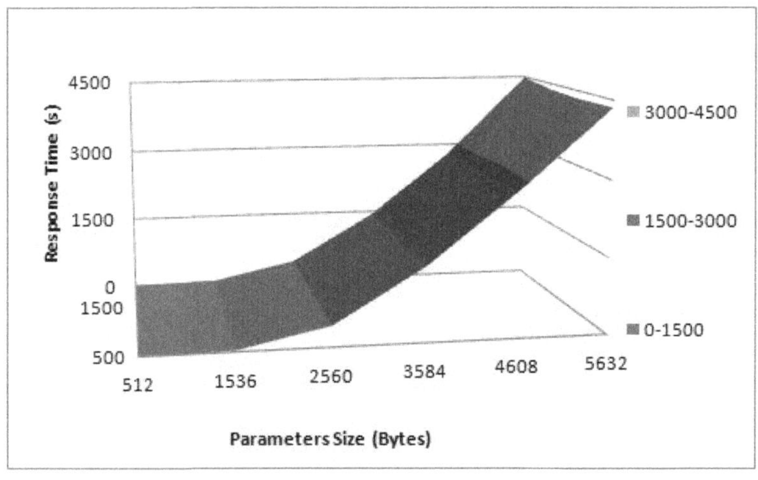

Fig. 4. Sort web service response time while hosted in the cloud

3.3 On-Premise vs. Cloud Performance Comparison

The performance of both web services (hosted on-premise and in the cloud) are compared with different payload depending on different message sizes and different number of concurrent messages.

Figure 5 depicts the Cloud vs on-premise relative response time comparison for Concat web service. The results show that cloud environment provides worse response time than traditional on-premise environment for each message size and for each number of concurrent messages. An interesting conclusion is that

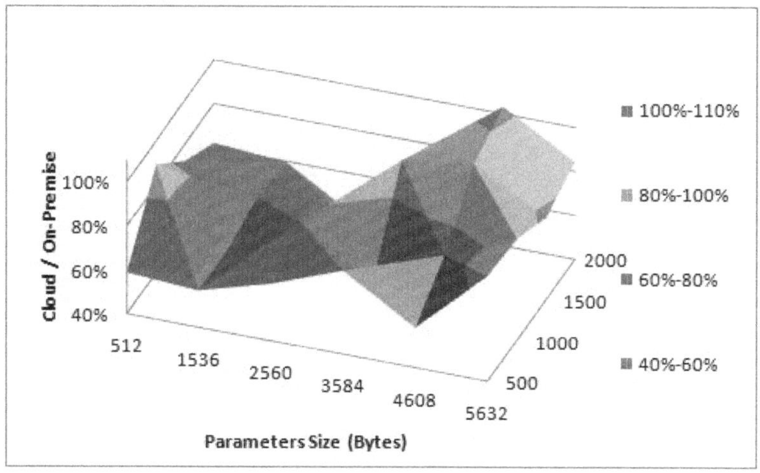

Fig. 5. Cloud vs on-premise relative response time for Concat web service

the cloud provides smaller penalties for greater messages. However, we found a local extreme. We believe that it appears due to communication time impact for small response time and the effect of the virtualization and cloud software which is part our further research.

Table 1 presents the relative performance for each test case and average for Concat web service. As in previous conclusions, the cloud performance average penalties depend only on the message size. The cloud provides smaller performance penalties for greater message sizes achieving an average performance of 82.42% and 83.99% correspondingly for parameters sizes of 4608 and 5632 bytes.

Table 1. Cloud relative performance compared to on-premise for Concat web service

Number / Size	512C	1536C	2560C	3584C	4608C	5632C	**AVG**
500	58.71%	56.71%	65.89%	77.96%	58.14%	87.91%	**67.55%**
1000	89.39%	56.58%	75.57%	53.44%	72.56%	86.19%	**72.29%**
1500	72.41%	62.16%	64.84%	91.17%	95.75%	77.45%	**77.30%**
2000	62.16%	59.55%	37.40%	56.88%	103.24%	84.41%	**67.27%**
AVG	**70.67%**	**58.75%**	**60.93%**	**69.86%**	**82.42%**	**83.99%**	**71.10%**

The relative response time comparison for Sort web service for cloud vs on-premise is shown in Figure 6. The results also show that cloud provides worse response time than on-premise for each message size and for each number of concurrent messages. The cloud provides smaller penalties for greater number of messages regardless of number of concurrent messages. The number of concurrent messages impacts the cloud performance for smaller messages.

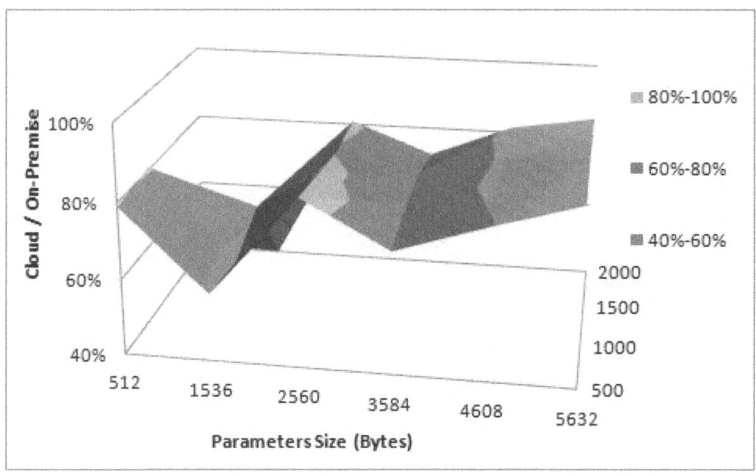

Fig. 6. Cloud vs on-premise relative response time for Sort web service

The relative performance of cloud vs. on-premises for Sort web service is shown numerically in Table 2. The worst performance the cloud provides for smaller parameters size, i.e. total $512B$ and $1536B$ for average 66.58% and 53.92% from on-premise performance. The cloud provides smaller performance penalties for greater messages for average 86.67% and for smaller number of concurrent messages, i.e. for 500 messages/sec. It provides on average 76.41% from on-premise performance. For huge number of concurrent messages it provides greater performance penalties, i.e. for 2000 messages/sec. it provides average 66.61% from on-premise performance.

Table 2. Cloud relative performance compared to on-premise for Sort web service

Number / Size	512C	1536C	2560C	3584C	4608C	5632C	AVG
500	78.90%	57.80%	84.07%	72.00%	79.40%	86.29%	**76.41%**
1000	80.93%	56.14%	81.88%	75.21%	81.29%	89.39%	**77.47%**
1500	65.26%	62.87%	79.70%	75.21%	80.66%	86.01%	**74.95%**
2000	41.22%	38.88%	80.87%	72.37%	81.35%	84.99%	**66.61%**
AVG	**66.58%**	**53.92%**	**81.63%**	**73.70%**	**80.68%**	**86.67%**	**73.86%**

4 Conclusion and Future Work

In this paper we realized performance analysis and comparison of web server performance by analyzing the response time. Two web services were tested for different loads varying the main input factors: the message size and the number of messages. The experiments are realized on the same web services hosted on-premise and in the cloud on the same hardware and runtime environment.

The results of the experiments show that the performance directly depends on input message size especially for both memory demand and compute intensive web service regardless of the platform as depicted in Figure 7. This is not emphasized for memory only demand web service.

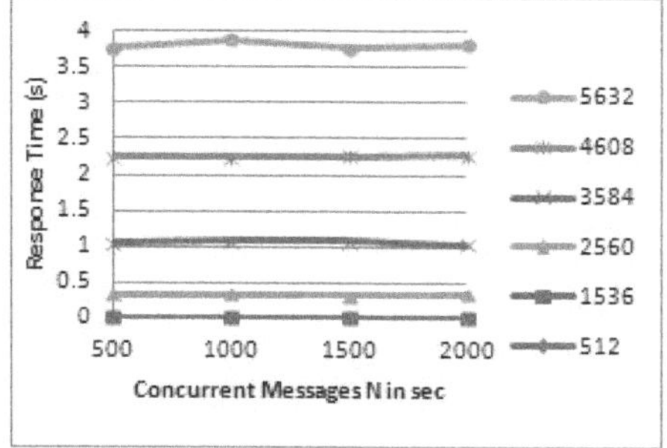

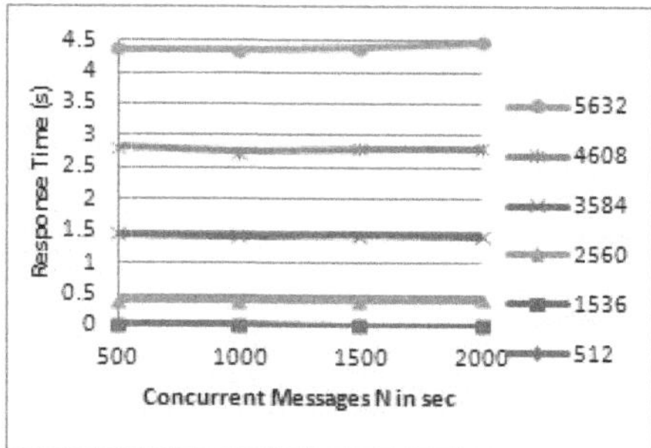

Fig. 7. Response time for constant message size but different number of concurrent messages for Sort web service hosted on-premise (left) and in the cloud (right)

In this paper we also defined quantitative performance indicators to determine the risk of migrating the services in the cloud for various message size and number of concurrent messages. The conclusion is that the performance is decreased to 71.10% of on-premise for memory demand and to 73.86% for both memory demand and compute intensive web service if it is migrated on the cloud. The cloud provides the smallest penalties for greater message sizes regardless of number of concurrent messages for memory demand web service. However, the smallest penalties for both memory demand and compute intensive web service migrated in the cloud are provided for smaller number of concurrent messages and for greater message sizes.

Our plan for further research is to continue with performance analysis of cloud computing on different hardware and cloud platforms with different hypervisors, web servers and runtime environment to analyze if cloud platform can provide better performance than on-premise.

References

1. Bentley, J.L., McIlroy, M.D.: Engineering a sort function. Softw. Pract. Exper. 23(11), 1249–1265 (1993)
2. Buyya, R., Sukumar, K.: Platforms for building and deploying applications for cloud computing. CoRR abs/1104.4379 (2011)
3. Castillo, P.A., Bernier, J.L., Arenas, M.G., Guervós, J.J.M., García-Sánchez, P.: Soap vs rest: Comparing a master-slave ga implementation. CoRR abs/1105.4978 (2011)
4. Curbera, F., Duftler, M., Khalaf, R., Nagy, W., Mukhi, N., Weerawarana, S.: Unraveling the web services web: An introduction to soap, wsdl, and uddi. IEEE Internet Computing 6(2), 86–93 (2002)

5. Gusev, M., Ristov, S.: The optimal resource allocation among virtual machines in cloud computing. In: Proc. of 3rd Int. Conf. on Cloud Computing, GRIDs, and Virtualization (CLOUD COMPUTING 2012), pp. 36–42 (2012)
6. IBM Web Services Arch. Team: Web services architecture overview (2012), https://www.ibm.com/developerworks/webservices/library/w-ovr/
7. Iosup, A., Ostermann, S., Yigitbasi, M.N., Prodan, R., Fahringer, T., Epema, D.: Performance analysis of cloud computing services for many-tasks scientific computing. IEEE Trans. on Par. and Dist. Syst. 22(6), 931–945 (2011)
8. Jackson, K.R., Ramakrishnan, L., Muriki, K., Canon, S., Cholia, S., Shalf, J., Wasserman, H.J., Wright, N.J.: Performance analysis of high performance computing applications on the amazon web services cloud. In: Proc. of the IEEE CLOUDCOM 2010, pp. 159–168. IEEE Computer Society, USA (2010)
9. Juric, M.B., Rozman, I., Brumen, B., Colnaric, M., Hericko, M.: Comparison of performance of web services, ws-security, rmi, and rmi-ssl. J. Syst. Softw. 79(5), 689–700 (2006)
10. Mizouni, R., Serhani, M., Dssouli, R., Benharref, A., Taleb, I.: Performance evaluation of mobile web services. In: ECOWS 2011, pp. 184–191 (2011)
11. Openstack: Openstack dual node (Febuary 2012), http://docs.stackops.org/display/documentation/Dual+node+deployment
12. Ristov, S., Gusev, M., Kostoska, M., Kjiroski, K.: Virtualized environments in cloud can have superlinear speedup. In: ACM Proceedings of 5th Balkan Conference of Informatics, BCI 2012 (2012)
13. Sasko, Tentov, A.: Performance impact correlation of message size vs. Concurrent users implementing web service security on linux platform. In: Kocarev, L. (ed.) ICT Innovations 2011. AISC, vol. 150, pp. 367–377. Springer, Heidelberg (2012)
14. SoapUI: Functional testing tool (January 2012), http://www.soapui.org/
15. Srirama, S.N., Jarke, M., Prinz, W.: A performance evaluation of mobile web services security. CoRR abs/1007.3644 (2010)
16. Suzumura, T., Takase, T., Tatsubori, M.: Optimizing web services performance by differential deserialization. In: Proc. of the IEEE Int. Conf. on Web Services, ICWS 2005, pp. 185–192. IEEE Computer Society, USA (2005)
17. Tripathi, S., Abbas, S.Q.: Performance comparison of web services under simulated and actual hosted environments. Int. J. of Computer Applications 11(5), 20–23 (2010)

Diatom Indicating Property Discovery with Rule Induction Algorithm

Andreja Naumoski and Kosta Mitreski

Ss. Cyril and Methodius University in Skopje,
Faculty of Computers Science and Engineering, Skopje, Macedonia
{andreja.naumoski,kosta.mitreski}@finki.ukim.mk

Abstract. In the relevant literature the diatoms have ecological preference organized using rule, which takes into account the important influencing physical-chemical parameters on the diatom abundance. Influencing parameters group typically consist from parameters like: conductivity, saturated oxygen, pH, Secchi Disk, Total Phosphorus and etc. In this direction, this paper aims in process of building diatom classification models using two proposed dissimilarity metrics with predictive clustering rules to discover the diatom indicating properties. The proposed metrics play important rule in this direction as it is in every aspects of the estimating quality of the rules, from dispersion to prototype distance and thus lead to increasing the classification descriptive/predictive accuracy. We compare the proposed metrics by classification and rule quality metrics and based on the results, several set of rules for each WQ and TSI category classes are presented, discussed and verified with the known ecological reference found in the diatom literature.

Keywords: Predictive clustering rules, accuracy, water quality category classes, diatoms.

1 Introduction

Predictive models that address environmental modelling can take many different forms that range from linear equations to logic programs. Two commonly used types of models are decision trees [1] and rules [2]. Their benefit is that they are comprehensible and can easily be analysed by a human. Unlike some representations, (e.g., linear regression equations that treat the entire example space simultaneously), the trees and the rules divide the space of examples into subspaces, and for each subspace provide a simple predictive model. Decision trees partition the set of examples into subsets in which examples have similar values of the target variable, while clustering produces subsets in which examples have similar values of all descriptive variables [2]. Here, every cluster has a symbolic description formed by the conjunction of conditions on the path from the root of the tree to the given node. On the other side is an area of machine learning in which formal rules are extracted from a set of examples mainly from measured ecological data. The rules extracted may

S. Markovski and M. Gusev (Eds.): *ICT Innovations 2012*, AISC 207, pp. 225–234.
DOI: 10.1007/978-3-642-37169-1_22

represent a full scientific model of the data, or merely represent local patterns in this measured data. Each rule represents an independent piece of knowledge that can be interpreted separately without other rules from the rule set. Typical rule set consists of rules in a form 'IF condition THEN prediction'. These rules are used for prediction of examples that do not satisfy the condition of any other rule.

The rule learning algorithms as a research topic has been investigated in many areas, and recently in the ecological modelling. Because the ecologist and biologist typically use statistical tools to analyse the data, the influence of the machine learning algorithm gradually increases. Most of these methods are direct descendant from the AQ series of methods [3,4]. The most commonly used rule induction algorithm is known as CN2 method [5,6], which produces classification rules in a "IF THEN" form. In every iteration a single rule is constructed using a heuristic beam search. After searching the space of the covered examples, if good rule is found it is added to the rule set, the rule search continues. The examples that are covered by this rule are removed from the measured dataset. In order to combine the advantages of the decision trees division of the set of the examples into subsets analogous to clustering, a new approach named predictive clustering [7,8] had been developed. Then a novel method as a combination of these two seemingly different methods, is made by [9], named predictive clustering rules. This method focuses on constructing clusters of examples that are similar to each other, but in general taking both the descriptive and the target variables into account [9]. Then the predictive model is associated with each cluster which describes the cluster, and, based on the values of the descriptive examples, predicts the values of the target examples and thus obtaining the rule model. In his research Zenko [9], defines several quality criteria's to estimate the quality of the learned rules. In all these quality criteria one of the important factors that determinates the quality of the rule is the dissimilarity metric. The Manhattan metric is used by [9], because of its simplicity and generality. However, in the process of revealing the diatom-indicator property, this metric has been show as weak metric. In this direction the paper proposes two dissimilarity metrics to be used in the process of the diatom classification. The Euclid and Sorensen metrics are introduces and experimentally evaluated. These metrics should produce higher classification accuracy due to the particular nature of the dataset.

The diatom dataset used for diatom classification consist from several important parameters that reflect the physical, chemical and biological aspects of the water quality and the trophic state of the ecosystem. These category classes are vital to understand how these systems function, which in turn helps to identify the sources of reducing the ecological health status of the lake. Diatoms are ideal bio-indicators, because they have narrow tolerance ranges for many environmental variables and respond rapidly to environmental change [10]. There is many known ecological preference for the diatoms, but still the biologist discovery many new diatoms which have unknown ecological indicating property. In this direction, in this paper we used the proposed metrics with the predictive clustering rules; to build diatoms models and the verified with the known ecological known references.

The remainder of this paper is organized as follows. In Section 2, we describe the machine learning methodology of the predictive clustering rule algorithm together with the proposed dissimilarity metrics used for diatom classification. Section 3 describes the data and explains the experimental design that is used to analyse the data at hand. In Section 4, we present and discuss several of the obtained evaluation results together with the rules for each WQ and TSI category class the obtained diatom models. The Section 5 concludes the paper and gives direction for future research.

2 Rule Learning for Diatom Classification

The rule induction process from examples has established itself as a basic component of many machine learning systems and it has been used in many research areas. By building predictive models, these methods are one the most used for interpretation of the results. Before we evaluate the results obtained by the diatom classification, we need to establish the quality criteria that each single rule or the rule set as a whole has to satisfy as it is done in [9]. In practice, these criteria can be hard to be achieved, and thus we need to find compromise between them. What is important to point is that, the rule set should be small and to have a small error rate. However very small rule sets tend to have high error rates.

2.1 Quality Criteria of Rules

Metrics that measure error in machine learning are specifically design to evaluate only classification or regression tasks and most of them can be used to evaluate single target models. In order to measure the error for both classification and regression tasks, and also the multi target models for the predictive clustering rules the author in [9] uses different metrics. This means that the metric that is used in [9] called dispersion takes one or more nominal numeric examples and return combined numeric value, within the [0, 1] interval. Because the distance between the two prototypes can be evaluated using any distance metric measure; we have decide to compare the already proposed Manhattan metric [9] with two other metrics. The two metrics in the process of inducing the models, should obtain more general rules that covers more examples and the final rule set will have fewer rule and be more understandable. However, we must be carefully, because the generalization also means higher error in the predictive model and a compromise between this two contrary measured must be found. The definition of relative coverage is straightforward and its definition is given in [9].

It is important to know that we should learn rules that cover examples with a prototype that is different than the prototype of the entire learning set of examples [9]. This means that the rule with a prototype equal to the default prototype would give the same predictions as the default rule. Predictions of such a rule would not be useful since they would not provide any new information, i.e., their information score would be zero [11]. Measuring prototype dissimilarity is in principle comparable to the measuring of significance of rules as applied in the CN2 algorithm [5,6]. In CN2, the likelihood ratio statistic is used to measure the difference between the class probability distribution in the set of examples covered by the rule, and the class probability distribution of the entire learning set [5,6]. If the value of this statistic suggests that the class distribution of the rule is not significantly different, the rule is not included in the output model. The prototype dissimilarity measure presented in [9] can be viewed as a generality of this approach since it can take into account more than one attribute.

2.2 Proposed Dissimilarity Metrics

In the literature there are several dissimilarity metrics which are suitable for different type of datasets. Concerning the diatom classification, the previously used Manhattan metric represents distance between the diatoms that more diverse, which is not the case of indicating organisms for a specific parameter. This metric examines the absolute differences among the coordinates of a pair of objects defined with equation (1). The formula for this distance between a point $x = (x_1, x_2, \ldots, x_n)$ and a point $y = (y_1, y_2, \ldots, y_n)$ is:

$$d = \sum_{i=1}^{n} |x_i - y_i| \tag{1}$$

where n is the number of variables, and x_i and y_i are the values of the i-th variable, at points x and y respectively.

The first proposed metric which is more suitable for diatom classification is Euclid distance. The Euclidean distance function measures the 'as-the-crow-flies' distance. The formula for this distance between a point $x = (x_1, x_2, \ldots, x_n)$ and a point $y = (y_1, y_2, \ldots, y_n)$ is:

$$d = \sqrt{\sum_{i=1}^{n} (x_i - x_i)^2} \tag{2}$$

where n is the number of variables, and x_i and y_i are the values of the i-th variable, at points x and y respectively. The second proposed metric is Sorensen metric. The Sorensen distance is a normalization method that common used in botany, ecology and environmental science field. It views the space as grid similar to the city block distance.

$$d = \frac{\sum_{i=1}^{n} |x_i - y_i|}{\sum_{i=1}^{n} x_i + \sum_{i=1}^{n} y_i} \tag{3}$$

The Sorensen distance metric has nice properties that if all coordinates are positive; its value is between zero and one. Sorensen metric represent exact similar coordinate. The normalization is done using absolute difference divided by the summation. Again

n is the number of variables, and x_i and y_i are the values of the i-th variable, at points x and y respectively.

The predictive clustering rules algorithm can induce two types of rules: order and unordered. When order rules are induced, each rule is tried in order, and the first rule that covers the example is used for classification. If no learned rule covers the example, the final default rule assigns the most common class in the entire learning set to the example. Alternatively, unordered rules can be induced with this algorithm. In this case, rules are learned iteratively for each possible class value in turn. When a new rule is found, only the examples covered by this rule which belong to the specified class are removed from the learning set. In our experimental setup we will generate only unordered rules because each rule is in depended one from another. The Euclid and Sorensen metrics by experimental evaluation is compared with the previously used Manhattan dissimilarity metric.

3 Data Description and Experimental Setup

The dataset used in the experiments consist from 13 input parameters representing the TOP10 diatoms species (diatom species that exist in Lake Prespa [12]) with their abundance per sample, plus the three WQ category classes for conductivity, pH and Saturated Oxygen. The one dataset is created for each WQ category class as output class. These measurements were made as a part of the TRABOREMA project [13]. The WQ category classes were defined according to the three physical-chemical parameters: Saturated Oxygen [14], Conductivity [14] and pH [14,15] and they are given in Table 1.

Table 1. Water quality classes for the physical-chemical parameters

Physico-chemical parameters	Name of the WQC	Parameter range
Saturated Oxygen	*oligosaprobous*	> 85 %
	β-mesosaprobous	70-85 %
	α-mesosaprobous	25-70 %
	α-meso / polysaprobous	10-25 %
pH	*acidobiontic*	pH < 5.5
	acidophilous	pH > 5.5
	circumneutral	pH > 6.5
	alkaliphilous	pH > 7.5
	alkalibiontic	pH > 8
	Indifferent	pH > 9
Conductivity	*fresh*	< 20 ($\mu S \cdot cm^{-1}$)
	fresh-brackish	< 90 ($\mu S \cdot cm^{-1}$)
	brackish-fresh	90 – 180
	brackish	180 - 900

For the proposed predictive clustering rules algorithm for diatom classification we have applied different parameter strategies to obtained the best descriptive and predictive performance of the models and in the same time the obtained models should be very easy interpretable and verified. We have used only Standard procedure for building rule models, the depth was set to 10, and maximum number of rules was set to 100 in the CLUS system, that can be obtained at http://dtai.cs.kuleuven.be/clus/. The range of the dispersion was set to default. The heuristic function was also set to adding as default setting in the predictive clustering rules algorithm. For each dataset, we have estimate the descriptive and predictive accuracy of the proposed space metrics with the Manhattan metric. We have used 10-fold cross validation to estimate the predictive performance of the models, while the descriptive performances of the models were estimate using the whole diatom dataset. To assess the quality of the learned rules of knowledge that we have gain from the models, we have compared also the coverage that each rule has over the whole or the 10-fold cross validation set.

4 Experimental Results

To assess the quality of the learned rules of knowledge that we have gain from the data, we have compared the descriptive and predictive performance of the models. We have also made a ranking system based on the ranking procedure proposed by [16] in order to estimate the best space metric for diatom classification.

4.1 Performance Evaluation

The performances of the experiments are given in the Table 2. Using standard 10-fold cross validation we have estimated the predictive performance of the two proposed metrics against the previously used Manhattan metric.

According to the table, the Manhattan metric have achieved only equal results with the proposed metrics for both descriptive and predictive classification accuracy. The Euclid metric is better for pH and Secchi Disk diatom classification for descriptive, but 5 out of 6 datasets the predictive performance are batter compared than the Manhattan metric. The second proposed metric – Sorensen, has achieve better descriptive performance for 5 from 6 datasets, while concerning the predictive performance has achieve highest predictive classification accuracy for Conductivity and Total Phosphorus datasets.

The same conclusion can be made from the ranking system, made with the proposed ranking procedure by [16]. According to the results, the Sorensen metric is the best metric for building diatom classification models to achieve best descriptive accuracy, while for predictive accuracy Euclid is the best space metric. Based on these results, we built many models and some of them are presented and discussed in the next section.

Table 2. Performance evaluation of the predictive clustering rules algorithm based for diatom classification

Dataset	Train			Test		
	Manhattan	Euclid	Sorensen	Manhattan	Euclid	Sorensen
Conductivity	97.25	97.35	**97.40**	**74.77**	**74.77**	**74.77**
pH	85.32	**87.00**	86.80	58.72	**60.55**	58.72
Saturated Oxygen	89.05	88.67	**89.11**	56.72	**58.21**	56.72
Secchi Disk	**84.74**	**84.74**	**84.74**	83.16	**83.68**	83.16
Total Phosphorus	**90.37**	90.06	**90.37**	35.32	33.49	**38.99**
Ranking		Quade			Friedman	
TRAIN	2.36	2.13	**1.5**	2.3	2.2	**1.5**
TEST	2.3	**1.73**	1.96	2.3	**1.6**	2.1

4.2 Classification Models for the Water Quality Classes

In this section we present several rules of the many induced rule models from the described datasets based on the highest classification accuracy using test procedure. Many of the rules are small, some of the rules are large in length and some of them cover more general knowledge. The space metric was chosen based on the highest predictive performance for each dataset.

The rules presented below reveals that APED and DMAU diatoms can be used for indicating *brackish* waters, while the STPNN and NSROT diatoms cannot be used as bio-indicators of *brackish* WQ Conductivity category class. Based on the rules the CSCU diatom can be indicator of *brackish-fresh* waters. According to the evaluation results given in brackets for every rule, the rules have relatively medium coverage for test procedure. For example, the first rule, in train procedure covers 48 data pointes from 196, and in test procedure 10 of 22 instances in the diatom dataset.

Rule 1: IF STPNN <= 1 AND APED > 2.0
THEN [*brackish*] [196/48: 22/10]
Rule 2: IF CSCU <= 1
THEN [*brackish-fresh*] [196/59: 22/6]
Rule 3: IF DMAU > 1.0 AND DMAU <= 4
THEN [*brackish*] [196/49: 22/8]
Rule 4: IF NSROT <= 0 AND APED > 0.0
THEN [*brackish*] [196/23: 22/6]

The other set of rules obtained by the algorithm, shows the indicating properties for several diatoms in pH WQ category classes.

Rule 5: IF CJUR > 0.0
THEN [*alkalibiontic*] [196/136: 22/13]
Rule 6: IF COCE > 13.0
THEN [*alkaliphilous*] [196/124: 22/14]
Rule 7: IF NROT > 3.0 AND COCE > 26.0 AND APED <= 1
THEN [*alkaliphilous*] [196/8: 22/8]
Rule 8: IF CSCU > 14.0 AND APED <= 2
THEN [*alkaliphilous*] [196/18: 22/4]

The Rules 5 has relatively high coverage in training procedure, compared with the rest of the rule set. Regarding the obtained knowledge, it is easy to note that CJUR diatom can be used for indicating *alkalibiontic* waters. Concerning the *alkaliphilous* water quality class, the COCE, CSCU and APED diatoms can be used for indicating such waters. The Rule 7 and Rule 8 have low coverage, but verification of the results will confirm the obtained knowledge. The last set of rules for WQ category classes is based on the Saturated Oxygen parameter.

Rule 9: IF CSCU > 5.0 AND APED > 0.0
THEN [*β-mesosaprobous*] [196/89: 22/9]
Rule 10: IF DMAU > 1.0 AND APED > 0.0 AND COCE > 3.0
THEN [*β-mesosaprobous*] [196/56: 22/8]
Rule 11: IF CSCU > 14.0 AND CJUR <= 10
THEN [*oligosaprobous*] [196/38: 22/5]
Rule 12: IF NROT <= 19 AND CSCU > 19.0 AND CPLA <= 0
THEN [*oligosaprobous*] [196/16: 22/3]

According to rules, the APED, COCE, DMAU and CSCU diatoms can be used for indicating *β-mesosaprobous* waters, while CSCU, CJUR and NROT diatoms can be used for *oligosaprobous* waters. The rule 12 concludes that CPLA diatom cannot be used for indicating *oligosaprobous* waters. Compared with the previous sets of rules, the rules have relatively medium coverage in test procedure.

4.3 Verification of the Diatom Model Results

Ecological references for the examined diatoms by the proposed algorithm are taken from the latest diatom ecology publications [17]. In the relevant literature, the APED diatom is known to be *alkaliphilous*, *fresh-brackish*, *nitrogen-autotrophic* (tolerates elevated concentrations of organically bound nitrogen), has high oxygen saturation (>75%), *β-mesosaprobic* and *eutrophic* (because of Organic N tolerance) diatom indicator [17]. The presented rule models have indicated that the APED diatom can be used as an indicator of *brackish*, *alkaliphilous*, and *β-mesosaprobous* waters. The β-*mesosaprobous* and the *brackish* indicating properties are verified by the models. In the relevant literature that CSCU is known as *alkalibiontic*, *fresh-brackish* water taxon, being *oligosaprobic* indicators with *eutrophic* preferences [17]. According to the models the CSCU diatom is *brackish-fresh* and *alkaliphilous* water diatom, which verifies the known ecological reference. COCE is known as *meso-eutro* taxon [17]. Based on the rule models, the COCE diatom is indicator for *alkaliphilous*, *β-mesosaprobous*. Regarding the DMAU diatom, this diatom is best developed in slightly *brackish* waters [17]. From the rule models the DMAU diatom van live in *brackish* waters (see Rule 3). Concerning the unknown ecological references the NSROT, NROT and CJUR diatom has no ecological references in the literature, so the results of the model are the first to be known. The NSROT diatom cannot be used for indicating *brackish* water. The NROT diatom according the rule models is indicator of *alkaliphilous* waters. The last CJUR diatom, the rule models stated that this diatom is indicator for *alkalibiontic* waters. Further investigation is needed before to any conclusion is made for the newly discovered diatom indicators.

5 Conclusion

In this paper, we applied machine learning methodology; in particular predictive clustering rules algorithm with two proposed dissimilarity metrics to build diatom models from diatoms abundance based several WQ and TSI category classes. We have made several experiments from the diatoms community that has different settings and different environmental physical-chemical and biological preference, using the proposed Euclid and Sorensen metric.

According to the experimental evaluation the proposed metrics have achieved better descriptive and predictive classification accuracy. Comparation with the previously used metric showed that in both cases this metric is superior as it was shown with the ranking system. The obtained rules based on this evaluation procedure later were discussed and verified with the known ecological references in the literature. Since the geographical location of the diatoms is not the limiting factor in the distribution of diatom species and the composition of communities; rather, the specific environmental variables prevailing at a particular location [18]. Several of the models have positively verified the obtained knowledge, and some of the even have added several new rules that are needed to be experimentally proven before are taken as indicating diatom preferences. An expert system aimed for the decision makers with advance rule induction algorithms could be based on this paper work.

We do strongly believe that these models will help explaining the very complex environmental patterns of influence within every water ecosystem. With this conclusion on mind, we plan to investigate more dissimilarity metrics modification in future, even to implement and modify the weighted algorithm, and adopt for regression problems and multi-target algorithm procedure for rule induction.

References

1. Quinlan, J.: C4.5: Programs for Machine Learning. Morgan Kaufmann, San Mateo (1993)
2. Flach, P., Lavrač, N.: Rule induction. In: Intelligent Data Analysis, 2nd edn., pp. 229–267. Springer, Berlin (2003)

3. Michalski, R.: On the quasi-minimal solution of the general covering problem. In: Proceedings of the Fifth International Symposium on Information Processing (FCIP 1969), Switching Circuits, Bled, Yugoslavia, vol. A3, pp. 125–128 (1969)
4. Michalski, R.S.: An Artificial Intelligence Approach. In: Understanding the Nature of Learning, vol. II, pp. 3–26. Morgan Kaufmann (1986)
5. Niblett, P., Clark, T.: The CN2 induction algorithm. Machine Learning 3(4), 261–283 (1989)
6. Boswell, P.: Rule induction with CN2: Some recent improvements. In: Proceedings of the Fifth European Working Session on Learning, pp. 151–163 (1991)
7. Blockeel, H.: Top-down Induction of First Order Logical Decision Trees. PhD thesis. Katholieke Universiteit Leuven, Department of Computer Science, Leuven, Belgium (1998)
8. Blockeel, H., De Raedt, L., Ramon, J.: Top-down induction of clustering trees. In: Proceedings of the Fifteenth International Conference on Machine Learning (ICML 1998), San Francisco, CA, USA, pp. 55–63 (1998)
9. Zenko, B.: Learning predictive clustering rules. University of Ljubljana, Faculty of Computer and Information Science. Ljubljana, Slovenia (2007)
10. Reid, M., Tibby, J., Penny, D., Gell, P.: The use of diatoms to assess past and present water quality, pp. 57–64 (1995)
11. Kononenko, I., Bratko, I.: Information-based evaluation criterion for classifier's performance. Machine Learning 6(1), 67–80 (1991)
12. Levkov, Z., Krstic, S.: Diatoms of Lakes Prespa and Ohrid (Macedonia). Iconographia Diatomologica, vol. 16, p. 603 (2006)
13. Krstić, S.: Description of sampling sites (2005)
14. Krammer, K., Lange-Bertalot, H.: Bacillariophyceae. 1. Teil: Naviculaceae. [Bacillariophyceae. 1. Part: Naviculaceae]. In: Ettl, H., Gerloff, J., Heynig, H., Mollenhauer, D. (eds.) Süsswasser Flora von Mitteleuropa, Band 2/1, p. 876. Gustav Fischer Verlag, Stuttgart (1986)
15. Van Der Werff, A., Huls, H.: Diatomeanflora van Nederland [Diatom flora in Netherlands]. Abcoude - De Hoef (1957, 1974)
16. García, S., Molina, D., Lozano, M., Herrera, F.: A study on the use of nonparametric tests for analyzing the evolutionary algorithms behaviour: A case study on the CEC'2005 special session on real parameter optimization. Journal of Heuristics 15, 617–644 (2009)
17. Van Dam, H., Mertens, A.: A coded checklist and ecological indicator values of freshwater diatoms from the Netherlands. Netherlands Journal of Aquatic Ecology 28(1), 117–133 (1994)
18. Gold, C.: Field transfer of periphytic diatom communities to assess shortterm structural effects of metals (Cd Zn) in rivers. Water Research 36, 3654–3664 (2002)

Cryptographic Properties of Parastrophic Quasigroup Transformation

Vesna Dimitrova[1], Verica Bakeva[1],
Aleksandra Popovska-Mitrovikj[1], and Aleksandar Krapež[2,⋆]

[1] Faculty of Computer Science and Engineering,
Ss. Cyril and Methodius University, Skopje, Macedonia
[2] Mathematical Institute of the Serbian Academy of Sciences and Arts,
Belgrade, Serbia
{vesna.dimitrova,verica.bakeva,
aleksandra.popovska.mitrovikj}@finki.ukim.mk,
sasa@mi.sanu.ac.rs

Abstract. We consider cryptographic properties of parastrophic quasigroup transformation defined elsewhere. Using this transformation we classify the quasigroups of order 4 into three classes: 1) parastrophic fractal; 2) fractal and parastrophic non-fractal; and 3) non-fractal. We investigate the algebraic properties of above classes and present a relationship between fractal and algebraic properties of quasigroups of order 4. We also find a number of different parastrophes of each quasigroup of order 4 and use it to divide the set of all quasigroups of order 4 into four classes. Using these classifications the number of quasigroups of order 4 which are suitable for designing of cryptographic primitives is increased compared to the case where parastrophes are not used.

Keywords: quasigroup, parastrophic quasigroup transformations, cryptographic properties, experimental mathematics.

1 Introduction

Quasigroups and quasigroup transformations are very useful for construction of cryptographic primitives, error detecting and error correcting codes. The reasons for that are: the structure of quasigroups, their large number, the properties of quasigroup transformations and others. The quasigroup string transformations and their properties were considered in several papers.

A quasigroup $(Q, *)$ is a groupoid (i.e. algebra with one binary operation $*$ on the finite set Q) satisfying the property:

$$(\forall u, v \in Q)(\exists! x, y \in Q) \quad (x * u = v \ \& \ u * y = v) \tag{1}$$

In fact, (1) says that a groupoid $(Q, *)$ is a quasigroup if and only if the equations $x * u = v$ and $u * y = v$ have unique solutions x and y for each given $u, v \in Q$. It

⋆ A.Krapež is supported by the Ministry of Education, Science and Technological Development of Serbia through projects ON 174008 and ON 174026.

S. Markovski and M. Gusev (Eds.): *ICT Innovations 2012*, AISC 207, pp. 235–243.
DOI: 10.1007/978-3-642-37169-1_23

Table 1. Parastrophes of quasigroup operations $*$

Parastrophes operation
$x\backslash y = z \iff x * z = y$
$x/y = z \iff z * y = x$
$x \cdot y = z \iff y * x = z$
$x/\!/y = z \iff y/x = z \iff z * x = y$
$x\backslash\!\backslash y = z \iff y\backslash x = z \iff y * z = x$

has been noted that every quasigroup $(Q, *)$ has a set of five quasigroups, called *parastrophes*, denoted by $/, \backslash, \cdot, /\!/, \backslash\!\backslash$ which are defined in Table 1.

In this paper we use the following notation for parastrophic operations:

$$f_1(x,y) = x * y,\ f_2(x,y) = x\backslash y,\ f_3(x,y) = x/y,$$
$$f_4(x,y) = x \cdot y,\ f_5(x,y) = x/\!/y,\ f_6(x,y) = x\backslash\!\backslash y.$$

Let $A = \{1,\ldots,s\}$ $(s \geq 2)$ be an alphabet and denote by $A^+ = \{x_1 \ldots x_k |\ x_i \in A,\ k \geq 1\}$ the set of all nonempty finite strings over A.

Note that $A^+ = \bigcup_{k\geq 1} A^k$, where $A^k = \{x_1 \ldots x_k |\ x_i \in A\}$. Assuming that (A, f_i) is a given quasigroup, for a fixed letter $l \in A$ (called leader) we define transformation $E = E_{f_i,l} : A^+ \to A^+$ by

$$E_{f_i,l}(x_1 \ldots x_k) = y_1 \ldots y_k \Leftrightarrow \begin{cases} y_1 = f_i(l, x_1), \\ y_j = f_i(y_{j-1}, x_j), \quad j = 2,\ldots,k. \end{cases} \tag{2}$$

In section 2 we give the new parastrophic quasigroup transformation and we analyze the application of that transformation in cryptography. Using the new transformation and the number of different parastrophes we make several classifications of quasigroups of order 4 presented in section 3. In addition, in section 4 we consider some algebraic properties of parastrophic fractal quasigroups and propose a mathematical model for parastrophic fractality.

2 Parastrophic Transformation and Its Cryptographic Properties

Using quasigroup parastrophes, Krapež gave in [4] an idea for quasigroup string transformation which can be applied in cryptography. A modified quasigroup transformation we defined in [2]. Here we describe that quasigroup transformation called parastrophic quasigroup transformation and further on, we consider its cryptographic properties.

Let p be a positive integer and $x_1x_2 \ldots x_n$ be an input message. Using previous transformation E, we define a parastrophic transformation $PE = PE_{l,p} : A^+ \to A^+$ as follows.

At first, let $d_1 = p$, $q_1 = d_1$, $s_1 = (d_1 \bmod 6) + 1$ and $A_1 = x_1x_2 \dots x_{q_1}$. Applying the transformation $E_{f_{s_1},l}$ on the block A_1, we obtain the encrypted block

$$B_1 = y_1y_2 \dots y_{q_1-2}y_{q_1-1}y_{q_1} = E_{f_{s_1},l}(x_1x_2 \dots x_{q_1}).$$

Further on, using last two symbols in B_1 we calculate the number $d_2 = 4y_{q_1-1} + y_{q_1}$ which determines the length of the next block. Let $q_2 = q_1 + d_2$, $s_2 = (d_2 \bmod 6) + 1$ and $A_2 = x_{q_1+1} \dots x_{q_2-1}x_{q_2}$. After applying $E_{f_{s_2},y_{q_1}}$, the encrypted block B_2 is

$$\begin{aligned} B_2 &= y_{q_1+1} \dots y_{q_2-2}y_{q_2-1}y_{q_2} = \\ &= E_{f_{s_2},y_{q_1}}(x_{q_1+1} \dots x_{q_2-2}x_{q_2-1}x_{q_2}). \end{aligned}$$

In general case, for given i, let the encrypted blocks $B_1, \dots, B_{i-1}$ be obtained and d_i be calculated using the last two symbols in B_{i-1} as previous. Let $q_i = q_{i-1}+d_i$, $s_i = (d_i \bmod 6) + 1$ and $A_i = x_{q_{i-1}+1} \dots x_{q_i-1}x_{q_i}$. We apply the transformation $E_{f_{s_i},y_{q_{i-1}}}$ on the block A_i and obtain the encrypted block

$$B_i = E_{f_{s_i},y_{q_{i-1}}}(x_{q_{i-1}+1} \dots x_{q_i}).$$

Now, the parastrophic transformation is defined as

$$PE_{l,p}(x_1x_2 \dots x_n) = B_1||B_2|| \dots ||B_r. \tag{3}$$

Note that the length of the last block A_r may be shorter than d_r (depends on the number of letters in input message). The transformation PE is schematically presented in Figure 1.

For arbitrary quasigroup on a set A and for given $l_1, \dots l_n$ and $p_1, \dots, p_n$, we define mappings PE_1, PE_2, $\dots$, PE_n as in (3) such that PE_i is corresponding to p_i and l_i. Using them, we define the transformation $PE^{(n)}$ as follows:

$$PE^{(n)} = PE^{(n)}_{(l_n,p_n),\dots,(l_1,p_1)} = PE_n \circ PE_{n-1} \circ \dots \circ PE_1,$$

where $\circ$ is the usual composition of mappings.

We made experiments with PE-transformation using different ways to determine the length of the next block and the quasigroup operation, in each iteration. Analyzing the output, we had the best results if we use the last two symbols to compute the length of the next block and corresponding quasigroup operation. Namely, if we take one symbol then we can obtain only 4 values to choose the quasigroup operations, but we have 6 parastrophes. On the other hand, when we take more then 2 symbols to compute the length of the next block and corresponding quasigroup operation, we conclude that the parastrophes are not changing very often. Therefore, in these cases we obtain worse results in terms of fractality, i.e., we have not increased the number of quasigroups suitable for cryptography.

An important property of one transformation for application in cryptography is the uniform distribution of the substrings in the output message. This property

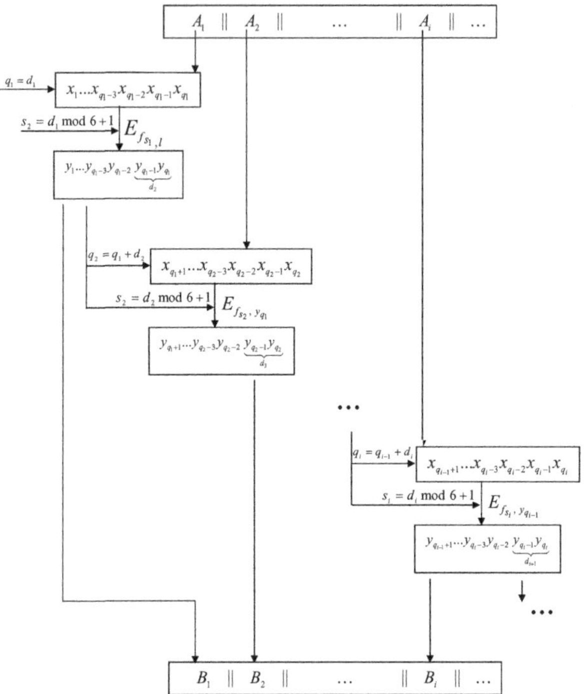

Fig. 1. Parastrophic transformation PE

is useful in proving the resistance to statistical attack. Therefore, we investigate experimentally the uniformity of the output message obtained after using PE-transformation and give the following hypothesis.

Hypothesis 1. *Let $\alpha \in A^+$ be an arbitrary string and $\beta = PE^{(n)}(\alpha)$. Then the m-tuples in β are uniformly distributed for $m \leq n$.*

The theoretical proof of this result is in progress.

3 Classifications of Quasigroups of Order 4 Useful in Cryptography

Using image pattern, Dimitrova and Markovski give a classification of quasigroups of order 4 in [3] as fractal and non-fractal. This classification is made in the following way. The authors start with a periodical sequence 123412341... of length 100 and apply 100 times the E-transformation given in (2) with given leaders. They present the transformed sequences visually using different color for each symbol 1,2,3,4. In this way, they obtain an image pattern for each quasigroup and then analyze the structure of this patterns. If the pattern has a fractal structure, the related quasigroup is called fractal. In the opposite case, the quasigroup is called non-fractal. The number of fractal quasigroups of order 4 is 192 and the number of non-fractal quasigroups is 384. Fractal quasigroups are not good for designing cryptographic primitives since they produce regular structures.

In order to increase the number of quasigroups suitable for application in cryptography, we give new classification of quasigroups of order 4:

- the classification by number of different parastrophes;
- the classification by PE-transformation.

3.1 Classification by Number of Different Parastrophes

We consider all 576 quasigroups of order 4 and for each quasigroup we find the set of all parastrophes. The cardinality of each of these sets is less than or equal to 6, i.e., not all parastrophes of a quasigroup are different.

Using the number of different parastrophes of each quasigroup we divide the set of all quasigroups of order 4 into 4 classes. The number of quasigroups in each of these classes is given in Table 2.

Table 2. Cardinality of classes by number of different parastrophes

No. parastrophes	No. quasigroups
1	16
2	2
3	240
6	318
Total	576

In Table 3 we give a sub-classification of the previous one. Namely, according to number of different parastrophes we classify separately, the fractal and non-fractal quasigroups of order 4.

Table 3. Cardinality of subclasses by number of different parastrophes

No. parastrophes	No. fractal quasigroups	No. non-fractal quasigroups
1	16	0
2	2	0
3	96	144
6	78	240
Total	192	384

From Table 3, we can see that the class of fractal quasigroups of order 4 is divided in 4 subclasses, and the class of non-fractal quasigroups is divided in just 2 subclasses. Comparing Table 2 and Table 3 we can conclude that all quasigroups with 1 or 2 parastrophes are fractal. Quasigroups with 3 and 6 parastrophes can be fractal or non-fractal.

The following proposition is proved by exhaustive verification.

Proposition 1. *Parastrophes of fractal quasigroups of order 4 are fractal as well.*

Consequently, all non-fractal quasigroups of order 4 have non-fractal parastrophes.

3.2 Classification Using PE-Transformation

Using image pattern as in [3], we make here a similar classification using new PE-transformation instead of E-transformation. We apply the new transformation $PE^{(n)}$ to the sequence 123412341234... as before and consider the fractal structure of the obtained image. Depending on that structure, we introduce new types of fractal quasigroups.

Definition 1. *Quasigroups with fractal structure obtained after applying of PE-transformation are called* ***parastrophic fractal quasigroups****. In the opposite case, the quasigroup is called* ***parastrophic non-fractal quasigroups****.*

We made experiments with image pattern for all 576 quasigroups of order 4 and found that 88 are parastrophic fractal and 488 are parastrophic non-fractal. We conclude that all parastrophic fractal quasigroups are in the class of fractal quasigroups, but not all fractal quasigroups are parastrophic fractal. This means that the class of all 192 fractal quasigroups is divided in 2 subclasses: 1) parastrophic fractal (88 quasigroups) and 2) fractal, but parastrophic non-fractal (104 quasigroups), called ***fractal parastrophic non-fractal quasigroups***.

In Figure 2, we give the quasigroup with lexicographic number 40 which is fractal (a), but is not parastrophic fractal (b).

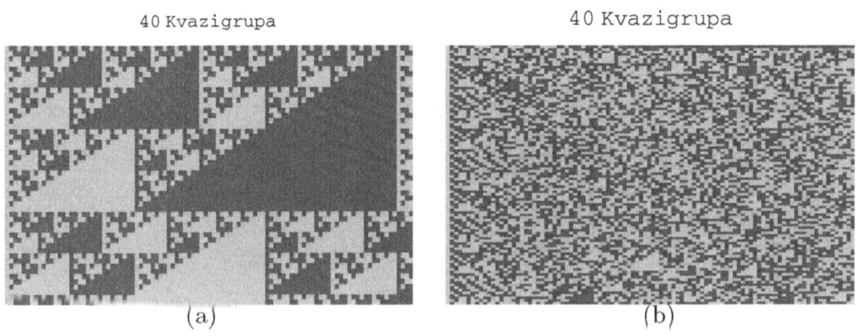

(a) (b)

Fig. 2. Fractal, but parastrophic non-fractal quasigroup

On the other hand, the class of all 384 non-fractal quasigroups is completely contained in the class of parastrophic non-fractal quasigroups.

According to the above, we make a new classification of quasigroups of order 4:

1) class of parastrophic fractal quasigroups (88 quasigroups);
2) class of fractal parastrophic non-fractal quasigroups (104 quasigroups);
3) class of non-fractal quasigroups (384 quasigroups).

With this classification the number of quasigroups useful in cryptography is increased. Namely, for cryptographic purposes we can use not only quasigroups from class 3, but the quasigroups from class 2, too.

Further on, according to the number of different parastrophes, we find the cardinality of subclasses of parastrophic fractal and fractal parastrophic non-fractal quasigroups, separately. The results are given in Table 4.

Table 4. Cardinality of subclass of fractal quasigroups by number of different parastrophes

No. parastrophes	No. parastrophic fractal	No. fractal parastrophic non-fractal
1	16	0
2	0	2
3	72	24
6	0	78
Total	88	104

Comparing Table 3 and Table 4 we can conclude that all fractal quasigroups with 1 parastrophe are parastrophic fractal, all fractal quasigroups with 2 and 6 different parastrophes are fractal parastrophic non-fractal quasigroups. Only the subclass of fractal quasigroups with 3 different parastrophes contains both parastrophic fractal and fractal parastrophic non-fractal quasigroups.

4 Algebraic Properties of Parastrophic Fractal Quasigroups

The authors of [6] gave there a mathematical model of fractality of quasigroups using some identities. Our goal is to find similar model, but for parastrophic fractality of quasigroups. For this purpose we investigate the algebraic properties of parastrophic fractal quasigroups of order 4. In order to find suitable identities to separate parastrophic fractal quasigroups, we investigated many identities, especially symmetric ones. Essential for our model are the following:

- commutativity ($ab = ba$),
- skew symmetry ($(ab)(ba) = const$),
- left loops ($ex = x$),

- right loops ($xe = x$),
- right symmetry ($(ab)b = a$),
- left symmetry ($b(ba) = a$),
- total symmetry (commutativity and left symmetry).

Using exhaustive verification we find that each parastrophic fractal quasigroup of order 4 belongs to the set of quasigroups I that satisfies the identity $x(x(x(xy))) = y$ and belongs to one of the following sets of quasigroups:

- Loops (L)
- Totally symmetric quasigroups (TS)
- Left Loops (LL) and Right symmetric quasigroups (RS)
- Right Loops (RL) and Left symmetric quasigroups (LS)
- Left Loops (LL) and Skew symmetric quasigroups (SS)
- Right Loops (RL) and Skew symmetric quasigroups (SS)
- Commutative quasigroups (C) and Skew symmetric quasigroups (SS).

Using the above notation, we give the following proposition:

Proposition 2. *The class of all parastrophic fractal quasigroups PFQ of order 4 can be presented as:*

$$PFQ = I \cap [L \cup TS \cup (LL \cap RS) \cup (RL \cap LS) \cup (SS \cap (LL \cup RL \cup C))].$$

In this way, we have the mathematical model for parastrophic fractality and even without using image pattern we can check if a given quasigroup is parastrophic fractal.

5 Conclusion

We consider here the parastrophic quasigroup transformation defined in [2] and using this transformation we define new types of quasigroups depending on parastrophic fractality. We give two new classifications of quasigroups of order 4: 1) classification by the number of different parastrophes; and 2) classification by PE-transformation. In this way, we increase the number of quasigroups of order 4 which are suitable to design cryptographic primitives. Also, we investigate the algebraic properties of parastrophic fractal quasigroups and propose a mathematical model for parastrophic fractality.

We give several propositions suitable for application in cryptography. These propositions are experimentally proved and for some of them the theoretical proof is in progress. At the end, if we summarize the obtained results we can conclude that:

- parastrophic fractal quasigroups should not be used for cryptographic primitives, since they have fractal structure, properties of symmetry and shape;
- we have mathematical model for separating the parastrophic fractal quasigroups;
- the parastrophic transformation is more suitable to design cryptographic primitives, since it increases the number of quasigroups useful in cryptography.

References

1. Bakeva, V., Dimitrova, V.: Some Probabilistic Properties of Quasigroup Processed Strings useful in Cryptanalysis. In: Gusev, M., Mitrevski, P. (eds.) ICT Innovations 2010. CCIS, vol. 83, pp. 61–70. Springer, Heidelberg (2011)
2. Bakeva, V., Dimitrova, V., Popovska-Mitrovikj, A.: Parastrophic Quasigrouop String Processing. In: Proc. of the 8th Conference on Informatics and Information Technology with International Participants, Macedonia, pp. 19–21 (2011)
3. Dimitrova, V., Markovski, S.: Classification of quasigroups by image patterns. In: Proc. of the Fifth International Conference for Informatics and Information Technology, Macedonia, pp. 152–160 (2007)
4. Krapež, A.: An Application of Quasigroups in Cryptology. Math. Maced. 8, 47–52 (2010)
5. Markovski, S., Gligoroski, D., Bakeva, V.: Quasigrouop string processing: Part 1. Contributions. Sec. Math. Tech. Sci., MANU XX(1-2), 13–28 (1999)
6. Markovski, S., Dimitrova, V., Samardziska, S.: Identities sieves for quasigroups. Quasigroups and Related Systems 18(2), 149–164 (2010)
7. Denes, J., Keedwell, A.D.: Latin Squares and their Applications. The English Universities Press Ltd. (1974)

Software Engineering Practices and Principles to Increase Quality of Scientific Applications

Bojana Koteska and Anastas Mishev

Faculty of Computer Science and Engineering,
Rugjer Boshkovikj 16, P.O. Box 393 1000 Skopje, Macedonia
bojana.koteska@gmail.com,anastas.mishev@finki.ukim.mk

Abstract. The goal of this paper is to propose some software engineering practices and principles that could increase the quality of scientific applications. Since standard principles of software engineering cannot fully engage in enhancing the development process of such applications, finding the right principles and their combination that will improve the quality is real challenge for software engineers. In order to provide more realistic representation of problems in the field of scientific high-performance computing, we conducted a survey where developers of scientific applications in the HP-SEE project answered some key questions about testing methods and conventions they used. Analysis of the results of the responses was major indicator of quality deficiencies in the high-performance scientific software development and that helped us to discern the possible improvements that need to be added in the planning, development and particularly in the verification phases of the software life cycle.

Keywords: Scientific Applications Quality, Software Engineering Principles, HP-SEE project.

1 Introduction

Solving complex problems in the field of natural sciences, especially in the last decade, strongly implies the need for inclusion of software engineering as an area that can certainly make a significant contribution in terms of improving performance, organization and quality of scientific applications. Since scientific computing is a multidisciplinary field, the need for computer modeling of scientific processes in order to get quantitative results is inevitable [5]. Thus confirming that the design of software which simulate any scientific phenomena is crucial for science and it could provide great support when it comes to automation of certain parts of the development process such as data analysis, management of workflows [7], testing, etc.

Practices and principles of software engineering impose new views and perceptions when it comes to raising the quality of high-performance scientific applications. Software engineering can be defined as a formal set of procedures and tools ensuring efficient project development [14] that could be used and adapted

S. Markovski and M. Gusev (Eds.): *ICT Innovations 2012*, AISC 207, pp. 245–254.
DOI: 10.1007/978-3-642-37169-1_24

to support the entire life cycle of scientific applications. The complexity of the domain which is the basic difference between the development of commercial and scientific applications [16] shows that the principles of software engineering cannot be fully applied in same way as in the development of commercial applications, so finding the appropriate combination and modification for their adjustment is a real challenge for software engineers.

Simulations of problems in science require maximum commitment and correct design that provide high processing performance and certainly better response time [8]. This shows that it is very important to pay attention to the applications quality and to make improvement efforts in order to create efficient development process and to detect and eliminate errors as soon as possible.

Hence, the main focus of this paper is aimed on identifying and proposing practices of software engineering that can significantly contribute to key improvements in the quality of scientific applications. The goal is, through research of already accepted quality recommendations, as well as practical understanding of the real problems faced by scientists, to define a quality model which will form the basis of what standards and criteria should be included in the quality review of scientific applications. Given the fact that standard principles of software engineering cannot fully apply in the development of scientific applications, the research involves the selection and adaptive combination of practices based on real answers from scientists who are part of ongoing development of this type of applications in the HP-SEE project. The results of detailed research helped to establish criteria and quality requirements whose application would significantly improve the planning and implementation practices for testing the quality of scientific applications.

2 Different Cultures of Software Engineers and Scientists

Seeing the potential problems that may occur as a result of different views of software engineers and scientists on the development process of scientific applications, especially from the vantage point of the quality of the applications, in particular, is the most important thing for this research. The obtained results will have a large impact in proposing recommendations that will remove some of the deficiencies and will solve future problems quickly.

One of the main reasons contributing to poor quality high performance computing applications is the lack of used software engineering formal methods. Since scientists have little or not knowledge of software engineering principles [16] and they learn programming techniques independently or by other scientists and not by software engineers [2], it is very difficult for them to adapt to the practices offered by software engineering.

Some scientists believe that the software development is only the process of coding and that it is not important for improving their scientific career because they think that many people can do it also [16]. They often create the code alone or they develop in small teams without previous formal training [14] [15] or they took the code from a source that is part of their research group [2], so the code refactoring and optimization process that will enable parallelism becomes even harder later. The first version of the code is usually created by scientists [15].

The first problem that software engineers could face when they are part of a scientific application development process is the meager list of requirements created by the scientists. That happens because creating requirements at the beginning is very difficult process due to the requirement changes, changes in models and algorithms changes [14] [15]. Software engineers expect a formal list of requirements, while scientists believe that many of the requirements should be discussed directly with engineers and that there is no need of written form [15]. Scientists do not use a formal approach for writing requirements and they do not create requirements document, so when performing software testing they do not know which requirements are completed and which are not. On the other hand, requirements could be written in high level language by using specific keywords in the scientific domain that software engineer cannot understand because he has no knowledge of the problem nature and science community practices [16][15].

Applications created by the scientists are mostly used in their research group and the most important thing for them is to get scientifically accurate result supported by well-simulated models [2] [15]. This clearly shows that they do not care how the processes of making calculations and optimizations are performed. Scientists are often guided by the thought that if a hypothesis is proven true, there is no longer need for reprogramming that part of the code [3]. The process of optimization becomes a matter of discussion among scientists only when they are unable to get results by using the available resources. If the code optimization is made, they want to use the free space to add new functionalities. They prefer readability of complex mathematical formulas rather than optimization which will be accepted only if necessary. It is very difficult to convince scientists that it is easier to use the existing frameworks, rather than creating new ones. Scientists do not accept changes in technology easy because they believe that making long term code and doing things from scratch are very difficult processes [2].

The processes of code verification and validation are differently accepted by scientists. They primarily give importance to the algorithm quality and much less to the implementation [2]. The verification process of high performance computing software is difficult [9] and it requires software engineers to be familiar with the scope of the problem because that science field is unknown for them [16]. However, providing the confidence that the solution is exactly what scientists in the community are looking to accomplish, is engineer's task [3].

Simulations of problems in the scientific world have no precise solution, but they are approximated with numerical algorithm. The process of validation usually means comparison with experimental results that sometimes are impossible to be obtained, so the problem of verification and validation is always present [14]. It's almost impossible to achieve 100% reliability because software engineers can not fully understand the theory and scientists the code [1]. Some errors could be found within the software code, but errors in the implementation logic could occur also [12]. Scientists validate software according to whether there is some progress in scientific research, so they are not very interested in the validation process until an error that affects scientific results is not found [16]. Therefore the creation of tests for scientific applications is a real challenge for software engineers.

3 Background Work

Software engineering can greatly affect the quality and productivity of scientific software in two ways, through introduction of new engineering practices that have proved as good in large projects by monitoring and publication of those already adapted [2]. The application of software engineering principles in the development process of high performance scientific software, can basically be the same with the application of software engineering in the development process of any application, but it requires only small additional changes in terms of planning, eliciting requirements, testing and development process [3]. Including software engineering practices and recommendations in the development of scientific applications requires making careful choices because their improper combination can decrease the quality of the applications which is contrary to our expectations.

The standard definitions of verification and validation cannot be fully applied in the scientific software development due to the facts that process of defining requirements is difficult and no external users exist. In the scientific world, verification can be defined as a process that ensures that the equations are solved properly without coding errors, i.e. in some way it is connected with mathematical models. Validation ensures that the results obtained from the code are accurate reproduction of the problem and that models that simulate the phenomena are correct also. Validation is confirmed by comparing the results with experimentally obtained data [13].

The difference in testing between scientific and commercial applications is only in the tests design because scientific application tests are more focused on numerical precision and proper use of mathematical libraries and their implementations, while the concepts of regression and automated testing remain the same [3]. Scientific applications testing means mainly testing the simulation models. The hardest task in this process is the definition of test cases, since the results that have to be obtained are usually not known from start. A good practice is to make a collection of test problems in the scientific community, i.e. to create benchmark tests that have already analyzed solutions which can be used as expected results [18] [3]. Good recommendations for generating test cases are to generate test cases based on code specification documents, to take into account the analysis of the limit values, to create test cases based on assumptions about errors and certainly at least 90 % of code statements to be included [4]. Due to frequent requirements and source code changes, regression testing that is performed after each made change can detect possible errors. The tests will help also to detect whether the change in the code caused another change that will affect the accuracy of the calculations ant it will help to find if the mistakes are in the range of tolerance [10]. The following testing methods can be also applied in the process of testing: module testing (testing smaller isolated sections of code), system testing (testing combination of modules), mutation testing (testing mutated code) and static testing (testing the code by checking each line manually) [18]. An additional challenge during the testing of scientific applications is to separate the errors in terms of whether they belong to the software or to the model and mathematical approximation [6].

Checking the accuracy of the results obtained from scientific applications cannot be done in the same way as in commercial software applications, so testing is conducted in a different way, primarily because most of the requirements are non-functional [3]. There is an evidence that shows that even 40% of the reasons for the decline of the software can be eliminated with the static analysis of errors [14]. The process of detecting errors can be enhanced by using predictive models that based on mathematical models and number and types of errors that occurred in previous versions could predict the errors that may occur in the next software version. There are some tools that automate process of writing tests, some are used for displaying the results, automation and tests management [4].

4 The Survey

4.1 HP-SEE

The HP-SEE project (High-Performance Computing Infrastructure for South East Europe's Research Communities) provides a common electronic infrastructure that connects HPC centres in South East Europe and enables new countries from this region to be included in the project also. This project creates opportunities for scientists and research communities in South East Europe to include their projects in any of the three available research communities: computational physics, computational chemistry and live sciences. In addition, it increases the process of collaboration between scientists and thus contributes to better research quality and achievements.

The success of the project is confirmed by 26 running applications in the three scientific communities and involvement of 14 countries from Southeast Europe region. There are 12 applications from 7 countries in the computational physics community, 7 applications from 6 countries in the computational chemistry community and 7 applications from 5 countries in life sciences community. Furthermore, it is motivation for other research centers and scientists to engage their projects and become part of this research infrastructure [17].

4.2 Results

In our quest to become more relevant, we conduct a survey where 12 scientists from the closed circle of people involved in the HP-SEE project gave answers to 29 questions, especially in the area of testing. These questions are written in accordance with the offered literature and research about software testing. It contains multiple choice and one choice questions about testing activities, testing models, ways of conducting the tests, testing tools, formal testing approaches, etc. The question short descriptions and answers from the survey are shown below.

Q1: Importance of testing process: 7 answered *Very important*, 4 *Extremely important*, 1 *Neutral*, 0 *Not important*, 0 *Not at all important*.

Q2: People responsible for testing process: 11 answered *Original developer*, 4 *Special software testing team*. This means that 3 of them chose the both options.

Q3: Defects in existing software: 6 answered *Not sure*, 4 *Yes*, 2 *No*.

Q4: Features that have to be tested specified in advance: 10 answered *Yes*, 2 *No*.

Q5: Features that have not to be tested specified in advance: 10 answered *No*, 2 *Yes*.

Q6: Use of automated testing: 12 answered *No*, 0 *Yes*.

Q7: Types of used testing methods: 11 answered *White box testing methods*, 7 *Black box testing methods*. It means that 6 of them chose the both options.

Q8: Types of used test activities: 7 answered *Test-last Development*, 3 *Not specified*, 2 *Test-first Development*.

Q9: Description of test cases: 6 answered *Freely*, 3 *In a formal text-based language*, 3 *No test cases described*, 0 *With help of standardized forms*.

Q10: Test cases generation technique: 6 answered *No explicit test case generation technique was used*, 5 *Model-based techniques*, 5 *Source code analysis*, 4 *Boundary value analysis*, 1 *Category partitioning*.

Q11: Used testing types: 8 answered *Integration testing*, 8 *Functional testing*, 7 *System testing*, 6 *adHoc testing*, 5 *Unit testing*, 2 *Regression testing*, 1 *Recovery testing*, 0 *None of the Above*.

Q12: Used static analysis techniques with tool support: 6 answered *Integration testing*, 3 *Control flow analyses*, 3 *Data flow analyses*, 3 *Static analyses of models*, 2 *Checking of coding standards*, 1 *Static analysis of texts/documents*, 1 *Other tool-supported static analyses*.

Q13: Used test design techniques based on experience: 5 answered *Other test techniques based on experience*, 4 *Exploratory testing*, 3 *No test techniques based on experience*, 2 *Error guessing*, 1 *Fault attack with defect checklists*.

Q14: Used test design techniques based on structure: 8 answered *No test techniques based on structures*, 2 *Statement coverage*, 2 *Branch coverage*, 2 *Condition coverage*, 1 *Path coverage*, 1 *Other test techniques based on structure*, 0 *Multiple condition coverage*.

Q15: Use of testing tools to perform testing: 8 answered *No*, 4 *Yes*.

Q16: Type of used testing tool in the testing process (if any): 4 answered *Source code testing tool*, 3 *Functional testing tool*.

Q17: Importance of requirement coverage: 4 answered *Very important*, 4 *Neutral*, 3 *Not important*, 1 *Not at all important*, 0 *Extremely important*.

Q18: Tracking the requirements coverage: 8 answered *No*, 4 *Yes*.

Q19: Importance of release version tracking: 4 answered *Very important*, 4 *Not important*, 3 *Neutral*, 1 *Not at all important*, 0 *Extremely important*.

Q20: Release version tracking implementation: 9 answered *No*, 3 *Yes*.

Q21: Importance of testing documents generation: 6 answered *Very important*, 4 *Neutral*, 2 *Not important*, 0 *Not at all important*, 0 *Extremely important*.

Q22: Testing documents generation: 7 answered *No*, 5 *Yes*.

Q23: Created test deliverables: 6 answered *Test plan*, 5 *Test cases*, 5 *Test reports*, 4 *Test scripts*, 3 *No test deliverables*, 0 *Defect/enhancement logs* .

Q24: Linking requirements to the tests that verify them: 6 answered *Yes*, 6 *No*.

Q25: Benefit of creating graphs and reports which outlines the state of system testing: 10 answered *Yes*, 2 *No*.

Q26: Validation of code by comparing simulation outputs with the results of physical experiments: 10 answered *Yes*, 2 *No*.

Q27: Barriers for testing: 8 answered *Time*, 3 *Costs*, 4 *No barriers for testing*, 1 *Available corresponding experiment data*, 0 *No support from the high-level management*.

Q28: Criteria to decide that the testing is completed: 10 answered *All test cases are executed without finding more bugs*, 8 *No bugs are found any more*, 2 *When code coverage analysis are reached*, 1 *Informal*, 1 *Coverage of tested models*, 1 *When results are validated using analytical or other solutions*.

Q29: Found defects: 11 answered *Corrected in accordance with its priority*, 1 *Allocated to a defect class*, 0 *Allocated to a priority*, 0 *Other*.

5 Software Engineering Practices for Creating Higher Quality Scientific Applications

The survey results helped us to determine in detail the shortcomings of software engineering formal practices and methods in the development of scientific applications and to propose specific quality improvement suggestions. Responses from the survey were also used to create a correlation between the questions and to discover potential dependencies that will contribute to improving the perceptions and practices in the development process of scientific applications. Correlations among questions refer primarily to related testing activities applied by development teams, but more specifically to the techniques and models that are used to generate test cases, error detection, types of testing, analysis, etc.

Some of the correlations are: Teams that perform unit testing also perform integration testing; Teams that perform release version tracking perform regression testing; Teams that track the requirements coverage perform functional testing; Teams that do not have testing plan and documentation perform adHoc testing; Teams that do not have any testing documents describe the test cases freely and do not link the requirements to tests that verify them; Teams where only original developer is responsible for testing do not use structure-based test design techniques and testing tools often.

The results showed that the process of testing is very important for scientists but the fact that 50% of them are not sure if their software has bugs indicates that the process of testing has major drawbacks. Given that most of the scientists answered that only software developers are responsible for the testing process, the need for inclusion of software engineers in the process of testing is imminent.

To improve the quality of scientific applications we propose a model that covers the following three aspects in the testing process: using well-defined practices, using formal methods and generating documents, certainly with the possibility of minor changes in order to adapt to the specific requirements of the application.

In terms of the practices for defining requirements first thing that needs to be created is a requirements formal specification. Although some of the requirements may change or occur in any of the later stages of development, all of them should have appropriate description using templates proposed by software engineering. That will provide a better overview of the features to be tested later. Irregularity in the process of requirements defining is confirmed by the survey where scientists are mostly neutral in terms of coverage requirements in the testing process and most of them do not practice that.

Well-specified requirements can significantly facilitate the process of defining test cases by providing greater visibility and coverage of the features that have to be tested. The survey results confirm that the features that have to be tested are written in advance in most applications, but the test cases are not created by linking them to the specified requirements, but test cases are mostly written in free form. Even some of them are not documented and the process of their generation does not use any specific technique. These principles must be changed with some software engineering practices like using formal language to define test cases, generating test cases by using specific techniques based on models, based on analysis of the boundary values or analyzes of the source code, etc.

The process of testing automation and use of tools that allow source code or functionality testing are very important and can greatly improve and accelerate the process of testing. To improve the quality of applications, the current practice of manual testing must be changed. Generally, today there are many modern tools, that support and perform various actions such as control and data flows analysis, static analysis of code and documents, checking coding standards etc.

Testing process must include various methods in order to provide complete assurance that the software is error free. According to the survey, most of the tests are performed on the source code and mostly after the software is developed. Thus, the integration testing is applied to check the functioning of the system after the integration of modules and functional testing which is based on requirements. However, to implement integrations testing, first the testing of each module should be performed (unit testing) and to implement functional testing all requirements should be specified and linked to test cases. Applying the regression testing or testing performed after each change is a good approach due to the frequent changes that occur. Agile software testing could be very useful testing method, especially when incremental and iterative development approaches are used. Testing the entire system and testing after system crashes or hardware error are very important also.

Generating documents across the entire development process of scientific applications will contribute to raising the quality of applications because it will allow better planning and scheduling activities, greater visibility and insight about what has been done and what should be done in the future. The results showed that there are no generated test documents in the development of some scientific applications, which means that the risk of errors is much higher because no detailed inspection of the features and no reports exist. The lack of practice to create documents must be replaced by using templates for test documents generations such as test plans, test cases, test reports, error logs, etc.

6 Conclusion

In this paper we have presented the software engineering principles and practices that can improve the scientific applications quality. The research has shown that many formal and standardized software practices and principles can be adapted to scientific software development process. The paper especially outlines the practices and principles that can be used in the testing process in order to ensure that errors have been removed and that the complex numerical calculations have been performed with great accuracy and precision. The survey we conducted helped us to understand the basic deficiencies in the scientific applications development process and that give us guidelines for better suggestions of software engineering practices that would improve not only the process of testing, but the overall quality of the applications. The proposed practices can be adjusted to the development process of any scientific application. A further expansion of these recommendations can be creation of a unified model that will be able to perform a detailed examination of the characteristics of a scientific application and therefore will be able to classify the problem in a given category and propose more specific practices.

References

1. Andersen, P.B., Prange, F., Serritzlew, S.: Software engineering as a part of scientific practice (April 2011), http://imv.au.dk/~pba/Homepagematerial/publicationfolder/softwareengineering.pdf
2. Basili, V.R., Carver, J., Cruzes, D., Hochstein, L., Hollingsworth, J., Shull, F., Zelkowitz, M.V.: Understanding The High Performance Computing Community: A Software Engineer's Perspective. IEEE Software 25, 29–36 (2008)
3. Baxter, R.: Software Engineering Is Software Engineering. In: Proceedings of the First International Workshop on Software Engineering for High Performance Computing System Applications, pp. 14–18. IEEE Computer Society, Washington, DC (2004)
4. Chin, L.S., Worth, D.J., Greenough, C.: A Survey of Software Testing Tools for Computational Science. Technical Report RAL-TR-2007-010, Software Engineering Group, Computational Science & Engineering Department, Rutherford Appleton Laboratory, Oxfordshire, UK (2007)

5. Eijkhout, V., Chow, E., Geijn, R.: Introduction to High-Performance Scientific Computing. Public Draft (July 2010), http://www.math-cs.gordon.edu/courses/cps371/doc/scicompbook.pdf
6. Hannay, J.E., MacLeod, C., Singer, J., Langtangen, H.P., Pfahl, D., Wilson, G.: How Do Scientists Develop and Use Scientific Software. In: Proceedings of the 2009 ICSE Workshop on Software Engineering for Computational Science and Engineering, pp. 1–8. IEEE Computer Society, Washington, DC (2009)
7. Howison, J., Herbsleb, J.D.: Scientific software production: incentives and collaboration. In: Proceedings of the ACM 2011 Conference on Computer Supported Cooperative Work, pp. 513–522. ACM, New York (2011)
8. Jurgens, D.: Survey on Software Engineering for Scientific Applications- Reusable Software, Grid Computing and Application. Institute of Scientific Computing. Technische Universitat Braunschweig, Germany (2009), http://rzbl04.biblio.etc.tu-bs.de:8080/docportal/servlets/MCRFileNodeServlet/DocPortal_derivate_00006306/Juergens-Survey-Software-Eng-Scientific-Applications.pdf
9. Loh, E., Van De Vanter, M.L., Votta, L.G.: Can Software Engineering Solve the HPCS Problem? In: Proceedings of the Second International Workshop on Software Engineering for High Performance Computing System Applications, pp. 27–31. ACM, New York (2005)
10. Neely, R.: Practical Software Quality Engineering on a Large MultiDisciplinary HPC Development Team. In: Proceedings of the First International Workshop on Software Engineering for High Performance Computing System Applications, pp. 19–23. IEE, Stevenage (2004)
11. Phadke, A.A., Allen, E.B.: Predicting Risky Modules in Open-Source Software for High-Performance Computing. In: Proceedings of the Second International Workshop on Software Engineering for High Performance Computing System Applications, pp. 60–64. ACM, New York (2005)
12. Post, D.E.: The Challenge for Computational Science. In: Proceedings of the First International Workshop on Software Engineering For High Performance Computing System Applications, pp. 8–13. IEEE Computer Society, Washington, DC (2004)
13. Post, D.E., Kendall, R.P.: Software Project Management and Quality Engineering Practices for Complex, Coupled Multiphysics, Massively Parallel Computational Simulations: Lessons Learned From ASCI. International Journal of High Performance Computing Applications 18, 399–416 (2004)
14. Roy, C.J.: Practical Software Engineering Strategies for Scientific Computing. In: Proceedings of the 19th AIAA Computational Fluid Dynamics Conference, pp. 1473–1485. American Institute of Aeronautics and Astronautics, Inc., Reston (2009)
15. Segal, J.: Models of scientific software development. In: First International Workshop on Software Engineering in Computational Science and Engineering, SECSE 2008, Leipzig, Germany (2008)
16. Segal, J.: Scientists and Software Engineers: A Tale of Two Cultures. In: Proceedings of the Psychology of Programming Interest Group, pp. 44–51. University of Lancaster, UK (2008)
17. The official HP-SEE site, http://www.hp-see.eu/
18. Zheng, B.: Documentation Driven Testing of Scientific Computing Software. Master Thesis, McMaster University, Ontario, Canada (2009), http://digitalcommons.mcmaster.ca/cgi/viewcontent.cgi?article=5421&context=opendissertations

Performance Evaluation of Computational Phylogeny Software in Parallel Computing Environment

Luka Filipović[1], Danilo Mrdak[2], and Božo Krstajić[3]

[1] University of Montenegro, Center of Information Systems,
Cetinjska 2, Podgorica, Montenegro
lukaf@ac.me

[2] University of Montenegro, Faculty of Natural Sciences,
Mihaila Lalića 1, Podgorica, Montenegro
danilomrdak@gmail.com

[3] University of Montenegro, Faculty of Electrical Engineering,
Džordža Vašingtona bb, Podgorica, Montenegro
bozok@ac.me

Abstract. Computational phylogeny is a challenging application even for the most powerful supercomputers. One of significant application in this are is Randomized Axelerated Maximum Likelihood (RAxML) which is used for sequential and parallel Maximum Likelihood based inference of large phylogenetic trees. This paper covers scalability testing results on high-performance computers on up to 256 cores, for coarse and fine grained parallelization using MPI, Pthreads and hybrid version and comparison between results of traditional and SSE3 version of RAxML.

Keywords: parallel computing, scalability testing, computational phylogeny, RAxML.

1 Introduction

As a result of fast development of different genetic molecular techniques in last 20 years we have huge data sets related to genome of different organisms. Those datasets in beginning were mainly used in identification of different populations of organisms with high significance but lately, due to rapid growth of such information data sets, DNA structure information are used in evolutionary biology purpose. Thus, DNA rows of nucleic base pars scientists are using in reconstruction of evolution history of different group of organisms. In beginning such phylogenetic trees were constructed on one gene for few species data sets but with rapid increasing of sequence data those phylogenetic analyses become more and more demanding in terms of numbers of species as well as in terms of number of genes in use. With aim to get closer to most realistic reconstruction of evolution processes, scientists use data sets of more than one gene for input data sets (multi-gene approach) in order to perform phylogenetic analysis. No matter if they use huge number of OUT (Operational Taxonomic Unit,

S. Markovski and M. Gusev (Eds.): *ICT Innovations 2012*, AISC 207, pp. 255–264.
DOI: 10.1007/978-3-642-37169-1_25

sometimes more than 1000 of them) or they use 5-10 genes for 20-50 OTUs, such approach demand strong computing resources and software for analysis execution. Moreover, since DNA sequence continuously grow in the number of organisms and in sequence length, there exists an increasing need for efficient programs for such analysis and process of software parallelization is adequate answers for such demands.

One of significant open source application from bioinformatics and DNA analysis area is Random Axelerated Maximum Likelikhood (RAxML) [1], program for sequential and parallel Maximum Likelihood based inference of large phylogenetic trees. RaxML exists in several versions; one is a pure sequential program lacking of any parallel code, the other is a parallel version which supports three parallelization techniques: MPI, Pthreads and hybrid parallelization. We analyzed all of them, compared using execution time, speedup, efficiency and CPU consumption on specified dataset. In our test model we used 123 different DNA sequences of Salmo trutta (Linnaeus, 1758) from Eurasian geographical region, 552 base pairs per DNA sequence. Those DNA data sets are available in gene bank [2]. We downloaded it in FASTA format and prepared them performing multiple DNA alignment in CLUSTAL_X [3].

2 The RAXML Software

The RAXML (Random Axelerated Maximum Likelikhood) [1] program has been developed to perform both sequential (on a single processor) and parallel (on multiple processors) phylogenetic analysis using the maximum likelihood optimality criterion. Historically, RAXML stems from the FASTDNAML program, which in turn stems from the DNAML program Although RAXML's design emphasis is on computationally efficient and biologically accurate analysis of very large data sets, it is also appropriate for and amenable to the analysis of data sets of any size. RAXML can use a variety of different character sets, including nucleotide, amino acid, binary, and multi-state character state data. Versions of the RAXML program are available for the Unix/Linux, Mac, and Windows operating systems [4][5].

The first step of the RAxML search strategy is the generation of a starting tree. This starting tree is constructed by adding the sequences one by one in random order, and identifying their optimal location on the tree under the parsimony optimality criterion [6]. The random order in which sequences are added is likely to generate several different starting trees every time a new analysis is run (especially for data sets with more than a few sequences), which allows better exploration of the tree space. If multiple analyses using different starting trees all converge on the same best tree, then confidence that this is the true best tree increases. The second step of the search strategy involves a method known as lazy subtree rearrangement or LSR [6][7]. Briefly, under LSR, all possible subtrees of a tree are clipped and reinserted at all possible locations as long as the number of branches separating the clipped and insertion points is smaller than N branches. RAXML estimates the appropriate N value for a given data set automatically, but one can also run the program with any fixed value.

The LSR method is first applied on the starting tree, and subsequently multiple times on the currently best tree as the search continues, until no better tree is found.

2.1 The RAXML Parallelization

RAxML has been parallelized in various ways and at various levels of granularity as the capabilities of the code have evolved. They can be divided by parallelization methods as:

- Coarse grained parallelization – using MPI
- Fine grained parallelization – using OpenMP and later Pthreads
- Hybrid version, which combines coarse and fine grained parallelization. [8]

In addition, three experimental versions have been developed:

- Version for Cell Broadband Engine, which include Cell-specific code for the fine-grained parallelization as well as a Cell-specific system-level scheduler for efficiently exploiting the coarse-grain parallelism with MPI.
- Version for BlueGene/L; It was multi-grained, but used MPI at both levels of granularity.
- Version where developers compared fine-grained parallelization done with MPI, Pthreads, or OpenMP. Although each approach could be fastest depending upon the data set and computer, the Pthreads implementation was adopted in production from version 7.0.0 and replaced the earlier OpenMP implementation.

Hybrid parallelization of RAxML was enabled from version 7.2.4. The fine-grained Pthreads parallelization is the same as in recent versions and is over the number of patterns. The coarse-grained MPI parallelization is over the number of separate tree searches, similar to that in Version 7.0.0. However, the new MPI approach is simpler than the former master/worker approach and has minimal MPI communication. Thus a fast and expensive interconnect is not required. The new approach is also more efficient when there is reasonable load balance, which is often the case.

All results presented in this paper were done with RAxML version 7.2.8., which had minor changes from version 7.2.4.

3 Benchmark Infrastructure

For the purpose of testing and result analysis of RAxML we needed HPC cluster with support of MPI, Pthreads and ability of system to run hybrid programs. One of suitable cluster in HP-SEE project [9] is HPCG cluster [10] is located at Institute of Information and Communication Technologies (IICT-BAS) in Sofia, Bulgaria. It has 576 computing cores organized in a blade system (HP Cluster Platform Express 7000 enclosures with 36 blades BL 280c with dual Intel Xeon X5560 @ 2.8Ghz). Parallel programming paradigms supported by HPCG cluster are Message passing, supporting several implementations of MPI, OpenMPI and OpenMP. The nodes (36 nodes) have relatively high amount of RAM (24GB per node). Hybrid approach available through combining the two approaches listed above.

4 RAXML Scalability Testing

4.1 Performance Metrics

Three metrics are commonly used to measure the performance of MPI programs, execution time, speedup and efficiency. Several factors such as the number of processors used, the size of the data being processed and inter-processor communications influence parallel program's performance. [11]

Speedup is defined by the following formula S(p) = T(1)/T(p) where p is the number of cores, T(1) is the execution time of the sequential application and T(p) is the execution time of the parallel algorithm with p processors. Efficiency can be defined as E(p) = S(p) / p .

WC bootstrapping test [12] showed that 100 bootstrap runs is minimum for any analysis, 500-1200 bootstrap runs is required for any serious analysis and minimum 5.000 bootstraps is needed for more than 99.5 % precision in ML searches. We decided to have two bootstrap tests: 1000 as a smaller test and 5000 bootstraps as a larger test on our test dataset.

4.2 Sequential Application Results

RAxML, version 7.2.8, offers two versions: standard and SSE3 vectorized version [13]. We tested parallelized versions of RAxML on smaller and larger number of bootstraps. Serial results for comparison are listed in Table 1.

Table 1. Execution time of serial version

	1000 bootstraps	*5000 bootstraps*
Standard version	18 455 s	86 828 s
SSE3 vectorized version	14 407 s	75 759 s

SSE3-based SIMD version gave better results for sequential version, but we used both version for MPI experiments on a HPCG cluster.

4.3 MPI Scalability Results

Coarse grained parallelization of RAxML is done using MPI. We tested MPI parallelization using described dataset with 8 and 16 cores per node.

Time of execution, speedup and efficiency of standard and SSE3 vectorized version for 1000 and 5000 bootstraps are listed in Table 2 and 3. Application speedup and efficiency are graphically presented in Figures 1-4.

Table 2. Execution time for 1000 bootstraps

	MPI with SSE3		*MPI without SSE3*	
Number of Cores	8 cores / node	16 cores / node	8 cores / node	16 cores / node
8	2129,29		3849,65	
16	1095,03	1545,57	2260,62	2560,12
32	649,33	887,51	1114,89	1342,29
64	437,39	436,79	607,83	740,24
128	262,88	282,25	362,54	608,42
256		233,12		389,55

Table 3. Execution time for 5000 bootstraps

	MPI with SSE3		*MPI without SSE3*	
No of cores	8 cores / nodes	16 cores / node	8 cores / nodes	16 cores / node
8	12477,51		15066,57	
16	6584,08	7559,33	7589,08	11577,62
32	4187,13	3831,41	4080,20	6033,55
64	2403,53	1963,52	2607,92	3180,24
128	1176,38	1098,83	1611,18	1607,92
256		723,42		954,88

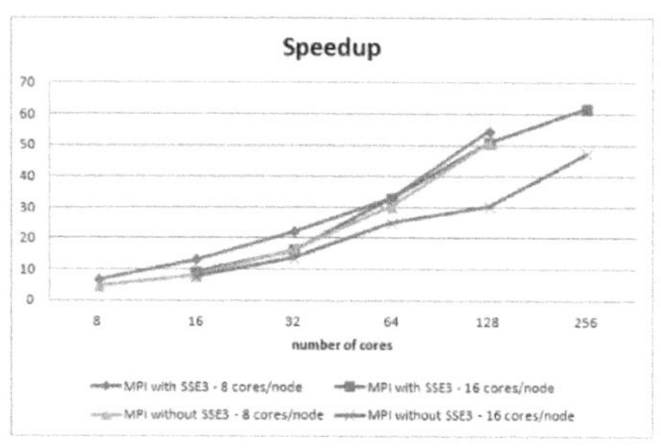

Fig. 1. Speedup for 1000 bootstraps

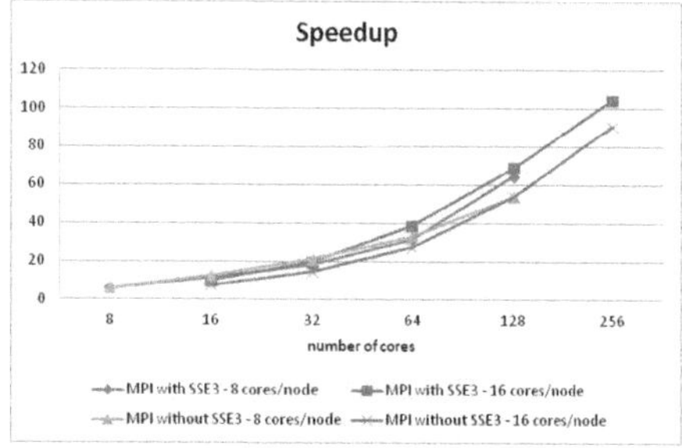

Fig. 2. Speedup for 5000 bootstraps

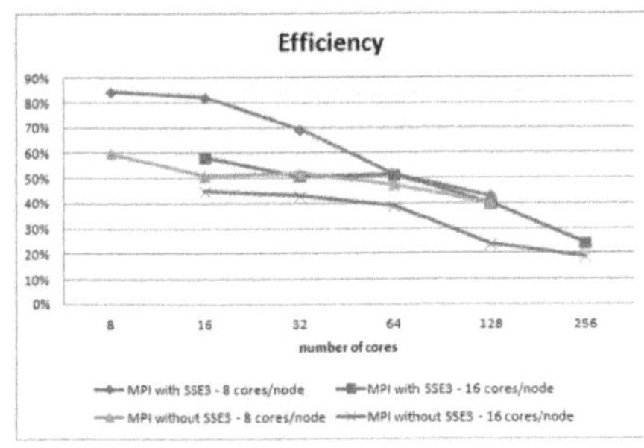

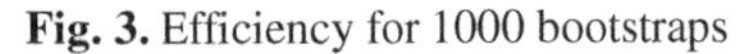

Fig. 3. Efficiency for 1000 bootstraps

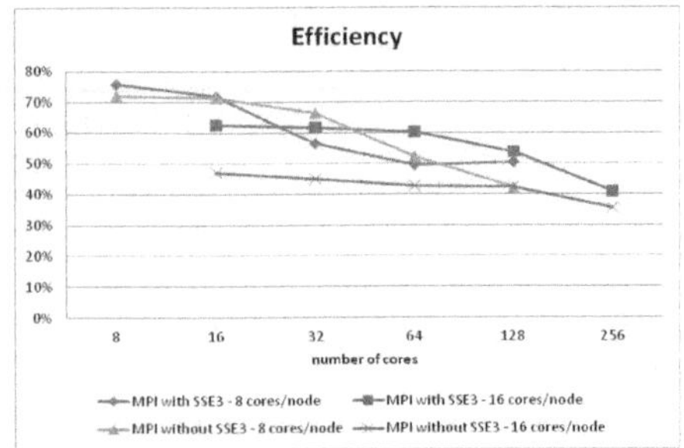

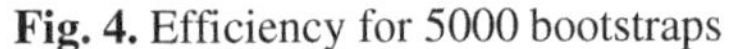

Fig. 4. Efficiency for 5000 bootstraps

After the analysis of execution time and the attached charts we conclude that MPI version with SSE3 instructions is significantly better than standard MPI version and should be used whenever possible.

As a general rule it is accepted that parallelization with efficiency lower than 50 % is not good parallelization. According to that fact standard MPI with 8 cores per node is acceptable for 32 cores for 1000 bootstraps (smaller test) and 64 cores for 5000 bootstraps (bigger test).

Figures 1 and 3 showed good scaling for MPI with SSE3 up to 128 cores, especially for run with 8 cores per node. Execution of application on 256 cores has not shown sufficient good because part tests executed on single core is too small and time for serial part of application is significant if we compare them with time of execution per core.

Figures 2 and 4 also showed good scaling for MPI with SEE3 up to 128 cores, but now version of 16 cores per node showed better results because it had larger amount of data to calculate. Execution of application on 256 cores was below 50%, but it showed 15% better efficiency than test with smaller number of bootstraps.

4.4 Pthreads Scalability Results

Pthreads tests are executing on single node. The optimal number of Pthreads increases with the number of distinct patterns in the columns of the multiple sequence alignment. It's limited with number of cores per node of the server being used.

Pthreads version is compiled with SSE3 support and performed on 1000 bootstraps. All results are compared with serial application. Results are shown in Table 4.

Table 4. Execution time for Pthreads version and comparison with serial version

	Serial application	Number of Threads				
		2	4	8	12	16
Time [s]	14407,02	9491,95	7679,72	5766,47	6685,62	6880,30
Speedup		1,51	1,87	2,49	2,15	2,09
CPU Time [CPUhours]	4,00	5,27	8,53	12,81	22,28	30,57

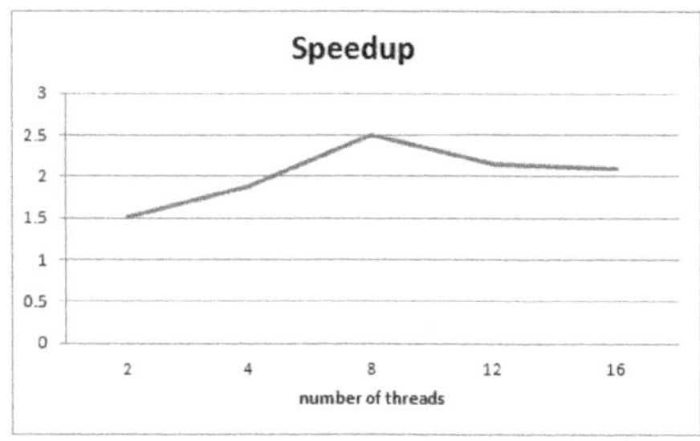

Fig. 5. Speedup for Pthreads parallelization

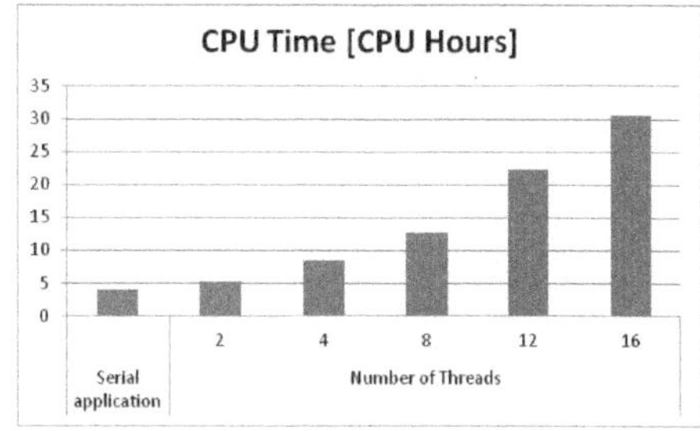

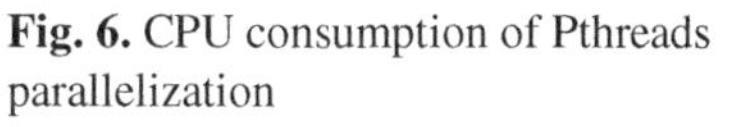

Fig. 6. CPU consumption of Pthreads parallelization

Figure 5 shows scaling up to 2.49 for 8 cores and negative return for more than 8 cores. Speedup rapidly drops down for more than 8 cores because set of 552 base pair is relatively small for splitting into more than 8 parts. CPU Time is shown in Figure 6 which shows us that Pthreads version takes up more CPU time than serial version for our dataset. We didn't executed larger Pthread test because smaller with 1000 bootstraps shown enormously using of resources.

4.5 Hybrid Scalability Results

The hybrid version of RAxML combines coarse and fine grained parallelization. It's especially useful for a comprehensive phylogenetic analysis, i.e., execution of many rapid bootstraps followed by a full maximum likelihood search. Multiple multi-core nodes can be used in a single run to speed up the computation and reduce the turnaround time. The hybrid code also allows more efficient utilization of a given number of processor cores. Moreover, it often returns a better solution than the stand-alone Pthreads code, because additional maximum likelihood searches are conducted in parallel using MPI. [9]

Since Pthreads version gave us bad results for larger number of cores we executed hybrid version with 2 and 4 threads per node and compared results with MPI version, 8 cores/node. Results are shown in Table 5.

Table 5. Execution times[s] for Hybrid parallelization and comparison with MPI and Pthreads version

	Hybrid version		*MPI version*	*Pthreads*
No of cores	2 Threads	4 Threads		*version*
8	2572,98	4646,24	2129,29	5766,47
16	1530,92	2433,89	1095,03	6880,30
32	978,75	1301,88	649,33	
64	518,62	836,98	437,39	

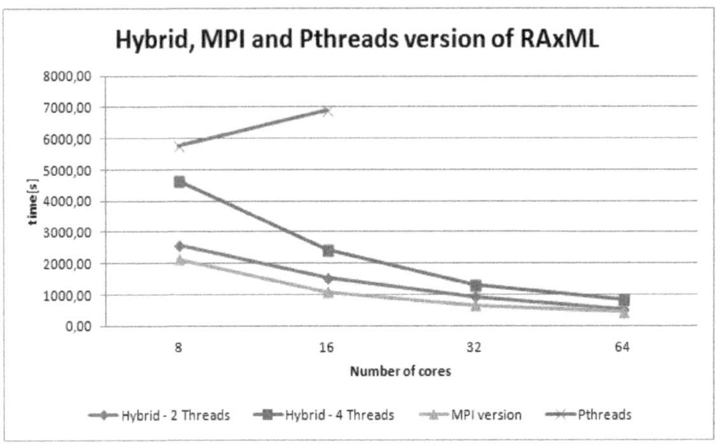

Fig. 7. Execution times of hybrid and MPI version

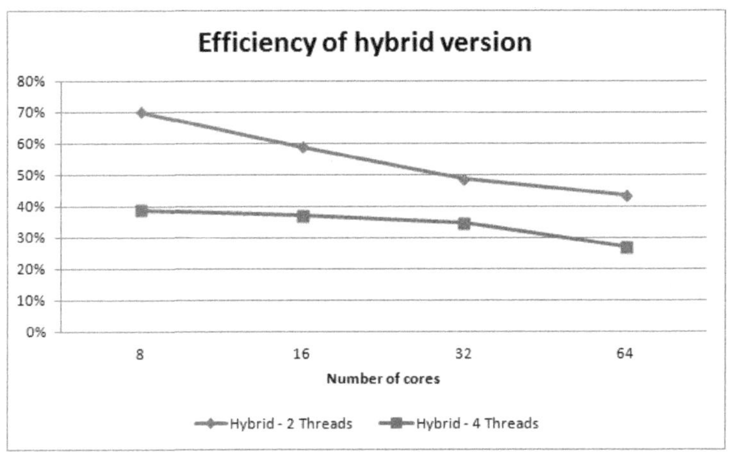

Fig. 8. Efficiency of RAxML hybrid version

Figure 7 shows comparison of execution times between Hybrid version for 2 and 4 threads for fine grained parallelization, MPI version (coarse grained parallelization) and Pthreads parallelization on 8 and 16 cores as a fine grained parallelization. MPI version showed best performance for our dataset. Pthreads version used up to 6.2 times more than MPI version and 4.5 times more than hybrid version for 16 cores.

Figure 8 shows efficiency of hybrid version where 2 thread version achieved significantly better results. Hybrid version with 2 Threads had efficiency over 50% on less than 30 cores, efficiency over 40 % on all tests up to 64 cores, while version with 4 threads had efficiency bellow 40% on all segment.

5 Conclusion

This paper has described a comparative analysis of RAxML parallelization results. MPI with SSE3 support version had best results, especially during execution on less than 128 cores on larger bootstrap test where every core had larger chunk of data to analyze. Pthread version had relatively small speedup up to 8 cores; it takes more CPU time than serial version for our dataset and had a negative return for more than 8 cores. Hybrid version showed worse results than MPI version, but only 20 % in some executions and it can be comparative for testing on some different sizes of input dataset. Hybrid version with 2 threads shown better results than version with larger number of threads because we used 552 base pairs per DNA sequence which can't be divided on larger number of cores without delay. We assume that Hybrid and Pthreads version can give better performance for different dataset with larger number of base pairs per DNA sequence.

Parallelized version of RAxML shown that results of analyses with big datasets can be received in shorter time using larger number of CPUs. From the point of efficiency and energy consumption every version shown that uses more energy than sequential, especially Pthreads. Users must find optimal number of cores, which is up to 128 in MPI version or less than 32 in Hybrid version in our example.

Acknowledgements. This work makes use of results produced by the High-Performance Computing Infrastructure for South East Europe's Research Communities (HP-SEE), a project co-funded by the European Commission (under contract number 261499) through the Seventh Framework Programme. HP-SEE involves and addresses specific needs of a number of new multi-disciplinary international scientific communities (computational physics, computational chemistry, life sciences, etc.) and thus stimulates the use and expansion of the emerging new regional HPC infrastructure and its services. Full information is available at: http://www.hp-see.eu/

References

1. RAxML - Scientific Computing Group, The Exelixis Lab, http://sco.h-its.org/exelixis/software.html
2. The National Center for Biotechnology Information, http://www.ncbi.nlm.nih.gov/
3. Thompson, J.D., Gibson, T.J., Plewniak, F., Jeanmougin, F., Higgins, D.G.: The Clustal_X windows interface: flexible strategies for multiple sequence alignment aided by quality analysis tools. Nucleic Acid Research 25, 24:4876–24:4882 (1997)
4. Rokas, A.: Phylogenetic Analysis of Protein Sequence Data Using the Randomized Axelerated Maximum Likelihood (RAXML) Program. Wiley Online Library, doi:10.1002/0471142727.mb1911s96
5. Stamatakis, A.: RAXML-VI-HPC: Maximum likelihood-based phylogenetic analyses with thousands of taxa andmixed models. Bioinformatics 22, 2688–2690 (2006)
6. Stamatakis, A., Ludwig, T., Meier, H.: RAXML-III: A fast program for maximum likelihood-based inference of large phylogenetic trees. Bioinformatics 21, 456–463 (2005)

7. Schmidt, H.A., von Haeseler, A.: Phylogenetic inference using maximum likelihood methods. In: Lemey, P., Salemi, M., Vandamme, A.M. (eds.) The Phylogenetic Handbook: A Practical Approach to Phylogenetic Analysis and Hypothesis Testing, pp. 181–209. Cambridge University Press, Cambridge (2009)
8. Pfeiffer, W., Stamatakis, A.: Hybrid MPI/Pthreads Parallelization of the RAxML Phylogenetics Code. In: Proc. Ninth IEEE Int'l Workshop High Performance Computational Biology (April 2010)
9. HP-SEE project, `http://www.hp-see.eu/`
10. Resource centre HPCG, HP SEE wiki repository, `http://hpseewiki.ipb.ac.rs/index.php/Resource_centre_HPCG`
11. El-Nashar, A.I.: To parallelize or not to parallelize, speed up issue. International Journal of Distributed and Parallel Systems (IJDPS) 2(2) (March 2011)
12. Pattengale, N.D., Alipour, M., Bininda-Emonds, O.R.P., Moret, B.M.E., Stamatakis, A.: How Many Bootstrap Replicates Are Necessary? In: Batzoglou, S. (ed.) RECOMB 2009. LNCS, vol. 5541, pp. 184–200. Springer, Heidelberg (2009)
13. Intel® 64 and IA-32 Architectures Software Developer's Manual

Cryptographically Suitable Quasigroups via Functional Equations

Aleksandar Krapež*

Mathematical Institute of the Serbian Academy of Sciences and Arts
Knez Mihailova 36, 11001 Belgrade, Serbia
sasa@mi.sanu.ac.rs

Abstract. The use of quasigroups in cryptography is increasingly popular. One method to find quasigroups suitable for cryptographic purposes is to use identity sieves, i.e. to find appropriate identities and check candidate quasigroups against them. We propose the use of functional equation approach to this problem. Namely, every identity can be considered as a functional equation and solutions to these equations as models of given identities. The identity i.e. functional equation can be transformed into related generalized functional equation which is suitable for algebraic treatment. A new method of solution of quadratic and parastrophically uncancelablle equations is given, using trees and dichotomies (a special equivalence relations). General solution is given by closed formulas. The quasigroups obtained can be further filtered using much simpler conditions.

Keywords: Identity sieve, quasigroup, quadratic functional equation, general solution.

1 Introduction

The use of quasigroups in cryptogaphy is increasingly popular. From the beginnings during the second world war, with the results of Schauffler, later published in [11], to the latest introduction of cryptographic systems based on quasigroups (Markovski, Gligoroski, Bakeva [7]), their parastrophes (Krapež [5]) and even n–ary quasigroups (Petrescu [9]), we see not only growing number of papers but also more versatile use of quasigroups. For an overview of this new field see Shcherbacov [12].

Here we consider the algebraic properties of quasigroups represented by equations. Namely, for the quasigroup to be usefull in cryptograhy, it has to satisfy two conflicting conditions – it should be computationally simple but cryptographically strong. The (relative) computational simplicity follows from the satisfaction of some identity (equation) Eq. The source of cryptographical strength is randomness. However, because of finitness of the quasigroup this is impossible

* The author is supported by the Ministry of Education, Science and Technological Development of Serbia through projects ON 174008 and ON 174026.

S. Markovski and M. Gusev (Eds.): *ICT Innovations 2012*, AISC 207, pp. 265–274.
DOI: 10.1007/978-3-642-37169-1_26

to achieve. So we use pseudorandomness instead and control it through the size q of the base set of the quasigroup and the number n ($n \gg q$) of variables in the equation Eq. This leads to identity sieves (Markovski, Dimitrova, Samardzijska [8]) of controlled complexity, i.e. properties represented by complicated identities (equations) with many variables.

We give here a method of solving equations of unbounded length n but somewhat bounded complexity. The strongest restriction is that equations are quadratic, so that every variable appears exactly twice in the equation. Equations are also generalized i.e. every operation appears exactly once in the equation. However, this restriction is negligible as we can index all occurrences of operations, solve resulting generalized equation and then try to find a solution that specializes to the case we started with. There is yet another property of equations – parastrophic uncancelabillity – which is too technical to be discussed in the introduction.

In the next section 'Quasigroups and Equations' we give necessary definitions and basic properties concerning these two notions. Section 'Equations and Dichotomies' present applications to equations of the facts given in the Appendix. Similarly, in the section 'Equations and Trees' we apply elementary properties of binary trees to trees of equations. Combining all these results, general solutions of equations are described. Finally, the way to use these solutions in cryptography is shown.

The Appendix defines a new notion of dichotomy, a very special type of equivalence relation and lists its basic properties important in formulation of solutions of equations.

The paper contains no proofs which will be published elsewhere.

2 Quasigroups and Equations

The standard references for quasigroups are books by V. D. Belousov [1], H. Pflugfelder [10] and O. Chein, H. Pflugfelder and J. D. H. Smith [2]. We present here only those facts which are the most important. For more on the subject of quasigroup functional equations see Krapež, Živković [6], where the general theory and relevant results are given.

We assume the following definition of a quasigroup:

Definition 1. *A triple groupoid* $(S; \cdot, \backslash, /)$ *is a* quasigroup *if it satisfies the following axioms:*

$$x\backslash xy = y \qquad\qquad xy/y = x$$
$$x(x\backslash y) = y \qquad\qquad (x/y)y = x\ .$$

The operation $\cdot$ *is* multiplication, $\backslash$ *is* left division *and* $/$ *is* right division. *We usualy shorten it saying that* $\cdot$ *is a quasigroup. These operations also satisfy* $xy = z$ *iff* $x\backslash z = y$ *iff* $z/y = x$.

A loop *is a quasigroup with an identity (i.e. an element* e *such that for all* $x \in S$ *the relations* $ex = xe = x$ *are true). The last condition can be replaced by* $x\backslash x = y/y$.

Definition 2. *We can define* dual operations *of* $\cdot, \backslash, /$:

$$x * y = yx \qquad x \backslash\backslash y = y \backslash x \qquad x // y = y/x \ .$$

They also give rise to quasigroups. The operations $\cdot, \backslash, /, *, \backslash\backslash, //$ are *parastrophes* of $\cdot$ (and of each other). Additionally, we say that $\cdot, \backslash\backslash$ and $//$ are *even*, while $\backslash, /$ and $*$ are *odd parastrophes* of $\cdot$. We use *notation* $x \circ y$, where $\circ$ is not an operation but just stands for one of $\cdot, *$.

Definition 3. *If* $(S; \cdot, \backslash, /)$ *and* $(T; \times, \nwarrow, \nearrow)$ *are quasigroups and* $f, g, h : S \longrightarrow T$ *bijections such that* $f(xy) = g(x) \times h(y)$ *then we say that* $(S; \cdot, \backslash, /)$ *and* $(T; \times, \nwarrow, \nearrow)$ *are* isotopic *and that* (f, g, h) *is an isotopy.* Dual isotopy *of two quasigroups is an isotopy of one of the quasigroups to the dual of the other. Two quasigroups are* isostrophic *if one of them is isotopic to a parastrophe of the other. Isostrophy is* even (odd) *if a quasigroup is isotopic to an even (odd) parastrophe of the other quasigroup.*

In the definition of isotopy, we do not need the relationships of all three pairs of operations. The formulas $h(x\backslash y) = g(x) \nwarrow f(y)$ and $g(x/y) = f(x) \nearrow h(y)$ follow automatically.

***

We consider here generalized quadratic functional equations on quasigroups. This means that all operation symbols represent quasigroup operations and that if a symbol F is used, we also have symbols for the parastrophes of F.

Definition 4. *Functional equation* $s = t$ *is* generalized *if every operational symbol* F *(including all parastrophes of* F*) appears only once in* $s = t$.

Definition 5. *Functional equation* $s = t$ *is* quadratic *if every variable appears exactly twice in* $s = t$. *Equation is* balanced *if every variable appears exactly once in* s *and once in* t.

A variable x *occurring in a quadratic functional equation* $s = t$ *is* linear *if it appears once in* s *and once in* t*; otherwise it is* quadratic.

Let us write $Eq[x_1, \ldots, x_n, F_1, \ldots, F_m]$ to emphasize that variables (operation symbols) of the equation Eq are exactly $x_1, \ldots, x_n$ $(F_1, \ldots, F_m)$. If we are focused just on variables (operations) we write $Eq[x_1, \ldots, x_n]$ $(Eq[F_1, \ldots, F_m])$.

Definition 6. *Let* $Eq[F_1, \ldots, F_m]$ *be a generalized quadratic functional equation on quasigroups. We write* $F_i \sim F_j$ $(1 \leq i, j \leq m)$ *and say that* F_i *and* F_j *are* necessarily isostrophic *if in every solution* $Q_1, \ldots, Q_m$ *of* $Eq[F_1, \ldots, F_m]$ *the operations* Q_i *and* Q_j *are mutually isostrophic. Moreover,* $F_i \approx F_j$ *if the isostrophy is even and* $F_i \leftrightarrow F_j$ *if it is odd.*

Sokhats'kyi [13,14] gave a definition of *parastrophic uncancellability* of functional equations. But this definition is very complicated. Krapež and Živković proved in [6] that a generalized quadratic quasigroup functional equation Eq is parastrophically uncancellable iff the relation $\sim$ is the full relation on the set of operation symbols of Eq. We shall use this characterization of parastrophic uncancellability instead of the original definition. Let us mention that this class of equations is used in Krapež [4] without giving it an explicit name.

3 Equations and Dichotomies

Let a generalized quadratic equation $Eq[x_1,\dots,x_n]$ be given. We choose an arbitrary but fixed sequence $a_1,\dots,a_n$ of elements of S with the understanding that whenever a variable x_i takes a value in S, it must be a_i. Let I be a subset of the set of indices $I_n = \{1,\dots,n\}$ and replace all variables x_j (j not in I) by appropriate elements a_j. The resulting equation $Eq[\dots,x_i,\dots,a_j,\dots]$ is called the *I–consequence* of Eq. Any I–consequence of Eq is an m–consequence provided I has exactly m elements.

Example 1. Let Eq be the equation of generalized associativity:

$$F_1(F_2(x_1,x_2),x_3) = F_3(x_1,F_4(x_2,x_3)) \ .$$

Then its $\{1,3\}$–consequence is $F_1(F_2(x_1,a_2),x_3) = F_3(x_1,F_4(a_2,x_3))$ and 2–consequences are $\{1,2\}$– , $\{1,3\}$– and $\{2,3\}$–consequences. There is only one 3–consequence of Eq and that is Eq itself. We also have three 1–consequences and one 0–consequence which is the $\varnothing$–consequence of Eq.

Let us define $R_2(x) = F_2(x,a_2)$ and $L_4(x) = F_4(a_2,x)$. Since F_2 and F_4 are quasigroups, both R_2 and L_4 are bijections, so called *translations* of quasigroups F_2,F_4, respectively. The $\{1,3\}$–consequence of Eq can be written now: $F_1(R_2x_1,x_3) = F_3(x_1,L_4x_3)$ and it implies that F_1 and F_3 are mutually isotopic.

In general, for any generalized quadratic equation Eq, the 2–consequences of Eq can be of one of the following 24 forms (where translations are ignored):

$F(x,x) = G(y,y)$	(0)	$E(x,F(G(x,y),y)) = e$	(1)
$F(x,y) = G(x,y)$	(2)	$E(x,F(G(y,x),y)) = e$	(2)
$F(x,y) = G(y,x)$	(1)	$E(x,F(G(y,y),x)) = e$	(0)
$F(x,G(x,y)) = y$	(1)	$E(F(x,x),G(y,y)) = e$	(0)
$F(x,G(y,x)) = y$	(2)	$E(F(x,y),G(x,y)) = e$	(2)
$F(x,G(y,y)) = x$	(0)	$E(F(x,y),G(y,x)) = e$	(1)
$F(G(x,x),y) = y$	(0)	$E(F(x,G(x,y)),y) = e$	(1)
$F(G(x,y),x) = y$	(2)	$E(F(x,G(y,x)),y) = e$	(2)
$F(G(x,y),y) = x$	(1)	$E(F(x,G(y,y)),x) = e$	(0)
$E(x,F(x,G(y,y))) = e$	(0)	$E(F(G(x,x),y),y) = e$	(0)
$E(x,F(y,G(x,y))) = e$	(2)	$E(F(G(x,y),x),y) = e$	(2)
$E(x,F(y,G(y,x))) = e$	(1)	$E(F(G(x,y),y),x) = e$	(1)

We see that in some cases quasigroups F and G are parastrophes or, since we ignored translations, isostrophes of each other. Some of these isostrophies are even and some are odd. Specifically, F and G are mutually even (odd) isostrophic iff they satisfy any of the 2–consequences denoted by (2) ((1)). In other cases, denoted by (0), we cannot draw conclusions about isostrophy of F and G. In no case we can infer isostrophy of E to either F or G.

Theorem 1. *Let Eq be a parastrophically uncancellable generalized quadratic quasigroup equation and let $\equiv$ ($\Leftrightarrow$) be the relation of even (odd) isostrophy obtained directly from 2–consequences of the equation Eq, as above. Then $(\approx, \leftrightarrow) = (\equiv_\infty, \Leftrightarrow_\infty)$ i.e. the pair $(\approx, \leftrightarrow)$ of even/odd isostrophies is the closure of the pair $(\equiv, \Leftrightarrow)$. In particular, the relation $\approx$ is a dichotomy on the set of operation symbols of Eq with a companion $\leftrightarrow$.*

For the *dichotomy* and other related notions see the Appendix. The Theorem 1 shows that from 2–consequences of Eq we get the complete information on relations $\approx$ and $\leftrightarrow$ for a given Eq. We call equation Eq *regular (singular)* iff the pair $(\approx, \leftrightarrow)$ for Eq is regular (singular).

4 Equations and Trees

A tree $(T; \leq)$ is an order on T such that the set $J(x)$ of elements of T smaller than $x \in T$ is well ordered by $\leq$. Since all of our trees will be finite, the above requirement reduces to $J(x)$ being a finite chain $x_1 < \ldots < x_m$ $(x_m < x)$.

The elements of T are *nodes*. The maximal elements of T are *leaves*. We require the existence of the smallest element 0 called the *root* (or *zero*), so our trees are *rooted*. They are also *binary* which means that for every nonleaf element a there are exactly two *children* (*immediate successors*).

We should know which of the children is the *first child* (or *left successor*) and which is the *second child* (or *right successor*). This makes our trees *ordered*.

Let us define a system $(T; \leq, 0, ^+, ^\sharp)$ based on a finite rooted binary ordered tree $(T; \leq)$. Unary operation $^+$ ($^\sharp$) is the first (second) successor function augmented by: $a^+ = a$ $(a^\sharp = a)$ iff a is a leaf.

There is a standard way to represent trees (see Kechris [3]). In our case of finite rooted binary ordered trees the corresponding standard tree is determined by a base set W, a subset of the set $\{0,1\}^*$ of all finite words over alphabet $\{0,1\}$ (of course, the concatenation closure operator $*$ has nothing to do with the dual operation of $\cdot$ from the Section 2).

We assume:

- The empty word $\varnothing \in W$ ($\varnothing$ is the root)
- For every word $w \in W$, all words made of initial segments of w are also in W
- If $w \in W$ is nonleaf, then $w0$ is its first and $w1$ is its second child
- For $w, w' \in W$, $w \sqsubseteq w'$ iff w is an initial segment of w'.

As suggested above, we have:

Theorem 2. *Every finite rooted binary ordered tree is (order) isomorphic to some standard finite rooted binary ordered tree.*

***

Let a generalized quadratic equation Eq (i.e. $s = t$) be given. Assume Eq contains variables $x_1, \ldots, x_n$ ($n > 2$). Since equation is generalized, there are exactly $2(n-1)$ operation symbols $F_1, \ldots, F_{2(n-1)}$ in Eq and we can write Eq as $Eq[x_1, \ldots, x_n, F_1, \ldots, F_{2(n-1)}]$. Let $X = \{x_1, \ldots, x_n\}$ and $\mathrm{Op} = \{F_1, \ldots, F_{2(n-1)}\}$. The equation is also quadratic, so every variable x_i appears twice in Eq. In order to distinguish between two *occurrences* of the same variable we replace the variable x_i either by $x_i^{'}$ (the leftmost occurrence of x_i) or by $x_i^{"}$ (the other occurrence of x_i). In such a way we get the equation (denoted also by Eq and $s = t$) with all symbols of operations and all occurrences of variables distinct. Let $X' = \{x_1^{'}, \ldots, x_n^{'}\}, X" = \{x_1^{"}, \ldots, x_n^{"}\}$ and let Δ be a new name for the equality symbol $=$. The tree of the formula Eq is determined by the following items:

- $M = \{\Delta\} \cup X^{'} \cup X^{"} \cup \mathrm{Op}$
- An order $\leq$ is defined on M:
    - Δ is the smallest element
    - $F_j < v$ iff the variable v occurs in the term $F_j(\ldots)$
    - $F_j < F_k$ iff the term $F_k(\ldots)$ is a proper subterm of the term $F_j(\ldots)$
    - all variables are maximal elements and they are mutually incomparable.
- The operations $^+$ and $^\sharp$ are defined by:
    - If $s = F_j(\ldots)$ and $t = F_k(\ldots)$ then $\Delta^+ = F_j$ and $\Delta^\sharp = F_k$
    - IF $F_p(F_q(\ldots), F_r(\ldots))$ is a subterm of Eq then $F_p^+ = F_q$ and $F_p^\sharp = F_r$
    - IF $F_p(F_q(\ldots), v))$ is a subterm of Eq then $F_p^+ = F_q$ and $F_p^\sharp = v$
    - IF $F_p(u, F_r(\ldots))$ is a subterm of Eq then $F_p^+ = u$ and $F_p^\sharp = F_r$
    - IF $F_p(u, v)$ is a subterm of Eq then $F_p^+ = u$ and $F_p^\sharp = v$
    - $v^+ = v^\sharp = v$ for all variables v.

The system $(M; \leq, \Delta, ^+, ^\sharp)$ is a finite rooted binary ordered tree dually (order) isomorphic to the set of all subexpressions of the equation $s = t$ with the order relation 'is a subexpression of'. It is also isomorphic to some standard finite rooted binary ordered tree $(W; \sqsubseteq, \varnothing, ^+, ^\sharp)$ where $w^+ = w0$ and $w^\sharp = w1$ for nonleaves $w \in W$ and $w^+ = w^\sharp = w$ if $w \in W$ is a leaf.

Definition 7. *If $\varphi : M \longrightarrow W$ is the above isomorphism, we define functions $f : I_{2(n-1)} \longrightarrow W$ and $g, h : I_n \longrightarrow W$ by:*

$$\begin{aligned} f(j) &= \varphi(F_j) && (1 \leq j \leq 2(n-1)) \\ g(i) &= \varphi(x_i^{'}) && (1 \leq i \leq n) \\ h(i) &= \varphi(x_i^{"}) && (1 \leq i \leq n) \ . \end{aligned}$$

We assume that *different* variables $g(i)$ and $h(i)$ (for all $i \in I_n$) represent just one variable since they are images of two occurrences of the same variable x_i. We need this schizophrenic role for variables as we move between two contexts – equations, where we need two occurrences of one variable, and trees, where we have two different nodes related to the same variable.

We also define a division of W into four disjoint subsets corresponding to subsets $\{\Delta\}, X^{'}, X^{"}$ and Op of M.

Definition 8. *Let Root* $= \{\varnothing\}$, $Y' = g(I_n) = \varphi(X')$, $Y" = h(I_n) = \varphi(X")$ *and* $\Omega = f(I_{2(n-1)}) = \varphi(Op)$.

Therefore $W = Root \cup Y' \cup Y" \cup \Omega$ and $Y' \cup Y"$ is a set of leaves of W. It is clear now that functions f, g, h represent isomorphism φ completely and we find them more convinient to use then φ itself. This is because the functions f, g, h connect linear structure of the equation (i.e. $I_n, I_{2(n-1)}$) and its tree structure (i.e. W).

Definition 9. *Let* λ_w, ϱ_w $(w \in \Omega)$ *be two families of permutations. We define*

- *Extensions of families* λ, ϱ *by the root index:* $\lambda_\varnothing = \varrho_\varnothing = \mathrm{Id}$
- *A new family* α_u $(u \in \Omega)$ *where* $\alpha_u = \begin{cases} \lambda_v^{-1}, & u = v0 \\ \varrho_v^{-1}, & u = v1 \end{cases}$
- *Yet another family* π_u $(u \in Y' \cup Y")$ *defined by* $\pi_u = \begin{cases} \lambda_v, & u = v0 \\ \varrho_v, & u = v1. \end{cases}$

Definition 10. *Condition* $(V_\bullet)$ *is defined as:*
For all i $(1 \leq i \leq n)$ $\begin{cases} \pi_{g(i)} x_i = \pi_{h(i)} x_i, & \text{for } x_i \text{ linear variable} \\ \pi_{g(i)} x_i \bullet \pi_{h(i)} x_i = e, & \text{for } x_i \text{ quadratic variable.} \end{cases}$

We are now ready for the final results.

5 Solutions

The equations under consideration were solved in [4]. However, the solution was not given as a closed formula but rather as an algorithm prescribing how to find that solution for any particular equation. Using trees and dichotomies we can augment old results with this missing piece.

Theorem 3. *Any solution of a parastrophically uncancelablle generalized quadratic quasigroup equation Eq (with* n *variables,* $n > 2$*) on a set* $S \neq \varnothing$ *satisfies:*

$$F_j(x, y) = \alpha_w(\lambda_w x \circ \varrho_w y) \qquad (1 \leq j \leq 2(n-1), w = f(j))$$

where:

- *operation* $\cdot$ *is a group on* S *with identity* e
- $\circ = \begin{cases} \cdot, & \text{if } F_j \approx F_1 \\ *, & \text{if } F_j \leftrightarrow F_1 \end{cases}$
- *functions* λ_w, ϱ_w $(w \in \Omega)$ *are arbitrary permutations of* S *such that condition* $(V.)$ *is satisfied.*

Theorem 4. *Let Eq be a parastrophically uncancelablle generalized quadratic quasigroup equation Eq (with* n *variables,* $n > 2$*) and* $+$ *an Abelian group on a set* $S \neq \varnothing$*. A family of operations:*

$$F_j(x, y) = \alpha_w(\lambda_w x + \varrho_w y) \qquad (1 \leq j \leq 2(n-1),\ w = f(j))$$

where functions λ_w, ϱ_w $(w \in \Omega)$ *are arbitrary permutations of* S *such that condition* (V_+) *is satisfied, is a solution of the equation Eq.*

Therefore the Theorem 3 gives necessary and the Theorem 4 sufficient conditions for a solution of Eq. A general solution, which lays on the borderline between them, is determined by the properties of the relation $\approx$ of Eq. Namely, we have:

Theorem 5. *If the equation Eq is regular then the family of operations, given by the formulas and conditions of the Theorem 3, is a solution of Eq.*

Theorem 6. *If the equation Eq is singular then the solution of Eq, given in the Theorem 4, is general.*

***

Assume now that a set of quasigroups F_j $(j \in I_m)$ of order q is used as a subset of cryptographic primitives for some cryptographic system. To ensure that the choice of quasigroups F_j is cryptographically suitable they should be computationaly simple but cryptographically strong. The (relative) computational simplicity follows from their being a solution of an equation Eq (with n variables, $n \gg q$) and the cryptographical strength follows from n being large. Let us transform Eq into the generalized equation $\text{Gen}(Eq)$ by giving a new index k $(k \in I_{2(n-1)} \setminus I_m)$ to every operation with multiple occurrence in Eq. Equation Eq is equivalent to the system:

$$\text{Gen}(Eq)$$

$$F_j = F_k \quad (j \in I_m, k \in I_{2(n-1)} \setminus I_m) \tag{1}$$

where F_k is the new name for F_j. From the Theorem 3 it follows that if $\text{Gen}(Eq)$ is parastrophically uncancelablle quadratic equation then all quasigroups F_j are (dually) isotopic to the same group. More precisely, they can be represented as

$$F_j(x, y) = \alpha_w(\lambda_w x \circ \varrho_w y) \qquad (1 \le j \le m,\ w = f(j))$$

for some permutations $\alpha_w, \lambda_w, \varrho_w$ $(w = f(j), 1 \le j \le m)$. Of course, they also satisfy further conditions which follow from (1) and Definitions 9,10.

References

1. Belousov, V.D.: Foundations of the Theory of Quasigroups and Loops, Nauka, Moscow (1967) (Russian) (MR 36-1569, Zbl 163:01801)
2. Chein, O., Pflugfelder, H.O., Smith, J.D.H.: Quasigroups and Loops: Theory and Applications. Sigma Series in Pure Math., vol. 9. Heldermann Verlag, Berlin (1990) (MR 93g:21033, Zbl 0719.20036)
3. Kechris, A.S.: Classical Descriptive Set Theory. Graduate Texts in Mathematics, vol. 156. Springer, New York (1995)
4. Krapež, A.: Strictly Quadratic Functional Equations on Quasigroups I. Publ. Inst. Math. N. S. 29(43), 125–138 (1981) (Belgrade) (Zbl 0489.39006)

5. Krapež, A.: An Application of Quasigroups in Cryptology. Math. Maced. 8, 47–52 (2010)
6. Krapež, A., Živković, D.: Parastrophically Equivalent Quasigroup Equations. Publ. Inst. Math. N. S 87(101), 39–58 (2010), doi:10.2298/PIM1001039K (Belgrade)
7. Markovski, S., Gligoroski, D., Bakeva, V.: Quasigrouop String Processing: Part 1. Contributions. Sec. Math. Tech. Sci., MANU XX(1-2), 13–28 (1999)
8. Markovski, S., Dimitrova, V., Samardziska, S.: Identities Sieves for Quasigroups. Quasigroups and Related Systems 18(2), 149–164 (2010)
9. Petrescu, A.: n–quasigroups Cryptographic Primitives: Stream Ciphers. Studia Univ. Babeş–Bolyai, Informatica LV (2), 27–34 (2010)
10. Pflugfelder, H.O.: Quasigroups and Loops: Introduction. Sigma Series in Pure Math., vol. 8. Heldermann Verlag, Berlin (1990) (MR 93g:20132, Zbl 719.20036)
11. Schauffler, R.: Die Associvität im Ganzen besonders bei Quasigruppen. Math. Zeitschr. 67(5), 428–435 (1957)
12. Shcherbacov, V.A.: Quasigroup Based Crypto–algorithms (2012) arXiv:1201.3572
13. Sokhats'kyi, F.M.: On the Classification of Functional Equations on Quasigroups. Ukrain. Mat. Zh. 56(9), 1259–1266 (2004) (Ukrainian)
14. Sokhats'kyi, F.M.: On the Classification of Functional Equations on Quasigroups. Ukrainian Math. J. 56(9), 1499–1508 (2004); English translation of [13]

Appendix: Dichotomies

Definition 11. *An equivalence relation δ on a set $A \neq \varnothing$ is a* dichotomy *if it has at most two equivalence classes i.e.*

$$\forall xyz(x\delta y \vee y\delta z \vee z\delta x)$$

A familiar example of a dichotomy is a congruence modulo 2 on the set of all integers.

The equality (full relation) on A will be denoted by $\Delta(\square)$. For a binary relation σ, we denote by σ^c the *complement* relation of σ (i.e. $\sigma^c = \square \setminus \sigma$).

Definition 12. *Let δ be an equivalence on $A \neq \varnothing$. A binary symmetric relation η on A is a* companion *to δ, if:*

$$\forall xy(x\delta y \vee x\eta y)$$

$$\forall xyz(x\delta y \wedge y\eta z \Rightarrow x\eta z)$$

$$\forall xyz(x\eta y \wedge y\eta z \Rightarrow x\delta z) \ .$$

A dichotomy δ and its companion η have the following properties:

Lemma 1. *Let $a_0, a_1, \ldots, a_n$ be a sequence of elements of A such that $a_{i-1}\sigma_i a_i$ $(1 \leq i \leq n)$ where σ_i is either δ or η. Then $a_0\delta a_n$ $(a_0\eta a_n)$ iff there is an even (odd) number of symbols η in the sequence $\sigma_1, \ldots, \sigma_n$.*

Lemma 2. *Let δ be a dichotomy on A with a companion η. If $\delta \cap \eta \neq \varnothing$ then $\delta = \eta = \square$.*

The following theorem is true:

Theorem 7. *An equivalence relation δ on A is a dichotomy iff it has a companion.*

If δ is not the full relation the companion is unique and equal to δ^c. If δ is the full relation it has exactly two possible companions: the empty relation ($\varnothing$) and the full relation ($\Box$).

The Theorem 7 states that a dichotomy δ with a companion η on A is either $\delta = \eta = \Box$ or $\delta = \Box, \eta = \varnothing$ or else $\delta = \sigma, \eta = \sigma^c$, where σ is a dual atom in the lattice of all equivalences on the set A.

Definition 13. *A* dichotomy pair (δ, η) *consisting of a dichotomy δ and its companion η is* regular *if $\eta \neq \Box$, otherwise it is* singular.

Note that (δ, η) is regular iff $\eta = \delta^c$ and singular iff $\delta = \eta = \Box$.

***

Definition 14. *Let δ and η be binary relations on $A \neq \varnothing$. Define sequences (δ_n) and (η_n) $(n \in \mathbb{N})$:*

$$\delta_1 = \Delta \cup \delta \cup \delta^{-1} \qquad\qquad \eta_1 = \eta \cup \eta^{-1}$$

and for $n > 1$:

$$\delta_{n+1} = (\delta_n \circ \delta_1) \cup (\eta_n \circ \eta_1) \qquad\qquad \eta_{n+1} = (\delta_n \circ \eta_1) \cup (\eta_n \circ \delta_1).$$

Let $\delta_\infty = \cup_{n=1}^{\infty} \delta_n$ and $\eta_\infty = \cup_{n=1}^{\infty} \eta_n$. A pair $(\delta_\infty, \eta_\infty)$ of relations $\delta_\infty, \eta_\infty$ is a *closure* of the pair (δ, η). We have:

Lemma 3. *Let $a, b \in A$. Then $a\delta_n b$ $(a\eta_n b)$ iff there are sequences $a_0, \ldots, a_n$ of elements of A and $\sigma_1, \ldots \sigma_n$ of relations $\Delta, \delta, \delta^{-1}, \eta, \eta^{-1}$ such that $a = a_0, b = a_n$ and $a_{i-1}\sigma_i a_i$ for all i $(1 \leq i \leq n)$, with an even (odd) number of relations η, η^{-1} in the sequence $\sigma_1, \ldots \sigma_n$.*

Theorem 8. *Let δ and η be binary relations on $A \neq \varnothing$ such that the equivalence generated by $\delta \cup \eta$ is $\Box$. Then the pair $(\delta_\infty, \eta_\infty)$ is a dichotomy pair.*

Ontology Supported Patent Search Architecture with Natural Language Analysis and Fuzzy Rules

Daniela Boshnakoska[1], Ivan Chorbev[2], and Danco Davcev[1]

[1] University for Information Science and Technology "St. Paul the Apostle", Building at ARM, Partizanska bb, 6000 Ohrid, R.Macedonia
{daniela.boshnakoska,dancho.davchev}@uist.edu.mk
[2] Faculty of Computer Science and Engineering, str. "Rugjer Boshkovikj" 16, P.O. Box 393, 1000 Skopje, Macedonia
ivan.chorbev@finki.ukim.mk

Abstract. We have recently witnessed a rapid growth in scientific information retrieval research related to patents. Retrieving relevant information from and about patents is a non-trivial task and poses many technical challenges. In this paper we present a new approach to patent search that combines semantic knowledge and ontologies used to annotate patents processed with natural language processing tools. The architecture uses fuzzy logic rules to organize the annotated patents and achieve more precise retrieval. Our approach to combine proven techniques in a composite architecture showed improved results compared to pure textual based indexing and retrieval. We also showed that results ranked using semantic annotation are better than results based on simple keyword frequencies.

Keywords: Ontology, semantic web, patent search, patent mining, semantic annotations, Natural Language Processing.

1 Introduction

Patent search and retrieval is a challenging area of research. The use of semantic web, ontologies and natural language processing proves itself interesting for supporting content mining of patents. In this paper, we describe our architecture for generating semantic annotations on patents and using ontologies for advanced search through the annotated patents. We use the textual contents of the patent documents processed by Natural Language Processing (NLP) tools. Combining multiple known techniques we aim to increase recall as well as precision in patent retrieval.

When applying for a new patent, old patents need to be searched to find any possible occurrence of copyright infringement. Also, during legal proceedings patents are scrutinized to find any repetition of similar ideas. Therefore results need to be precise and meaningful. Responses to queries need to be fast.

This paper is organized as follows. The next section presents a short overview of recent work in the area of patent search. In Section 3 the overall architecture of our

S. Markovski and M. Gusev (Eds.): *ICT Innovations 2012*, AISC 207, pp. 275–284.
DOI: 10.1007/978-3-642-37169-1_27

system is presented. Section 4 presents some experimental results and simulation with our search architecture. Section 5 explains plans for future work, and the conclusion is given in Section 6.

2 Overview of Recent Work

The objective of our prototype architecture is to encompass existing proven methods, approaches or present services in a novel composition.

BioPatentMiner [1] facilitates information retrieval from biomedical patents. It relies on the identification of biological terms and relations in patents. PATEXPERT [2] is focusing on patent analysis based on a semantic web approach.

The authors in [3] describe an approach for generating semantic annotations on patents, by relying on the structure and on a semantic representation of patent documents and their contents. Authors in [5] describe a different approach to constructing ontologies especially suited for patent search. They propose an ontology modeling method which combined UML (Unified Modeling Language) with OWL (Web Ontology Language) in order to put forward the improvement of existing modeling methods.

Among the challenges of patent search is the issue of terminological inconsistencies that are used in the documents [8]. The authors focus on the terminological inconsistency issue by exploring domain knowledge through the use of ontology standards.

Recent attempts in advancing patent search see the inclusion of fuzzy logic [6]. The authors describe a model for representing the patent request by a set of concepts related to existing knowledge ontology.

Ontology based semantic retrieval has been used in searching different kinds of data, like for instance literature [4]. Their SemPub system centers on semantic retrieval and user customization. They search for terms in a concept space (a graph of terms occurring in documents linked to each other by the frequency and relationships between them and with which they occur together).

Since ranking the found results is the most important task after retrieving the records, semantic relationships are used in this stage as well [7]. The goal of the work presented in their paper is to use semantic relationships for ranking documents without relying on the existence of any specific structure in a document or links between documents.

3 System Architecture

Our goal was to build a composite system that is based on improved methods that we have adapted for the searching process in the patent domain. The system we created is modular and in the future it can be extended with additional advanced techniques that provide innovative approach. Currently, we have integrated composite document indexing that is not solely based on plain text-based analysis, but also includes semantic extractions as well. Natural language processing analysis is used for parsing the user's search query. The document retrieval phase is enriched with ranking mechanisms based on fuzzy logic. All modules are developed using the Microsoft.NET platform.

3.1 Process Flow

The initial stages of the architecture (or "Structural Clustering") include the data preprocessing phase, the retrieval of experimental patents and their preparation for analysis. Before indexing, each patent document is transformed using shallow NLP techniques, like splitting and tokenizing sentences.

When an inquirer searches with a query through the web interface, the query is processed using Natural Language Processing tools and query chunks are produced. The query chunks are used to search the data clusters. A fuzzy ruler ranks the results and merges them as a complete set. The web interface provides the output linking every patent from the result set to its original document. Furthermore, the user can choose the preferred search mode: text, semantic or both and can switch on or off the fuzzy ranking functionality. The composite system architecture is presented graphically in the Figure 1.

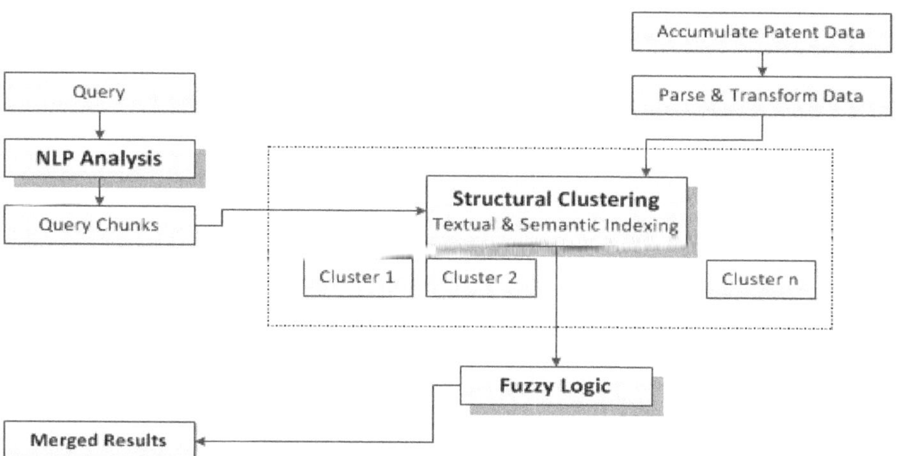

Fig. 1. Main architecture flow

3.2 Data Preprocessing

The data preprocessing phase is done by the "Parse and Transform Data" module. We used data from available online sources gathered by independent search with patent retrieval tools and online patent aimed searches. High portion of the patents available in the world are published and accessible online. This gives the opportunity to test recall by accessing patents focused on a designated topic. As main patent sources we have used The United States Patent and Trademark Office (USPTO) [10], European Patent Organization (EPO) [16] and World Intellectual Property Organization (WIPO) [17]. We used a set of patents (about 2000 patents) available on the EPO, USPTO and WIPO systems mainly on predefined enzymes topics. The preliminary step in our structuring phase was to transform the original data and extract meaningful and structured concepts from every document. Each patent document is processed, using sentence splitter and tokenizer to isolate words in a sentence. The parsed data content is the main input in the next stage in the process of multi-level indexing. Additionally the documents were stripped from excess formatting tags and prepared for clustering.

3.3 Structural Clustering

Patent documents and the systems they are organized in differ from other types of online web documents. They feature specifics unique to them and contain structural portions that are found in every patent. Patent components (inventors, abstract, claims, citations etc.) are not heterogeneous, and can be structured in a finite number of sets. We relied on the uniform structure among different patent sources and we build composite structure sets (clusters), based on the document modules. Each file is analyzed, segmented and textually indexed accordingly to the patent section. This structural preparation results in indexed clusters of titles, abstracts, authors, claims, references and content.

In patents, as in any technical text, some document sections (such as title, abstract or claims) are considered more significant since they contain the essence of the patent. The important parts are short and technically written, using terms found in ontologies, and the gain of using semantic annotation is significant. The less important parts of the patent are more descriptive, lengthy and often written in common language having the technical terms intermingled. In this case the semantic annotation would increase the computational burden, bringing small overall gain in the search quality. We used enzyme based ontologies - GlycO and EnzyO [13] to do semantic annotations that rely on the class trees available in these ontologies. During the file analysis each patent component is matched and indexed according to the classes available in the ontologies. The described method results in another level of indexing based on ontologies. Additionally, the annotated data is embedded as part of the original documents and can be visually displayed as in Figure 2. This process incorporates SharpNLP models to identify classes of basic entities within the text (such as person, location, date, organization), enriched with entities gathered from the ontology classes.

Although patents are mostly related to a predefined domain, searches with more general queries should not be misplaced. To support and back such search, we have included another pure textual indexing based on the entire document tokenized content without any prior segmentation.

All three modes of clustering and indexing (structural, textual and ontology-based) are implemented using the Apache Lucene .NET indexing and search library [14].

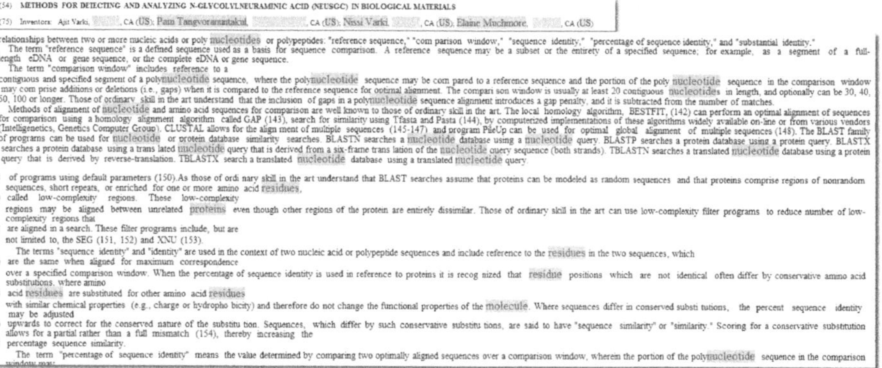

(54) METHODS FOR DETECTING AND ANALYZING N-GLYCOLYLNEURAMINIC ACID (NEU5GC) IN BIOLOGICAL MATERIALS

(75) Inventors: Ajit Varki, , CA (US); Pam Tangvoranuntakul, , CA (US); Nissi Varki, , CA (US); Elaine Muchmore, , CA (US)

relationships between two or more nucleic acids or poly nucleotides or polypeptides: "reference sequence," "com parison window," "sequence identity," "percentage of sequence identity," and "substantial identity."

The term "reference sequence" is a defined sequence used as a basis for sequence comparison. A reference sequence may be a subset or the entirety of a specified sequence; for example, as a segment of a full-length eDNA or gene sequence, or the complete eDNA or gene sequence.

The term "comparison window" includes reference to a contiguous and specified segment of a polynucleotide sequence, where the polynucleotide sequence may be com pared to a reference sequence and the portion of the poly nucleotide sequence in the comparison window may com prise additions or deletions (i.e., gaps) when it is compared to the reference sequence for optimal alignment. The compari son window is usually at least 20 contiguous nucleotides in length, and optionally can be 30, 40, 50, 100 or longer. Those of ordinary skill in the art understand that the inclusion of gaps in a polynucleotide sequence alignment introduces a gap penalty, and it is subtracted from the number of matches.

Methods of alignment of nucleotide and amino acid sequences for comparison are well known to those of ordinary skill in the art. The local homology algorithm, BESTFIT, (142) can perform an optimal alignment of sequences for comparison using a homology alignment algorithm called GAP (143), search for similarity using Tfasta and Pasta (144), by computerized implementations of these algorithms widely available on-line or from various vendors (Intelligenetics, Genetics Computer Group). CLUSTAL allows for the align ment of multiple sequences (145-147) and program PileUp can be used for optimal global alignment of multiple sequences (148). The BLAST family of programs can be used for nucleotide or protein database similarity searches. BLASTN searches a nucleotide database using a nucleotide query. BLASTP searches a protein database using a protein query. BLASTX searches a protein database using a trans lated nucleotide query that is derived from a six-frame trans lation of the nucleotide query sequence (both strands). TBLASTN searches a translated nucleotide database using a protein query that is derived by reverse-translation. TBLASTX search a translated nucleotide database using a translated nucleotide query.

of programs using default parameters (150).As those of ordi nary skill in the art understand that BLAST searches assume that proteins can be modeled as random sequences and that proteins comprise regions of nonrandom sequences, short repeats, or enriched for one or more amino acid residues, called low-complexity regions. These low-complexity regions may be aligned between unrelated proteins even though other regions of the protein are entirely dissimilar. Those of ordinary skill in the art can use low-complexity filter programs to reduce number of low-complexity regions that are aligned in a search. These filter programs include, but are not limited to, the SEG (151, 152) and XNU (153).

The terms "sequence identity" and "identity" are used in the context of two nucleic acid or polypeptide sequences and include reference to the residues in the two sequences, which are the same when aligned for maximum correspondence over a specified comparison window. When the percentage of sequence identity is used in reference to proteins it is recog nized that residue positions which are not identical often differ by conservative amino acid substitutions, where amino acid residues are substituted for other amino acid residues with similar chemical properties (e.g., charge or hydropho bicity) and therefore do not change the functional properties of the molecule. Where sequences differ in conserved substi tutions, the percent sequence identity may be adjusted upwards to correct for the conserved nature of the substitu tion. Sequences, which differ by such conservative substitu tions, are said to have "sequence similarity" or "similarity." Scoring for a conservative substitution allows for a partial rather than a full mismatch (154), thereby increasing the percentage sequence similarity.

The term "percentage of sequence identity" means the value determined by comparing two optimally aligned sequences over a comparison window, wherein the portion of the polynucleotide sequence in the comparison window may

Fig. 2. Portions of the semantically annotated document using entities from models and ontology classes

3.4 Retrieval Phase

When searching for patents, the diversity of users should not be neglected. Different personal or technical background of the users and the level of the abstractness that users practice must be taken in consideration. Technically trained users will usually have exact term as an input query. On the other hand, basic users might use more common, general and less specific input.

The system supports different levels of retrieval, based on the application mode selected by the user. Full text retrieval is the simplest method that uses the textual indexing of the entire document's tokenized contents, and Lucene search capability. Fuzzy Ranking retrieval uses the input query to create sub-result sets from document segments clustered and indexed in the previous phase. It uses fuzzy rules to order these sets in a complete result list. Fuzzy Ranking supports two modes of searching: textual – where each subset is gathered from textual search preformed on different patent segments (clusters); and semantic – where enhanced search is performed using nested classes from the ontology trees for each patent segment.

3.5 Fuzzy Ranking

Once the user makes a request, the input is analyzed using natural language processing approach. When the tagged input consists of general speech tags, our fuzzy ruler should mark content portion of documents as more relevant than other parts. The search query is submitted to all the clusters of indexed patents. Accordingly, a patent set is returned from every cluster (each patent segment, the ontology based indexed cluster and the pure textual cluster). The returned results are what we refer to as sub-result sets. Every returned patent sub-result set contains a list of patents ordered according to the correlation of the query and the indexing of the cluster itself (semantic, textual or both). Merging the multiple sub-result sets into a single result, ordered appropriately is essential for achieving satisfactory precision of the system. Therefore, aiming to merge the sub-results in a meaningful manner, we introduced the relevance formulas based on fuzzy logic. Patents that contain relevant data in high correlation with the query will be prioritized in the final resulting set.

The relevance ranking process starts with chunking the input query in part-of speech (POS) tags. We use the SharpNLP tool [15] (part of the Open NLP library) to chunk up the input query. Each word in the query is associated a POS tag (noun, adjective, verb, adverb, number, foreign word, etc.). Some insignificant words, like conjunctions, are ignored. POS tags from the query are used to calculate the weight of each sub-result set so that the order of the sub-result sets in the final result can be determined. Patents from the multiple sub-result sets are added in the final result set according to the weight of the initial sub-result set they belong to. The weight of each sub-result set is calculated according to the Formula 1.

$$RSetW = A_W * \frac{\sum (POSi_W * N_{[POSi]})}{\sum N_{[POSi]}} \tag{1}$$

where *RSetW* is the final weight of the sub-result set. *POSiW* is the specific weight for a given POS tag i depending on the cluster that the sub-result set originated from. *N[POSi]* is the occurrence frequency of the POS tag i in the query. A_W is a shifting factor that transfers the weight value in a predefined membership range [0-20].

Table 1. Coefficient values for A_w and $POSi_W$ for different cluster types

Cluster	A_W	NNP proper noun	NN noun	FW foreign word	VB verb	JJ adjective	RB adverb	SYM symbol
Title	0.3	4	10	4	9	7	0	0
Author	2.4	10	6	9	0	0	0	0
Abstract	2.64	4	10	4	9	9	9	8

The values of the coefficients *POSiW*, given in Table 1, are chosen according to the impact of the particular POS tag in the cluster type. We scaled each POS tag with a value between 0 and 10, depending on its influence in the type of document segment (cluster). For example, tags that represent proper nouns (personal names) have greater impact in clusters such as 'Author' or 'Citations'. Also, in the 'Title' or 'Claims' cluster, the weight formula should not have large values for personal nouns, since it is unlikely that a personal name is contained directly in the patent title. The coefficient values were derived empirically after several experiments and manual testing.

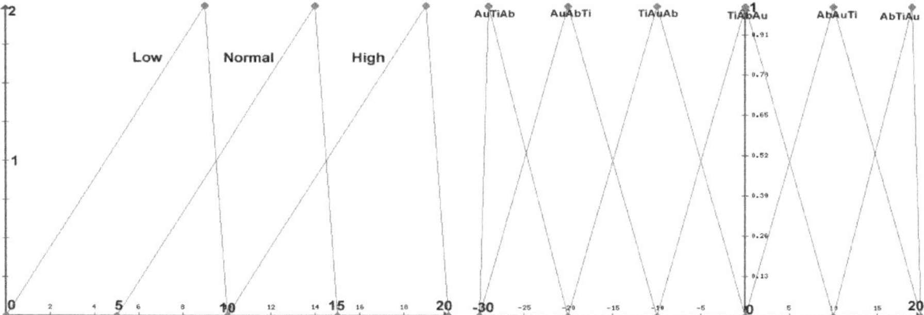

Fig. 3. Input and Output Membership Function for each document structure

Each cluster type yielding a sub-result set relates to a fuzzy set in our Fuzzy Ruler. We use Low, Normal and High as fuzzy parameters, as shown in Figure 3. The weights of the sub-result sets are combined in logical products. For mathematical simplicity, we use a triangular membership function and the same membership function for all sub-result sets. The Min-max method was chosen to include all contributing rules. Defuzzification into crisp output, gave us a reordered relevance set that is more significant than the predefined one.

4 Experimental Results

We have considered two strategies of evaluation experiments. In the first set of experiments, our focus was aimed at estimation of the benefits from the use of the fuzzy ruler in the ranking results. The second set of experiments was focused on the benefits of semantic search. The goal was to evaluate its advantages over results based solely on text based indexing.

In the first type of experiments, we used different variations of author (inventor) names combined with additional words or tags. More than 100 search queries were executed ranging from a single name request, to mixed queries consisting of several personal names mixed with nouns, adjectives, adverbs, verbs or other tags. Experiments with queries that contained mixed type of part of speech tags showed better results in ranked mode. For example, the search query 'progress Shawn Marth Herman Kreig immune' has personal names as a dominant part, followed with a noun and an adjective. With personal names as predominate terms, we expect that authors or citations that contain the given names should be higher in the resulting list. Assignees and references also contain personal names, so their presence cannot be fully removed. In the resulting chart on Figure 4, personal names are more notable (because of their higher importance) than the other type of tags who are weighted based on their input order. The axes represent word frequencies in both modes with Fuzzy Ruler and pure textual retrieval.

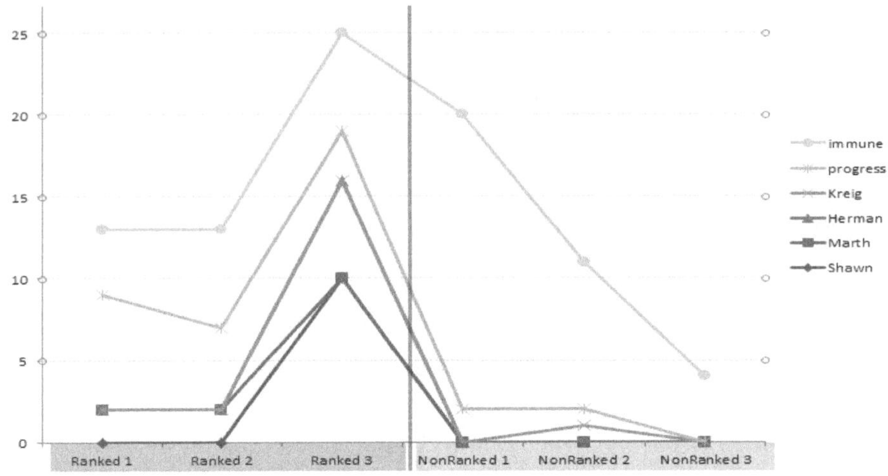

Fig. 4. Keyword frequency in top results with and without Ranking

It is evident in Figure 4 that the data ranked using the fuzzy ruler has proven better compared to the results produced without it.

In the second type of experiments, we have investigated more than 50 different variations of ontology based queries. They vary from a single item query, to mixed queries with different POS tags.

For instance, we can consider the search query: 'carbohydrate amino acid'. This input is dominated with noun POS tags, which increases the weights of semantically annotated clusters, pushing their results higher in the fuzzy controller. We examined the top result sets in the models with and without semantic annotation. Our results showed that given keywords had higher overall frequencies in the results when searching without the semantic component. However, after manual inspection of documents returned in the top result sets in both modes, a different conclusion can be derived. We measured which portions of the resulting patent document contained the keywords. We found that in documents searched with semantic annotation the keywords in question occurred in significant portions of the patent (abstract, claims or summary), as well as in the contents. In the results obtained without semantic annotation the keywords were spread throughout the patents, often not in the significant components of the patent. Therefore in this case the relevance of the top results was not as high. This lead us to the conclusion that semantically annotated search resulted in more relevant documents than pure textual annotation based on word frequencies only.

Patent amendments contain common inventors, sometimes even common title, but are published in different time period. Because of their similarity, when searching for keywords contained in them, we expect that both documents will be returned in results as equally relevant. During our inspection, we noticed that in the semantic indexing mode such amendment documents are successive in the list, while when using pure textual indexing, they returned diffused in the resulting set (Figure 5).

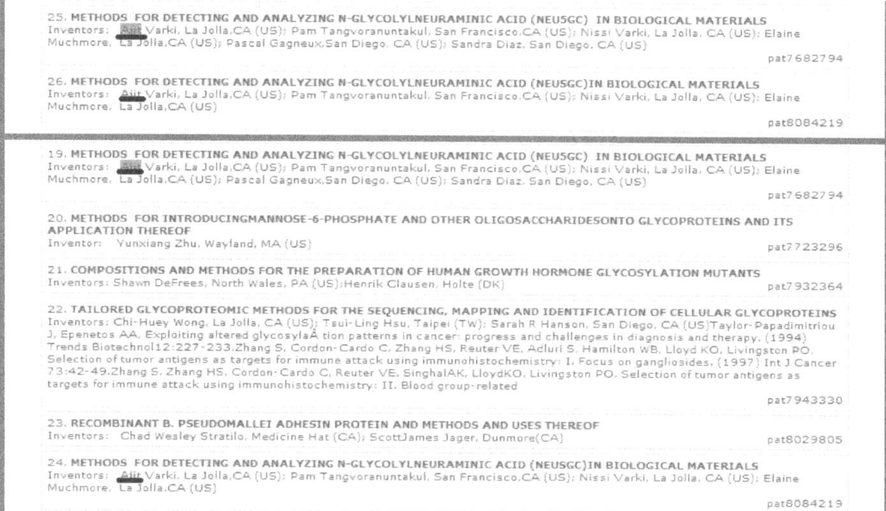

Fig. 5. Position in relevant patents with and without semantic indexing

5 Future Work

After the establishment of the basic architecture and experiments, our future focus is to enlarge the patent dataset, as well as including well new ontologies. Patent datasets will be collected from various sources and different areas, mainly from free data collections such as CLEF or TREC [18]. Also, the challenge of searching patents in different languages will be tackled. This will require an additional multilingual methodology and architecture component. Larger number of data clusters will be needed in order to comprise divergent sources and documents. Advanced query manipulation in prior art search can further increase retrievability [9].

Since patents almost regularly contain schematics and images, an image processing module should be involved. The image processing module will be used to index patents according to the images within.

Additional weighting filters for every patent in the sub-resulting sets will be developed to refine the relevance of the final result set.

6 Conclusion

We have witnessed new approaches and implementations of various technologies in all aspects of patent search and retrieval. Most of them have proven valuable in their strictly defined goal. However, many techniques are mostly focused on specific areas and fail when the entire patent search process is taken into consideration.

Our approach to combine proven techniques in a composite architecture showed improved results compared to pure textual based indexing and retrieval. We also showed that results ranked using semantic annotation are better than results based on simple keyword frequencies. When searching technical documents, such as patents in our case study, semantic annotation on important document sections are more adequate to produce relevant documents. We expect that finer filtering on topmost results will be even more beneficial. We plan for our future work to incorporate additional strategies and approaches and investigate their correlation and importance in the search and retrieval phase.

References

1. Mukherjea, S., Bamba, B., Kankar, P.: Information Retrieval and Knowledge Discovery Utilizing a BioMedical Patent Semantic Web. IEEE Transactions on Knowledge and Data Engineering 17(8), 1099–1110 (2005)
2. Giereth, M., Brügmann, S., Stäbler, A., Rotard, M., Ertl, T.: Application of semantic technologies for representing patent metadata. In: Proceedings of the First International Workshop on Applications of Semantic Technologies (2006)
3. Ghoula, N., Khelif, K., Dieng-Kuntz, R.: Supporting Patent Mining by using Ontology-based Semantic Annotations. In: Proceedings of the IEEE/WIC/ACM International Conference on Web Intelligence, Washington, DC, pp. 435–438 (2007)
4. Loganantharaj, R., et al.: An Ontology Based Semantic Literature Retrieval System. In: Proc. of the 19th IEEE Symp. on Computer-Based Medical Systems (2006)
5. Zhi, L., Wang, H.: A Construction Method of Ontology in Patent Domain Based on UML and OWL. In: International Conference on Information Management, Innovation Management and Industrial Engineering, December 26-27, pp. 224–227 (2009)
6. Segev, A., Kantola, J.: Patent Search Decision Support Service. In: Seventh International Conference on Information Technology, New Generations (ITNG), April 12-14, pp. 568–573 (2010)
7. Aleman-Meza, B., Arpinar, I.B., Nural, M.V., Sheth, A.P.: Ranking Documents Semantically Using Ontological Relationships. In: IEEE Fourth International Conference on Semantic Computing, Semantic Computing (ICSC), September 22-24, pp. 299–304 (2010)
8. Taduri, S., Lau, G.T., Law, K.H., Kesan, J.P.: Retrieval of Patent Documents from Heterogeneous Sources Using Ontologies and Similarity Analysis. In: Fifth IEEE International Conference on Semantic Computing, pp. 1–8 (2011)
9. Bashir, S., Rauber, A.: Improving Retrievability of Patents in Prior-Art Search. Information and Software Technology 52(6), 641–655 (2010)
10. The United States Patent and Trademark Office, http://www.uspto.gov
11. Patent Retriever, http://www.patentretriever.com
12. Patent Search Tool, http://www.pat2pdf.org
13. The Glycomics Ontology, http://lsdis.cs.uga.edu/projects/glycomics
14. Apache Lucene .NET, http://incubator.apache.org/lucene.net
15. Sharp NLP Project, http://sharpnlp.codeplex.com
16. European Patent Office, http://www.epo.org
17. The World Intellectual Property Organization, http://www.wipo.int
18. CLEF-IP, http://www.ir-facility.org/clef-ip

Superlinear Speedup for Matrix Multiplication in GPU Devices

Leonid Djinevski[1], Sasko Ristov[2], and Marjan Gusev[2]

[1] FON University, Av. Vojvodina, 1000 Skopje, Macedonia
leonid.djinevski@fon.edu.mk
[2] Ss. Cyril and Methodious University,
Faculty of Information Sciences and Computer Engineering,
Rugjer Boshkovikj 16, 1000 Skopje, Macedonia
{sashko.ristov,marjan.gushev}@finki.ukim.mk

Abstract. Speedup in parallel execution on SIMD architecture according to Amdahl's Law is finite. Further more, according to Gustrafson's Law, there are algorithms that can achieve almost linear speedup. However, researchers have found some examples of superlinear speedup for certain types of algorithms executed on specific multiprocessors.

In this paper we achieved superlinear speedup for GPU devices, which are also categorized as SIMD. We implement a structure persistent algorithm which efficiently exploits the shared cache memory and avoids cache misses as much as possible. Our theoretical analysis and experimental results show the existence of superlinear speedup for algorithms that run on existing GPU device.

Keywords: Cache memory, Matrix Multiplication, GPU, CUDA, HPC, Superlinear Speedup.

1 Introduction

Matrix-matrix multiplication (MM) is one of the most used algorithms in computing. There has been a lot of research done in optimizing MM in different directions. For example, decreasing the average number of operations improves the algorithms runtime. DeFlumere et al. [6] present that the optimal rectangular partitioning can be significantly outperformed by the optimal non-rectangular one on real-life heterogeneous HPC platforms. Introducing parallelization can also speedup the runtime of MM. Clarke et al. [5] proposed an algorithm that reduces the communication time, and thus total execution time, in a heterogeneous processors by arranging the partitioning so that all processors finish their work in the same time. Volkov and Demmel [21] present detailed benchmarks of the GPU memory system, kernel start-up costs, and arithmetic throughput on dense matrix operations. Several efficient implementations of sparse matrix-vector multiplication in CUDA are given in [3].

Other researchers contribute for fast implementations of the linear algebra support libraries like BLAS [4], LAPACK [2], which provide easy interface for people without big computational background.

S. Markovski and M. Gusev (Eds.): *ICT Innovations 2012*, AISC 207, pp. 285–294.
DOI: 10.1007/978-3-642-37169-1_28

Hierarchical cache memory levels increase over time. Modern CPUs have Level 1 to Level 3 cache memories (shared or dedicated per CPU core / socket) between the registers and the main memory. Modern GPUs have also multilevel memory hierarchy. Low-latency GPU memories are much more promising than general purpose LRU caches for the tiling and partitioning strategies [11]. New Fermi architecture [22] consists of configurable cache memory. The authors in [14] present an improved GEMM algorithm for efficiently use of Fermi's extended memory hierarchy and memory sizes.

Algorithm optimization for a certain architecture platform should achieve maximum speedup. Speedup defines the ratio of the execution time on a single processing element over the execution time of p processing elements. Amdahl's Law [1] states that the speedup is constrained by the algorithm, while according to Gustafson [10] speedup can achieve almost linear speedup proportional to the number of processors.

However, speedup can achieve even higher values than linear speedup (superlinear speedup), which in parallel computing this implementation on various CPU devices is known for certain time. It is possible even in the cloud [9, 20]. Additionally, superlinear speedup is also found in multi-GPU implementation on Fermi architecture GPU, due to configurable cache memory [18]. In this paper we have looked more in-depth the Fermi memory hierarchy, and found that superlinear speedup can be achieved in a single GPU. To our knowledge, we are the first to address superlinearity for matrix multiplication on a single GPU.

The remainder of this paper is organized as follows: A short overview of the GPU architecture and memory hierarchy is presented in Section 2. Theoretical analysis with cache memory occupancy analysis and expected superlinear regions is presented in Section 3. The testing methodology is presented in Section 4, followed by the obtained results analyzed in Section 5. Conclusions are presented in Section 6.

2 GPU Architecture

General-Purpose computing on Graphics Processing Units (GPGPU) origins from utilization of the raw performance of the GPU devices, using OpenGL or DirectX API calls [12]. This approach has proved to be very ineffective because it requires to cast the problems in native graphics operations. Therefore NVIDIA developed the CUDA programming model [16] which facilitates by tapping into the available computational resources. CUDA programs are accelerated by data-parallel computations of millions of threads, which in this context means instance of a kernel, where krenel is the program running on the GPU device. This approach characterizes the GPU architecture as a SIMD parallel machine [8] with convenient memory hierarchy.

NVIDIA GPUs have evolved into massively parallel, many-core architectures. These GPUs contain an array of Streaming Multiprocessors (SM), each containing 8 Scalar Processors (SP) for the Tesla architecture [13], 32 SPs for the Fermi architecture [7], and 192 SPs for the latest Kepler architecture [17]. However,

CUDA in particular is a Single Instruction Multiple Thread (SIMT) programming model [15], where all threads execute in step the same instruction, but within one SM. On the other hand, threads in different SMs are executing instructions independently from each other, thus providing scalability.

The memory hierarchy of NVIDIA Fermi GPU device is presented in Fig. 1. The GPU devices have off-chip memory, so called global memory where average single fetching of data takes at least 400 cycles. The first level in the memory hierarchy is the L1 and shared memory, which is shared by a number of threads organized in thread blocks. It can be accessed almost as fast as register memory and is called private memory which is exclusive to a single thread. L2 cache is off-chip memory and can be accessible from all threads in any SM.

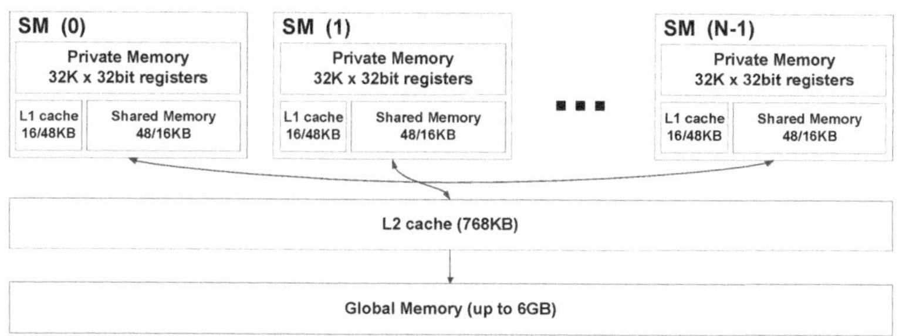

Fig. 1. Memory hierarchy of NVIDIA Fermi architecture

3 Theoretical Analysis

In this section we analyze the cache memory occupancy for sequential and parallel execution in GPUs and propose an algorithm which efficiently exploits configurable GPU's L1 cache memory.

3.1 Superlinear Regions

Cache memory occupancy for sequential execution is greater than the cache memory occupancy for parallel execution since, smaller chunks of data are stored, thus allowing larger size problems to be stored without generation of cache misses. As described in Section 2 if the cache memory requirements fit in L1 cache, then we expect the highest processor speed and call this region L1 region. In L2 region data does not fit in L1, but fits in the L2 cache generating cache misses in L1, but not for L2.

The theoretical expectations for performance are presented in Fig. 2. Cache occupancy for the sequential execution is defined by the points L1S and L2S and for the parallel execution by the points L1P and L2P. Due to different dedicated L1 cache occupancy in dedicated L1 cache there is difference between L1 regions for sequential and parallel execution. We expect better performance for parallel execution in case when the sequential execution generates cache misses for L1 and L2 still does not generate cache misses, and in case of parallel execution L1 does not generate cache misses. This region is called *superlinear region* since it leads to a possible superlinear speedup [19].

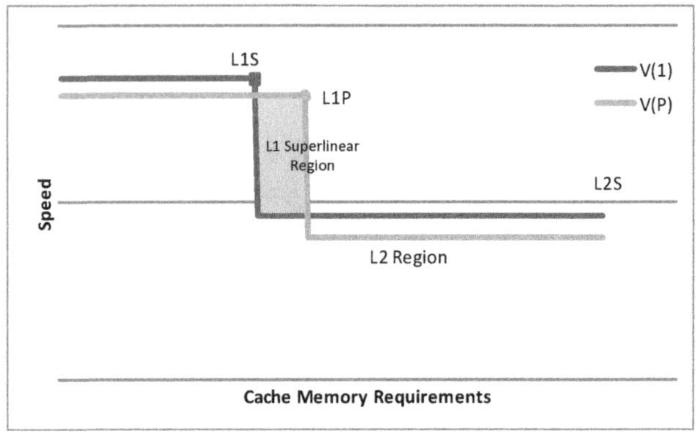

Fig. 2. Expected speed with real cache

3.2 Analysis of Memory Utilization

Based on the SIMD architecture as described in Section 2, the L1 cache memory is shared among all 32 threads that run per single SM. However, we are interested only in scenarios where each core has its own dedicated cache memory, in order to test our assumption presented in Section 3.1. Hence, L1 cache is dedicated only per SM. On one hand we can control how many threads to run per threadblock, but on the other hand, there is no specified way to control how many threadblocks can be run per SM. Nevertheless, there are limitations regarding active threadblocks per SM, so by allocating maximum shared memory we can ensure that only one threadblock is running per SM. Thus, if the threadblock is defined by one thread, we ensure that one thread is running per SM.

A sequential implementation of our algorithm that runs one thread per only one active SM, occupies the L1 cache memory as presented in Fig. 3, where the accessed memory blocks are denoted with gray color.

The cache memory occupancy for the parallel implementation is depicted in Fig. 4. In this implementation several SMs are used such that each SM uses only one thread and full L1 cache. L2 cache in this implementation is shared among all SMs.

3.3 The Algorithm

For simplification, we multiply square matrices of same sizes $N * N$. The basic MM algorithm is defined by $c_{ij} = \sum_{k=0}^{N-1} a_{ik} * b_{kj}$ where a_{ik}, b_{kj} and c_{ij} are correspondingly elements of matrices A, B and C, for all $i, j, k = 0, \ldots, N-1$.

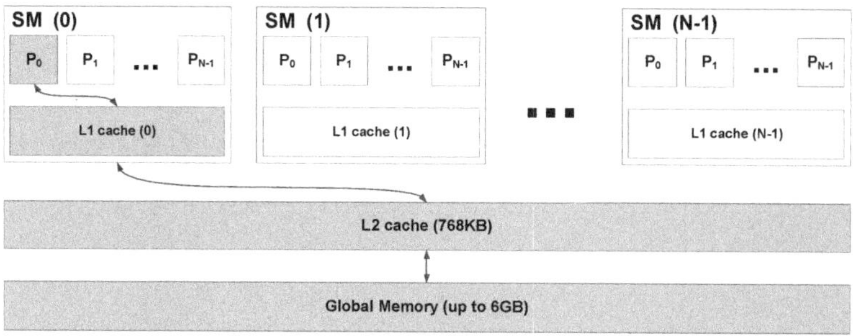

Fig. 3. Memory utilization of the sequential implementation

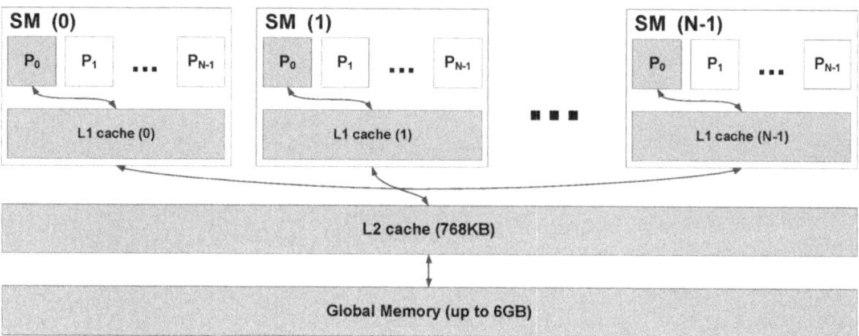

Fig. 4. Memory utilization of the parallel implementation

The authors in [19] achieved superlinear speedup in parallel execution on a multiprocessor device by partitioning the A and C into smaller P submatrices. The idea is to store greater part of B in the L2 cache and share it among all processes avoiding cache misses as much as possible. Based on the algorithm in [19], we have developed a parallel algorithm for a GPU device. Since multicore processors have larger cache memories it is easier to store the whole B matrix. In GPU the largest cache memory is L2 (736KB) and the matrix B cannot be fitted in L2 for larger problem sizes. However, we solve this, by partitioning the matrix B with the number of available processing elements. An example of two processing elements is presented in Fig. 5, where the horizontal and the vertical striped matrices (A and B respectively) are multiplied, and the unstriped matrix which is divided in four regions is the resulting matrix C, m is the size of the partitioned submatrix, am_ost is the residual of problem size and the number of processing elements, bx stands for the ID of the processing element and jj together with bx indicate which submatrix has to be processed.

The region 1 in matrix C is calculated for each problem size. However, for problem sizes which is a factor of the number of processing elements, there is no

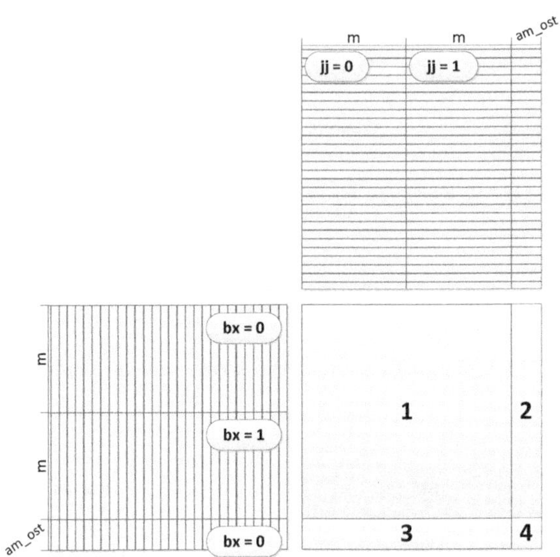

Fig. 5. GPU matrix multiplication algorithm

residual and there is perfect alignment with the number of divided submatrices, thus regions 2, 3 and 4 do not have data to calculate.

4 Testing Methodology

We have performed our tests on GPU NVIDIA Tesla C2070. All experiments were performed on the same hardware infrastructure (Intel i7 920 CPU at 2.67GHz, with 12GB RAM at 1333MHz) and Linux operating system Ubuntu 10.04 LTS. The implementations were compiled with the NVIDIA compiler nvcc from the CUDA 4.2 toolkit.

The testing methodology is based on 2 experiments which show speedup dependence on problem size (cache memory requirements) and on the number of processing elements.

Experiment 1 determines the speed dependence for different cache memory requirements. We increase the problem size and measure the speed for sequential execution and parallel execution with maximum SMs. Our hypothesis to be confirmed experimentally is to achieve a superlinear region.

In *Experiment 2* we'll choose particular problem size from superlinear region and L1 region, and execute the algorithm on 2, 4, 6, 8, 10, 12, and 14 SMs. Our hypothesis to be confirmed experimentally is to achieve superlinear speedup for each number of SMs for the problem size selected from superlinear region and sublinear speedup for the problem size selected from the L1 region.

5 Results

This section presents the results that show superlinear speedup existence in superlinear region.

5.1 Experiment 1 – Speedup vs. Cache Memory Requirements

Fig. 6 depicts the speed for different cache memory requirements, where V(1) stands for the speed of the sequential execution. In order to better depict the Fig. 6 for parallel execution on 14 cores, we have normalized the speed per core, therefore V(14) presents the average speed per processing element.

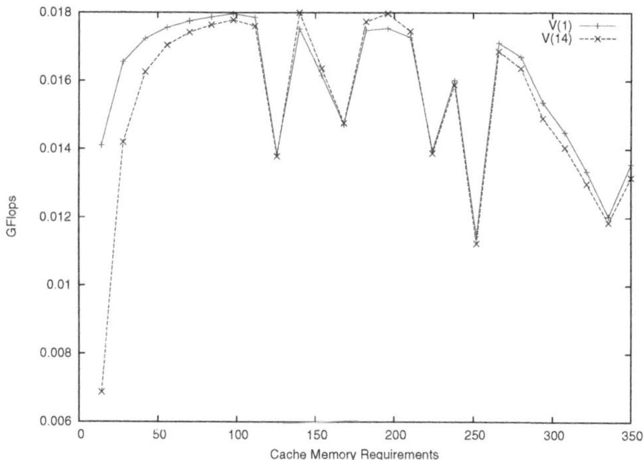

Fig. 6. Normalized speed for sequential (1SM) and parallel (14SMs) execution

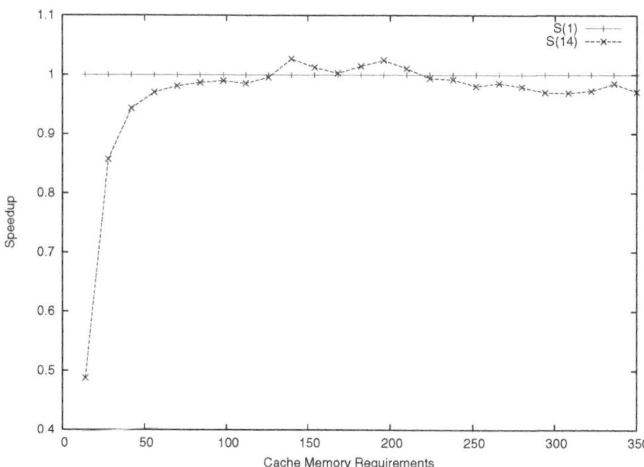

Fig. 7. GPU speedup for the parallel execution (14 active SMs)

Accordingly, the speedup S(14) is denoted for the parallel execution on 14 cores for the same experiment in Fig. 7. Thus, it is easy to notice that there is an existence of a superlinear region, and an appropriate superlinear speedup.

5.2 Experiment 2 – Speedup vs. Number of SMs

The results of the Experiment 2 are depicted in Fig. 8. We can conclude that superlinear speedup exists for each parallel execution with 2, 4, 6, 8, 10, 12 and 14 SM for particular problem size of the superlinear region. For the problem size of L1 region we achieved sublinear speedup for each number of SMs.

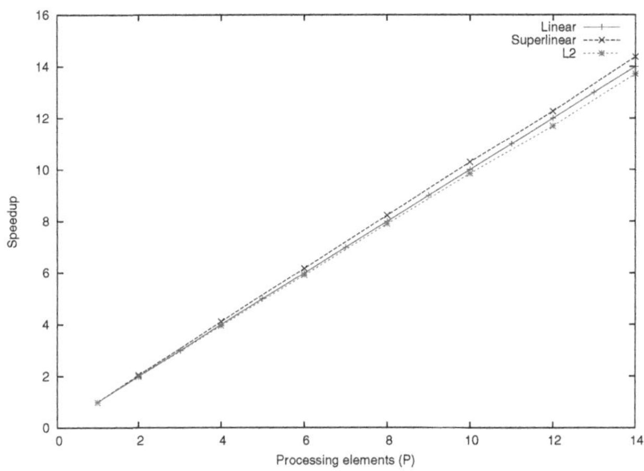

Fig. 8. GPU speedup for the second experiment

6 Conclusion

The experiments have confirmed the theoretical analysis about existence of superlinear regions of the problem size for matrix multiplication using GPU devices, where the normalized performance per processing element for parallel execution is better than in sequential execution. Thus, we have obtained a superlinear speedup beyond the limits specified in Gustafson's law [10]. However, we have only obtained superlinear speedup in the superlinear region. In this case speedup originates from the larger cache that we dedicated per core, therefore the cache memory requirements of the working set fit more in L1 for parallel execution and generate L1 cache misses for sequential execution.

Based on the experiments, we have presented further proof that there is existence of superlinear speedup for SIMD architecture processors with dedicated caches, more particular GPU devices. In our algorithm we have kept the same number of operations, therefore the obtained superlinear speedup from our analysis is justified. However, having a larger cache memory is sufficient, but the reuse of data is necessery condition for achiving superlinear speedup. Additionally, we found that the speedup performance is directly dependent on cache performance and generation of cache misses.

The results in this paper show that there has to be a revision of the speedup definition and cache memory requirements have to be added in order to make proper calculations. Presented analysis in this paper should contribute in the development of future architectures.

References

1. Amdahl, G.M.: Validity of the single-processor approach to achieving large scale computing capabilities. In: AFIPS Conference Proceedings, April 18-20, vol. 30, pp. 483–485. AFIPS Press, Reston (1967)
2. Anderson, E., Bai, Z., Bischof, C., Blackford, S., Demmel, J., Dongarra, J., Croz, J.D., Greenbaum, A., Hammarling, S., McKenney, A., Sorensen, D.: LAPACK Users' Guide. Soc. for Ind. and Appl. Math., 3rd edn., PA (1999)
3. Bell, N., Garland, M.: The impact of cache misses on the performance of matrix product algorithms on multicore platforms. Research Report NVR-2008-004 (December 2008), http://hal.inria.fr/inria-00537822/en/
4. Blackford, L.S., et al.: An updated set of basic linear algebra subprograms (blas). ACM Trans. Math. Softw. 28(2), 135–151 (2002)
5. Clarke, D., Lastovetsky, A., Rychkov, V.: Column-based matrix partitioning for parallel matrix multiplication on heterogeneous processors based on functional performance models. In: Alexander, M., D'Ambra, P., Belloum, A., Bosilca, G., Cannataro, M., Danelutto, M., Di Martino, B., Gerndt, M., Jeannot, E., Namyst, R., Roman, J., Scott, S.L., Traff, J.L., Vallée, G., Weidendorfer, J. (eds.) Euro-Par 2011, Part I. LNCS, vol. 7155, pp. 450–459. Springer, Heidelberg (2012)
6. DeFlumere, A., Lastovetsky, A., Becker, B.: Partitioning for parallel matrix-matrix multiplication with heterogeneous processors: The optimal solution. In: HCW 2012. IEEE Computer Society, Shanghai (2012)
7. Glaskowsky, P.: Nvidias fermi: the first complete gpu computing architecture. Tech. rep., NVIDIA (2009) (white Paper)
8. Grama, A., Karypis, G., Kumar, V., Gupta, A.: Introduction to Parallel Computing, 2nd edn. Addison-Wesley (January 2003)
9. Gusev, M., Ristov, S.: Superlinear speedup in windows azure cloud. Tech. Rep. IIT:06-12, University Ss Cyril and Methodius, Skopje, Macedonia, Faculty of Information Sciences and Computer Engineering (July 2012)
10. Gustafson, J.L.: Reevaluating amdahl's law. ACM 31(5), 532–533 (1988)
11. Jacquelin, M., Marchal, L., Robert, Y.: The impact of cache misses on the performance of matrix product algorithms on multicore platforms. Research Report RR-7456, INRIA (November 2010), http://hal.inria.fr/inria-00537822/en/
12. Kirk, D., Hwu, W.M.: Programming Massively Parallel Processors: A Hands-on Approach, 1st edn. Morgan Kaufmann Publishers Inc., USA (2010)
13. Lindholm, E., Nickolls, J., Oberman, S., Montrym, J.: Nvidia tesla: A unified graphics and computing architecture. IEEE Micro 28(2), 39–55 (2008)
14. Nath, R., Tomov, S., Dongarra, J.: An improved magma gemm for fermi graphics processing units. Int. J. High Perf. C. App. 24(4), 511–515 (2010)
15. Nickolls, J., Dally, W.: The gpu computing era. IEEE Micro 30(2), 56–69 (2010)

16. NVIDIA: Cuda programming guide (Auguest 2012), http://developer.download.nvidia.com/compute/DevZone/docs/html/C/doc/CUDA_C_Programming_Guide.pdf/
17. NVIDIA: Next generation cuda compute architecture: Kepler gk110 (2012)
18. Playne, D.P., Hawick, K.A.: Comparison of gpu architectures for asynchronous communication with finite-differencing applications. Concurrency and Computation: Practice and Experience 24(1), 73–83 (2012)
19. Ristov, S., Gusev, M.: Superlinear speedup for matrix multiplication. In: Proceedings of the ITI 2012 34th International Conference on Information Technology Interfaces, pp. 499–504 (2012)
20. Ristov, S., Gusev, M., Kostoska, M., Kjiroski, K.: Virtualized environments in cloud can have superlinear speedup. In: ACM Proceedings of 5th Balkan Conference of Informatics, BCI 2012 (2012)
21. Volkov, V., Demmel, J.W.: Benchmarking gpus to tune dense linear algebra. In: Proceedings of the 2008 ACM/IEEE Conference on Supercomputing, SC 2008, pp. 31:1–31:11. IEEE Press, Piscataway (2008)
22. Wittenbrink, C.M., Kilgariff, E., Prabhu, A.: Fermi gf100 gpu architecture. IEEE Micro 31(2), 50–59 (2011)

Verifying Liveness in Supervised Systems Using UPPAAL and mCRL2

Jasen Markovski* and M.A. Reniers

Eindhoven University of Technology,
PB 513, 5600MB, Eindhoven, The Netherlands
{j.markovski,m.a.reniers}@tue.nl

Abstract. Supervisory control ensures safe coordination of high-level discrete-event system behavior. Supervisory controllers observe discrete-event system behavior, make a decision on allowed activities, and communicate the control signals to the involved parties. Models of such controllers are automatically synthesized from the formal models of the unsupervised system and the specified safety requirements. Traditionally, the supervisory controllers do not ensure that intended behavior is preserved, but only ensure that undersired behavior is precluded. Recent work suggested that ensuring liveness properties during the synthesis procedure is a costly undertaking. Therefore, we augment state-of-the-art synthesis tools to provide for efficient post-synthesis verification. To this end, we interface a model-based systems engineering framework with the state-based model checker UPPAAL and the event-based tool suite mCRL2. We demonstrate the framework on an industrial case study involving coordination of maintenance procedures of a high-end printer. Based on our experiences, we discuss the advantages and disadvantages of the used tools. A comparison is given of the functionality offered by the tools and the extent to which these are useful in our proposed method.

Keywords: Supervisory control, model checking, μ-calculus, UPPAAL.

1 Introduction

Control software development is becoming a significant bottleneck in the development of high-tech complex systems due to ever-increasing system complexity and demands for better quality. Traditional approaches to software development employing (re)coding-testing loops have proven not entirely adequate to handle the challenge as control requirements frequently change during the design process [11]. This issue gave rise to supervisory control theory [16,4], where models of supervisory controllers, referred to as *supervisors*, are synthesized automatically based on formal models of the uncontrolled hardware, referred to as *plant*, and the model of the *control requirements*.

* Supported by Dutch NWO project ProThOS, no. 600.065.120.11N124. Also affiliated with Faculty of Computer Science and Engineering, Ss. Cyril and Methodius University, Skopje, Republic of Macedonia.

S. Markovski and M. Gusev (Eds.): *ICT Innovations 2012*, AISC 207, pp. 295–304.
DOI: 10.1007/978-3-642-37169-1_29

Supervisory controllers coordinate high-level system behavior by receiving sensor signals from ongoing activities, make a decision on allowed activities, and send back control signals to the hardware actuators. A standard assumption is that the controller reacts sufficiently fast on machine input, which enables the modeling of the supervisory control feedback loop as a pair of synchronizing processes [16,4]. The observed activities of the system are modeled by means of discrete events, which are split into *controllable events*, which model interaction with the actuators of the machine, and *uncontrollable events*, which model observation of sensors. The synchronization of the plant and a supervisor, referred to as *supervised plant*, models the supervisory control loop. The supervisor can disable controllable events by not synchronizing with them, but it must always enable available uncontrollable events by always synchronizing with them as the latter cannot be ignored or altered. In addition, supervised plants must satisfy the control requirements, which model allowed or safe system behavior.

There are, however, several issues that need to be addressed. Automated control software synthesis does not come with the guarantee that the supervised system has all intended functionalities. It only ascertains safe and nonblocking system behavior, i.e., it prevents the system of reaching dangerous or otherwise undesired states, or deadlocking [16,4]. More specifically, the state space of the uncontrolled system comprises both unsafe and useful behavior, whereas the desired safe functioning of the system is specified by the control requirements. The synthesis procedure ensures safety, which may require elimination of important 'live' states that may lead the execution of the system to unsafe states [9,4]. In most cases, the latter is not directly deducible from the control requirements and, moreover, it can only be observed during or following the synthesis procedure. In that case, either the control requirements are too strict, or the model of the uncontrolled system is not sufficiently detailed, or it is flawed [13].

To remedy these issues, extensions of the theory have been proposed to incorporate liveness requirements in the synthesis procedures. Extensions with the temporal logic CTL* are proposed and analyzed in [9], whereas a proposal is given in [20] to extend the NuSMV model checker for the purpose of synthesis employing CTL*. An alternative is given in [17], where a translation from temporal logic control requirements to a standard automata setting is given and standard synthesis tools are applied. In the domain of software synthesis, a variant of the temporal logic LTL precisely depicts fairness and liveness assumptions made by the supervisor in [6]. Unfortunately, these approaches suffer from high (doubly-exponential) complexity due to enforcing of complex liveness properties during the synthesis procedure [15]. Consequently, the proposed frameworks can handle systems with only $10^3 - 10^5$ states [20,9,17,6], which calls for post-synthesis verification. The work of [19] proposes verification of synthesized extended finite automata as means to validate process operations for resource allocation. In [3] counterexamples are exploited to guide synthesis and verification procedures, whereas [18] investigates structural restrictions with hierarchical interfaces.

In this paper, we compare two such extensions that rely on the state-of-the-art model checkers UPPAAL [10] and mCRL2 [7] for post-synthesis verification.

The model checker UPPAAL relies on a restricted variant of CTL [10], whereas mCRL2 supports μ-calculus that subsumes CTL [8]. The comparison is given in terms of the functionality offered by the tools, summarized in Table 3, and the extent to which these are useful in our proposed method. As an illustration, we revisit an industrial case study involving coordination by supervisory control of maintenance procedures of a printing process of a high-end Océ printer [13]. Due to confidentiality issues, we can only present an (obfuscated) part of the case study. The complete case study comprises 25 automata with 2–24 states resulting in more than 6 billion states of the uncontrolled system. The (coordination) control problem is to ensure that quality of printing is not compromised by performing maintenance procedures as required, while interrupting ongoing print jobs as little as possible. We present the modeling of a part of the case study using the synthesis tool Supremica [1], which is interfaced with the model checkers by the tools Supremica2{UPPAAL, mCRL2} [12], respectively. Afterwards, we specify and verify several progress and liveness properties of interest.

2 Supervisory Coordination of Maintenance Procedures

We are dealing with high-tech Océ printers of [13], the control architecture of which is abstractly depicted in the upper left corner of Fig. 1. The user initiates print jobs, which are assigned to the embedded software by the printer controller in order to actuate the hardware to realize them. The embedded software is organized in a distributed way, per functional aspect, such as, paper path, printing process, etc. Several managers communicate with the printer controller and each other to assign tasks to functions, which take care of the functional aspects.

We depict a printing process function comprising one maintenance operation in Fig. 1. Each function is hierarchically differentiated to (1) controllers: Target Power Mode and Maintenance Scheduling, which receive control and scheduling tasks from the managers; (2) procedures: Status Procedure, Current Power Mode, Maintenance Operation, and Page Counter, which handle specific tasks and actuate devices, and (3) devices as hardware interface. Status Procedure is responsible for coordinating the other procedures given the input from the controllers. The control problem is to synthesize a supervisory coordinator that ensures that quality of printing is not compromised by timely performing maintenance procedures, while interrupting ongoing print jobs as little as possible [13]. The coordination rules that ensure safe behavior are given below.

We briefly describe the procedures of which the Supremica models are depicted in Fig. 1. Automata and procedure names coincide, whereas state names hint on physical representation. Uncontrollable events are underscored, whereas variable assignments that trace the automaton state are placed below transitions labels. The plant model is formed by the parallel composition of these automata, which must synchronize on the labels with common names [1,16,4]. Current Power Mode sets the power mode to run or standby depending on the enabling signals (*Stb2Run* and *Run2Stb*) from Status Procedure, and sends back feedback by employing *_InRun* and *_InStb*, respectively. Maintenance Operation

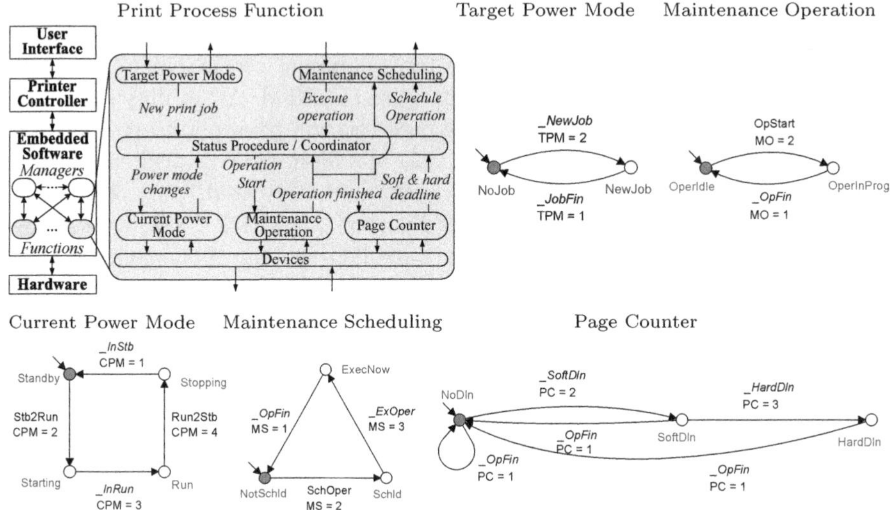

Fig. 1. Modeling of the printing process function

either carries out a maintenance operation, started by *OpStart* or it is idle. The confirmation is sent back by the event *_OpFin*, which synchronizes with Maintenance Scheduling and Page Counter. Page Counter counts the printed pages since the last maintenance and sends signals *_SoftDln* and *_HardDln*, when soft or hard deadlines are reached, respectively. A soft deadline signals that maintenance should be performed, but it is not yet compulsory if there are pending print jobs. A hard deadline is reached when maintenance of the printing process must be performed to ensure quality of the print. The page counter is reset, triggered by the synchronization on *_OpFin*, each time that maintenance is finished. The controller Target Power Mode sends signals regarding incoming print jobs to Status Procedure by *_NewJob*, which should set the printing process to run mode for printing and standby mode for maintenance and power saving. When the print job is finished, the signal *_NoJob* is sent. Maintenance Scheduling receives a request for maintenance with respect to expiration of Page Counter from Status Procedure, by the signal *SchOper* and forwards it to the manager. The manager confirms the scheduling with the other functions and sends a response back to the Status Procedure, using *_ExOper*. It also receives feedback from Maintenance Operation that the maintenance is finished in order to reset the scheduling, again triggered by *_OpFin*.

Status Procedure is restricted by several coordination rules: 1) Maintenance operations can be performed only when Printing Process Function is in standby; 2) Maintenance operations can be scheduled only if soft deadline has been reached and there are no print jobs in progress, or a hard deadline is passed; 3) Only scheduled maintenance operations can be started; and 4) The power mode of the printing process function must follow the power mode dictated by the managers, unless overridden by a pending maintenance operation. For

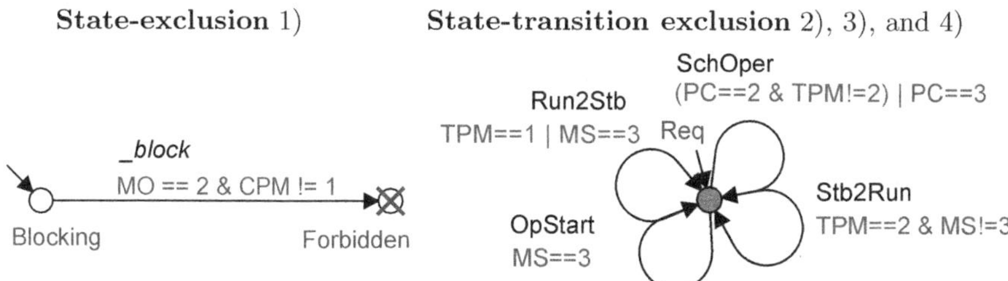

Fig. 2. Control requirements

a detailed account of the model-based systems engineering process and specification and formalization of the control requirements, we refer to [13,2].

1) To model this property in Supremica, we consider the states Standby and from Current Power Mode and OperInProg from Maintenance Operation, identified by $CPM == 1$ and $MO == 2$, respectively. We need a state-exclusion property [13] to model control requirements 1), i.e., we specify that no other combination of states is possible. To this end, we employ the notion of forbidden states, which force the supervisor to eliminate all states that inevitably reach them by traces of uncontrollable events [1]. We add a plant automaton that contains one uncontrollable transition with a unique label and let it target a forbidden state, as depicted in Fig. 2. The uncontrollable transition is guarded by $PM != 1 \,\&\, MO == 2$, where != denotes inequality and & denotes conjunction, so all states that do not conform to 1) are eliminated during supervisor synthesis.

2) States SoftDln and HardDln identify when a soft and hard deadline is reached, respectively. State NewJob of Target Power Mode states that there is a print job in progress. The event *SchOper* is responsible for scheduling maintenance procedures. Thus, it is enabled if $(PC == 2 \,\&\, TPM != 2) \mid PC == 3$, where | denotes disjunction. To restrict the occurrence of *SchOper*, we employ a control requirement automaton comprising a guarded self loop, given in Fig. 2.

3) Similarly to 2), we model control requirement 3) in Fig. 2 by a self-loop guarded with $MS == 3$ as this guard identifies the state ExecNow from Maintenance Scheduling.

4) We model this requirement separately for switching from Run to Standby mode, and vice versa. We can change from Run to Standby mode if this is required by the manager, i.e., there is a new print job identified by $TPM == 2$, and there is no need to start a maintenance operation, identified by $MS != 3$. Thus, we have a self loop labeled by *Stb2Run* and guarded by $TPM == 2 \,\&$ $MS != 3$, as depicted in Fig. 2. Contrariwise, when changing from Run to Standby power mode, the manager must be followed, unless it is overridden by a pending maintenance operation modeled by the guard $TPM == 1 \mid MS == 3$ of the self loop labeled with *Run2Stb*, again depicted in Fig. 2.

Employing the control requirements 1) – 4), depicted in Fig. 2, we synthesize a nonblocking supervisor for the plant depicted in Fig. 1, and we derive the supervised plant. Next, we validate the functionality of the supervised plant, to which end we employ the model checkers UPPAAL and mCRL2.

3 Verification of Desired Functionality with UPPAAL

In general, each system has more properties than the ones that can be handled by supervisor synthesis. Typically, properties that prescribe that the system is bound to make some progress, or is productive, are not easy to consider as control requirements. Therefore, it is necessary to use verification technology to establish whether or not these additional properties are satisfied by the supervised plant.

We first validate the functionality of Status Procedure, which we synthesized as a supervisory coordinator. We employ the model checker UPPAAL [10], which requires as input an UPPAAL automaton representation of the supervised plant. This automaton is obtained by translating the Supremica model, using the developed tool Supremica2UPPAAL [12]. Since UPPAAL supports the variable types of Supremica, we can employ the same variable assignments to identify states in UPPAAL. We treat the supervised plant as a closed system, i.e., it does not interact with the environment, so we translate all labeled transitions as outgoing broadcast channels [10]. Thus, the translation preserves the labeled transition structure of the supervised plant.

We model the verification properties using the temporal logic supported by the tool. The standard logical operators are $\neg$, $\wedge$, $\vee$, $\Rightarrow$, and *deadlock* denotes presence of deadlock in a system. By $\mathcal{B}$ we denote the set of propositional logic formulas constructed by these operators. The logic is a restricted variant of CTL [5], where the combinations of `A` and `E`, standing for all paths and there exists a path, respectively, and $\Box$ and $\Diamond$, standing for all states and there exists a state, respectively, are allowed, but without nesting. A useful progress property is given by $\phi \rightarrow \psi$, for $\phi, \psi \in \mathcal{B}$, which is equivalent to $A\Box(\phi \;\Rightarrow\; A\Diamond\, \psi)$.

We verify some representative verification properties that are commonly employed, cf. [14], which we denote by V_i for $i \in \mathbb{N}$. First, we verify that Status Procedure does not have a deadlock, by using $V_1\colon A\Box \neg\, \mathit{deadlock}$.

Next, we check that the state-exclusion requirement is satisfied. For this task, we employ variables with the same name and value as in the Supremica model, as our translation preserves the variables with their corresponding assignments. Thus, we verify that $V_2\colon A\Box\ MO == 2 \;\Rightarrow\; CPM == 1$.

Next, we check that if the system reaches a hard deadline and no maintenance operation is scheduled, then the maintenance operation becomes scheduled. This is specified as $V_3\colon PC == 3 \;\wedge\; MS == 1 \rightarrow MS == 2$.

To ascertain that the maintenance procedure can be scheduled when soft deadline has been reached, we employ $V_4\colon E\Diamond\ PC == 1 \;\wedge\; MO == 2$.

Finally, we can check that Status Procedure follows the commands from the Target Power Mode manager, by verifying that if the target power mode is run, then the printing process also eventually switches to run mode as well, given by $V_5\colon TPM == 2 \rightarrow CPM == 3$.

4 Verification of Desired Functionality with mCRL2

First, we give a condensed introduction to modal μ-calculus [8], the underlying temporal logic of mCRL2 [7] in which a large class of (event-based) properties of labeled transition systems can be specified. The syntax and semantics of the μ-calculus is more complicated and less intuitive than more prominent temporal logics, like CTL or LTL, so we spend a bit more space bringing it closer to the reader. The syntax of the modal μ-calculus, as used in this paper, is given by:

$$\phi ::= \text{true} \mid \text{false} \mid \neg\phi \mid \phi \wedge \phi \mid \phi \vee \phi \mid \phi \Rightarrow \phi \mid [pf]\phi \mid \langle pf \rangle\phi \mid \mu X.\phi \mid \nu X.\phi \mid X$$
$$pf ::= af \mid af^* \qquad af ::= \ell \mid !\ell \mid \text{true} \mid A,$$

where $\ell \in \mathcal{L}$ is a transition label for a given set of transition labels $\mathcal{L}$, $A \subseteq \mathcal{L}$ is a set of action labels, and X is a fixpoint variable.

Satisfiability of a state formula ϕ is defined with respect to a state in a labeled transition system. The formula true holds for every state, false does not hold for any state. The interpretation of the logical connectives is standard. The formula $[pf]\phi$ holds for every state for which the formula ϕ holds in all states that can be reached via a path described by pf. The formula $\langle pf \rangle\phi$ holds if there is some path described by pf such that ϕ holds in the state that is reached by that path.

A path formula pf describes a set of paths. The path formula af describes the set of paths consisting of just one transition taken from the set of actions described by the action formula af. The path formula af^* describes all paths over the actions from the set of actions described by the action formula af. An action formula af describes a set of actions. The action formula ℓ describes the set of actions that only contains $\ell \in \mathcal{L}$. The action formula $!\ell$ describes the set of all actions besides ℓ. The action formula true describes the set of all actions. The action formula A describes the set of actions $A \subseteq \mathcal{L}$.

The formula $\mu X.\phi$ is the least fixed point and $\nu X.\phi$ stands for the greatest fixed point. The formula $\mu X.\phi$ holds for the smallest set of states for which ϕ is valid. Similarly, $\nu X.\phi$ holds for the largest set of states for which ϕ holds. For a more intuitive explanation and an extensive discussion of the fixed point operators, we refer to [8].

The μ-calculus is an event-based temporal logic. This means that it cannot employ the variables to identify states as in the translation to UPPAAL. Therefore, to identify states, we employ the unique state labels of the automata depicted in Fig. 1. For each state, we insert a self loop labeled by the name of that state, and the set of all such labels is denoted by $\mathcal{S}$. Since, the name of the state is unique, the parallel composition of the automata will not synchronize on the state names, so we can employ these self loop to identify states of interest.

First, we describe how the state formulae that are used in the UPPAAL queries are transcribed into equivalent formulae in the modal μ-calculus. To ascertain whether state s (from the original Supremica model) has been reached in the labeled transition system of the supervised plant, we have to check whether $\langle s \rangle$true holds, for $s \in \mathcal{S}$. This means that we do not use the variables that were introduced for the purpose of verification in UPPAAL. All logical connectives that are employed in state formulae are also present in the modal μ-calculus, albeit a different syntax is used.

The only state formula that has not been discussed yet is the special formula *deadlock*. A state has deadlock (i.e., satisfies the property *deadlock*) if it has no

UPPAAL	Modal μ-calculus	UPPAAL	Modal μ-calculus
$A\Box\phi$	$[\mathcal{L}^*]\phi_\mu$	$A\Diamond\phi$	$\mu X.[\mathcal{L}]X \vee \phi_\mu$
$E\Diamond\phi$	$\langle\mathcal{L}^*\rangle\phi_\mu$	$E\Box\phi$	$\nu X.\langle\mathcal{L}\rangle X \wedge \phi_\mu$
		$\phi \rightarrow \psi$	$[\mathcal{L}^*](\phi_\mu \Rightarrow \mu X.([\mathcal{L}]X \vee \psi_\mu))$

Fig. 3. Transcription of UPPAAL path formulae into modal μ-calculus formulae; for a state formula ϕ, ϕ_μ denotes the modal μ-calculus formulation of ϕ

Property	Modal μ-calculus
V_1:	$[\mathcal{L}^*]\neg[\mathcal{L}]\text{false}$
V_2:	$[\mathcal{L}^*](\langle \mathit{OperInProg}\rangle\text{true} \Rightarrow \langle \mathit{Standby}\rangle\text{true})$
V_3:	$\mu X.([\mathcal{L}]X \vee ((\langle \mathit{HardDln}\text{true} \wedge \langle \mathit{NotSchld}\rangle\text{true}) \Rightarrow \langle \mathit{Schld}\rangle\text{true}))$
V_4:	$\langle\text{true}^*\rangle(\neg\langle \mathit{NoDln}\rangle\text{true} \wedge \langle \mathit{OperInProg}\rangle\text{true})$
V_5:	$\mu X.([\mathcal{L}]X \vee (\langle \mathit{NewJob}\rangle\text{true} \Rightarrow \langle \mathit{Run}\rangle\text{true}))$

Fig. 4. Transcription of UPPAAL properties into modal μ-calculus properties

outgoing transition. Deadlock is the property that no single event can be executed. In modal μ-calculus this is formulated as [true]false. In labeled transition systems where state information is encoded by means of self loops, deadlock must be interpreted as the property that no transition with event label in $\mathcal{L}$ can be executed. Thus, we describe a deadlock state by the formula $[\mathcal{L}]$false.

The transformation of path formulae into modal μ-calculus formulae takes placing using the patterns presented in Fig. 3. Making use of the transcription presented above one may readily translate the UPPAAL properties into their modal μ-calculus formulations as given in Fig. 4.

Sometimes, simpler μ-calculus formulae can be given for properties. For example, property V_1, absence of deadlock, may also be written $[\mathcal{L}^*]\langle\mathcal{L}\rangle$true. Using the mCRL2 toolset the transcribed versions of the properties V_1–V_5 are re-checked effortlessly.

Do to the highly open nature of our system specification it is impossible to ascertain the property that if the system reaches a hard deadline then the maintenance operation becomes scheduled, see [14] for a detailed discussion. The UPPAAL formulation of this property is $PC == 3 \wedge MS == 1 \rightarrow MS == 3$. In modal μ-calculus this property can be formalised as follows:

$$[\mathcal{L}^*]((\langle \mathit{HardDln}\rangle\text{true} \wedge \langle \mathit{NotSchld}\rangle\text{true}) \Rightarrow \mu X.([\mathcal{L}]X \vee \langle \mathit{Schld}\rangle\text{true}))$$

which expresses that in any state where a hard deadline is reached and no maintenance operation is scheduled eventually (and unavoidably) a state will be reached where the maintenance operation is scheduled. Obviously this property then also does not hold for the mCRL2 model (and the mCRL2 tooling confirms this). However, there are several ways to weaken the property slightly such that meaningfull properties about the system can be obtained. For example,

$$[\mathcal{L}^*]((\langle \mathit{HardDln}\rangle\text{true} \wedge \langle \mathit{NotSchld}\rangle\text{true}) \Rightarrow \nu X.([\mathcal{L}]X \vee \langle \mathit{Schld}\rangle\text{true}))$$

expresses the above property with the relaxation that it may be the case that there are infinite paths along which maintenance operation will not be scheduled. This property holds! Another way of weakening the property is by excluding looping through the states *NoJob* and *NewJob* by means of the events *_NewJob* and *_JobFin*. This is achieved by omitting either one of these labels from $\mathcal{L}$, e.g.:

$$[\mathcal{L}^*]((\langle \mathit{HardDln} \rangle \mathrm{true} \wedge \langle \mathit{NotSchld} \rangle \mathrm{true}) \Rightarrow \mu X.([\mathcal{L} \setminus \{\mathit{\_NewJob}\}]X \vee \langle \mathit{Schld} \rangle \mathrm{true}))$$

Also this property holds. Thus the modal μ-calculus allows us to express relevant properties of the system that are beyond the expressive power of the query language used by UPPAAL. For property V_5 a similar weakening can be applied.

5 Discussion

Models of supervised systems obtained by synthesis are usually formulated in terms of automata. Both UPPAAL and mCRL2 are very capable of dealing with automata-like structures, although UPPAAL here fits more naturally. A clear difference is the way the languages deal with state information. UPPAAL caters for the concept by allowing one to introduce names for states and by using state variables. mCRL2 is focussed entirely on the events that may occur. Several transformations that take place inside the tool (outside the scope of the user) may result in loss of state information. As we show here, one can circumvent this by putting relevant state information (such as state names) in special events.

Also in expressing properties of systems UPPAAL allows reference to state information, whereas mCRL2 does not. However, the class of properties that can be expressed for UPPAAL models is strictly smaller than the ones that can be expressed by modal μ-calculus formulae, provided that state names are displayed as done in this paper. In case UPPAAL properties refer to the values of certain variables a similar 'trick' needs to be applied in the mCRL2 model to make this information accessible through an event. We have shown that the modal μ-calculus is capable of expressing relevant properties that are beyond the expressive power of UPPAAL path formulae. However, this comes at the price of modeling convenience, as formulating path formulae specifications is much easier than mastering the involved constructs of the modal μ-calculus.

References

1. Akesson, K., Fabian, M., Flordal, H., Malik, R.: Supremica - an integrated environment for verification, synthesis and simulation of discrete event systems. In: Proceedings of WODES 2006, pp. 384–385. IEEE (2006)
2. Baeten, J.C.M., van de Mortel-Fronczak, J.M., Rooda, J.E.: Integration of Supervisory Control Synthesis in Model-Based Systems Engineering. In: Proceedings of ETAI/COSY 2011, pp. 167–178. IEEE (2011)
3. Brandin, B.A., Malik, R., Malik, P.: Incremental verification and synthesis of discrete-event systems guided by counter examples. IEEE Transactions on Control Systems Technology 12(3), 387–401 (2004)

4. Cassandras, C., Lafortune, S.: Introduction to discrete event systems. Kluwer Academic Publishers (2004)
5. Clarke, E.M., Emerson, E.A., Sistla, A.P.: Automatic verification of finite-state concurrent systems using temporal logic specifications. ACM Transactions on Programming Languages and System 8(2), 244–263 (1986)
6. D'Ippolito, N.R., Braberman, V., Piterman, N., Uchitel, S.: Synthesis of live behaviour models. In: Proceedings of SIGSOFT 2010, pp. 77–86. ACM (2010)
7. Groote, J.F., Mathijssen, A.H.J., Reniers, M.A., Usenko, Y.S., van Weerdenburg, M.J.: Analysis of distributed systems with mCRL2. In: Process Algebra for Parallel and Distributed Processing, pp. 99–128. Chapman & Hall (2009)
8. Groote, J.F., Reniers, M.A.: Algebraic process verification. In: Handbook of Process Algebra, ch. 17, pp. 1151–1208. Elsevier (2001)
9. Jiang, S., Kumar, R.: Supervisory control of discrete event systems with CTL* temporal logic specifications. SIAM Journal on Control and Optimization 44(6), 2079–2103 (2006)
10. Larsen, K.G., Pettersson, P., Yi, W.: Uppaal in a Nutshell. International Journal on Software Tools for Technology Transfer 1(1-2), 134–152 (1997)
11. Leveson, N.: The challenge of building process-control software. IEEE Software 7(6), 55–62 (1990)
12. Markovski, J.: Supremica2{UPPAAL, mCRL2} and demo models (2012), `http://sites.google.com/site/jasenmarkovski`
13. Markovski, J., Jacobs, K.G.M., van Beek, D.A., Somers, L.J.A.M., Rooda, J.E.: Coordination of resources using generalized state-based requirements. In: Proceedings of WODES 2010, pp. 300–305. IFAC (2010)
14. Markovski, J., Reniers, M.A.: An integrated state- and event-based framework for verifying liveness in supervised systems. In: Proceedings of ICARCV 2012. IEEE (2012) (to appear)
15. Piterman, N., Pnueli, A., Sa'ar, Y.: Synthesis of reactive(1) designs. In: Emerson, E.A., Namjoshi, K.S. (eds.) VMCAI 2006. LNCS, vol. 3855, pp. 364–380. Springer, Heidelberg (2006)
16. Ramadge, P.J., Wonham, W.M.: Supervisory control of a class of discrete-event processes. SIAM Journal on Control and Optimization 25(1), 206–230 (1987)
17. Seow, K.T.: Integrating temporal logic as a state-based specification language for discrete-event control design in finite automata. IEEE Transactions on Automation Science and Engineering 4(3), 451–464 (2007)
18. Song, R., Leduc, R.: Symbolic synthesis and verification of hierarchical interface-based supervisory control. In: Proceedings of WODES 2006, pp. 419–426. IEEE (2006)
19. Voronov, A., Akesson, K.: Verification of process operations using model checking. In: Proceedings of CASE 2009, pp. 415–420. IEEE (2009)
20. Ziller, R., Schneider, K.: Combining supervisor synthesis and model checking. ACM Transactions on Embedded Computing Systems 4(2), 331–362 (2005)

Recognition of Colorectal Carcinogenic Tissue with Gene Expression Analysis Using Bayesian Probability

Monika Simjanoska, Ana Madevska Bogdanova, and Zaneta Popeska

Ss. Cyril and Methodius University in Skopje, Faculty of Computer Science and Engineering
m_simjanoska@yahoo.com,
{ana.madevska.bogdanova,zaneta.popeska}@finki.ukim.mk

Abstract. According to the WHO research in 2008, colorectal cancer caused approximately 8% of all cancer deaths worldwide. Only particular set of genes is responsible for its occurrence. Their increased or decreased expression levels cause the cells in the colorectal region not to work properly, i.e. the processes they are associated with are disrupted. This research aims to unveil those genes and make a model which is going to determine whether one patient is carcinogenic. We propose a realistic modeling of the gene expression probability distribution and use it to calculate the Bayesian posterior probability for classification. We developed a new methodology for obtaining the best classification results. The gene expression profiling is done by using the DNA microarray technology. In this research, 24,526 genes were being monitored at carcinogenic and healthy tissues equally. We also used SVMs and Binary Decision Trees which resulted in very satisfying correctness.

Keywords: DNA microarray, machine learning, colorectal cancer, Bayes' theorem, posterior probability, Support Vector Machines, Binary Decision Trees.

1 Introduction

According to the World Health Organization and the GLOBOCAN project which provided research of the cancer incidence, mortality and prevalence worldwide in 2008, colorectal cancer is responsible for nearly 608,000 deaths, or, it causes 8% of total cancer deaths. This fact makes the colorectal cancer the fourth most common cause of death from cancer [1].

In this paper, the colorectal cancer is considered as a problem of particular genes which have increased or decreased expression levels in the colorectal region. The gene expression profiling is done by using the Illumina HumanRef-8 v3.0 Expression BeadChip microarray technology. This whole-genome expression array allows 24,526 transcript probes.

Gene expression data used in this paper is downloaded from the ArrayExpress, EMBL-EBI biological database [2]. It is collected according to the MIAME standard and can be accessed using the unique identity number E-GEOD-25070.

The paper is organized as follows: 2. Methods and methodology, where we give an overview of the related work and the developed original procedure for using the Bayes' theorem, 3. Experiments and results and 4. Summary and conclusions.

S. Markovski and M. Gusev (Eds.): *ICT Innovations 2012*, AISC 207, pp. 305–314.
DOI: 10.1007/978-3-642-37169-1_30

2 Methods and Methodology

2.1 Related Work

In this section we briefly review some of the research literature related to colorectal cancer statistical and discriminant analysis.

L. C. LaPointe [3] in his Ph.D. thesis describes the discovery and validation of biomarker candidates for colorectal neoplasia. Some genes exhibit gene expression patterns which correlate with the neoplastic phenotype and these results enable investigation of the central practical aim: the identification from the pool of differentially expressed genes those candidate biomarkers which could serve as leads for clinical assay research and development in the future. He has given an overview to discriminant analysis, Bayes' theorem, and machine learning algorithms in candidate biomarkers identification.

Gene expression data set used in our paper has also been used in other scientific researches. The experiment authors in [1] together with Christopher P.E. Lange et al. [4] used this expression data to perform analysis of the aberrant DNA methylation in colorectal cancer.

Another paper that used the same gene expression data and that can be helpful in comparing different methods for identification of colorectal cancer genes is the research done by Bi-Qing Li, et al. [5]. In order to identify colorectal cancer genes, they used method based on gene expression profiles and shortest path analysis of functional protein association networks.

In the Shizuko Muro's, et al. [6] research, when making the classification model they assumed that the gene expression data is distributed according to a mixture of Gaussian distributions.

2.2 The Methodology

In our research, we used Bayes' theorem to classify the colorectal carcinogenic tissue using the gene expression analysis. In order to achieve realistic results, we developed an original methodology that includes several steps – data preprocessing, statistical analysis, modeling the a priori probability for all significant genes and the classification process itself. Furthermore, we used the Support Vector Machines and Decision trees to compare the obtained classification results.

2.2.1 Data Preprocessing

Gene expression profiling of 26 colorectal tumors and matched adjacent 26 non-tumor colorectal tissues is retrieved for further analysis. The gene expression data consists of raw and processed data. Processed data is log2 transformed and normalized using Robust Spline Normalization (RSN) method.

Normalization Methods. Our research aims to unveil the differential expression of the genes expression level. We assume that only a small set of genes are differentially expressed. In such cases Quantile normalization is a suitable normalization method. Quantile normalization (QN) makes the distribution of the gene expression as similar as possible across all samples [7]. However, Quantile normalization forces same distribution for intensity values across different samples which can cause small differences among intensity values to be lost [8]. Therefore, we also analyzed processed data which is normalized using the RSN method. RSN method combines the good features from the Quantile and the Loess normalization. Rather, it combines the strength of Quantile normalization and the curve fitting [9].

Filtering Methods. Some genes may not be well distributed over their range of expression values, i.e. low expression values can be seen in all samples except one [10]. This can lead to incorrect conclusion about gene behavior. To remove such genes, we used an entropy filter. Entropy measures the amount of information (disorder) about the variable. Higher entropy for a gene means that its expression levels are more randomly distributed [11], while low entropy for a gene means that there is low variability [12] in its expression levels across the samples. Therefore, we used low entropy filter to remove the genes with almost ordered expression levels.

Statistical Tests. This model assumes that whole-genome gene expression follows normal distribution [13]; therefore, we used unpaired two-sample t-test for differential expression. The t-test is most commonly used method for finding marker genes that discriminate carcinogenic from healthy tissue. Here we have two independent groups, cancer vs. normal tissue. We expect that most of the genes are not differentially expressed. Thus, the null hypothesis states that there is no statistical difference between the cancer and the normal samples. The rejection of the null hypothesis depends on the significance level which we determine. In this paper we consider the genes as statistically significant for a p-value less than 0.01, which means that the chances of wrong rejection of the null hypothesis is less than 1 in 100.

Using the t-test only, we confront with the problem of false positives. The term false positive refers to genes which are considered statistically significant when in reality differential expression doesn't exist. To remove such genes from further analysis, we used False Discovery Rate (FDR) method. FDR method is defined as a measure of the balance between the number of the true positives and false positives [14]. For a threshold of 0.01 we expect 10 genes to be false positive in a set of 1000 positive genes. The significance in terms of false discovery rate is measured as a q-value. It can be described as a proportion of significant genes that turn out to be false positives [14]. This method is supposed to reduce the number of significant genes supplied from the t-test.

The t-test and the FDR method identified differential expression in accordance with statistical significance values. However, these methods do not consider biological significance. The biological significance is measured as a fold change which describes how much the expression level changed starting from the initial value. Fold change is measured as ratio between the two expression intensities and does not take into account the variance of the expression levels. Because of its simplicity it is usually used in combination with another statistical method [15]. Therefore, we used the volcano plot visual tool to display both statistically and biologically significant genes using a p-value threshold of 0.01 and a fold change threshold of 1.2. The genes that lie in the area cut off by the horizontal threshold, which implicates statistical significance, and the vertical thresholds, which implicate biological significance, are the genes that are up or down regulated depending on the right and the left corner of the plot respectively.

2.2.2 Modeling the a Priori Probability

Using the histogram[1] visual tool, we represented gene expressions at carcinogenic and healthy tissues. Observing the Fig. 1 and Fig. 2, which show the gene expression distribution in a carcinogenic and healthy tissue respectively, we perceived that their distribution substantially differs one from another. In order to confirm the visual assumption of difference, we used the Kolmogorov-Smirnov test at which all genes rejected the null hypothesis of having the same distribution. Having the prior knowledge about gene expressions distribution, we can use the Bayes' theorem to compute the posterior probability $p\ (C_i \mid \vec{x})$, where $\vec{x} = \{e_1, e_2, \ldots, e_n\}$ is a tissue vector containing the expression values for all significant genes, and C_i is one of the classes – carcinogenic or healthy. The posterior probability expresses how probable the class C_i is for a given tissue $\vec{x}$. According to the Bayes' theorem [16], in order to obtain this probability, we must determine the class-conditional densities $p\ (\vec{x} \mid C_i)$ for each class C_i individually, and the class prior probabilities $p\ (C_i)$.

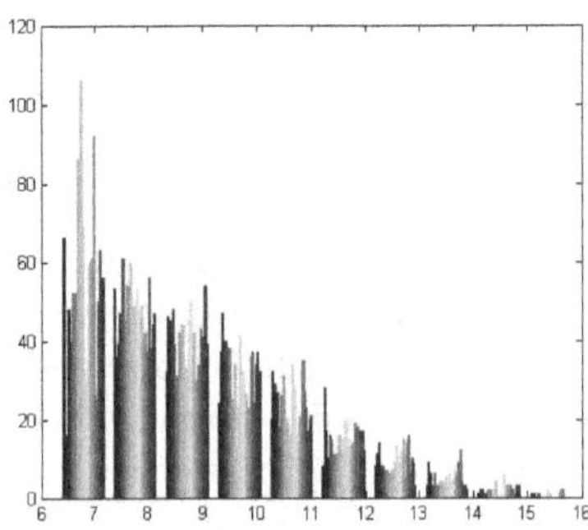

Fig. 1. Gene expression distribution of the carcinogenic tissue samples

[1] The histogram represents each sample (patient) with different color, putting its expression values on the x-axis and the number of genes on the y-axis.

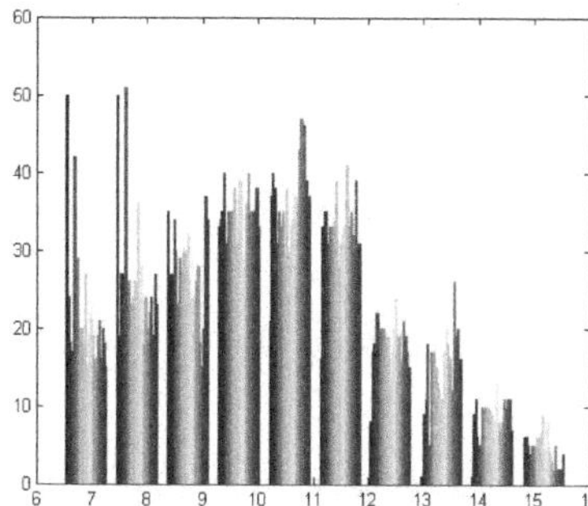

Fig. 2. Gene expression distribution of the healthy tissue samples

Since the probability $p\ (\vec{x})$ is calculated using the law of total probability and is the same for both cases it is usually ignored the Bayes' theorem takes the form

$$p\ (C_i \mid \vec{x}) = p\ (\vec{x} \mid C_i)\ p\ (C_i). \tag{1}$$

Estimating the class prior probabilities $p\ (C_i)$ is simple in this case, because we have equal number of samples into both of the classes - carcinogenic and healthy. Thus, the prior probabilities are $p\ (C_1) = p\ (C_2) = 0.5$ for the carcinogenic and the healthy class, respectively.

The class-conditional density $p\ (\vec{x} \mid C_i)$ is the probability density function for $\vec{x}$ given the particular class C_i. Unlike most of the models which assume Gaussian distribution, we followed generative approach and modeled class-conditional densities by ourselves. Thus, assuming independence of gene distribution we modified the class-conditional densities as

$$p\ (\vec{x} \mid C_i) = \Pi\ f_1 f_2 \ldots f_n, \tag{2}$$

where f_i is the continuous probability distribution of each gene distinctively.

In order to determine the distribution of each gene, we needed to observe a large quantity of data. Therefore, using the holdout cross-validation technique, we involved ¾ of the data in the training process. For each gene we performed statistical tests over the continuous and asymmetric Lognormal, Gamma and Extreme value distribution and we have chosen the one with the highest probability. Once we have modeled the class-conditional densities and the prior probabilities, we used the Bayes formulation to calculate the a posteriori probability to classify the tissues (1).

2.3 Classification Techniques

As we revealed the genes whose differential expression is significant for the colorectal cancer in the data preprocessing part, we can use supervised learning methods to diagnose whether the tissue is healthy or carcinogenic and choose the one that recognizes the carcinogenic tissues with highest precision.

Bayes' Theorem. Once we have modeled the class-conditional densities and the prior probabilities, we proceeded to calculate the posterior probability and to classify the tissues using (1), by the rule

$$\text{If } p\,(C_1 \mid \vec{x}\,) > p\,(C_2 \mid \vec{x}\,), \text{ then choose } C_1 \quad (3)$$

$$\text{If } p\,(C_2 \mid \vec{x}\,) > p\,(C_1 \mid \vec{x}\,), \text{ then choose } C_2.$$

Support Vector Machines. SVM is a method that can classify high-dimensional data as are multiple genes' expression levels. Given significant genes expression levels, we constructed tissue vectors $\vec{x}$ for each patient. This binary classifier is supposed to choose the maximum margin separating hyperplane among the many [17] that separates the carcinogenic from healthy samples in the m-dimensional expression space, where m is the number of significant genes. In order to investigate the expression data separability, we trained the classifier using three types of kernels: linear kernel, quadratic kernel and radial basis function. To avoid over-fitting, we used hold-out cross-validation technique which avoids the overlap between training data and test data, yielding a more accurate estimate for the generalization performance of the algorithm [18]. In addition, we also used bootstrapping method for accuracy improvement.

Decision Trees. Decision tree is a hierarchical data structure implementing the divide-and-conquer strategy. The tree codes directly the discriminants separating class instances without caring much for how those instances are distributed in the regions. The decision tree is a discriminant-based, whereas the statistical methods are likelihood-based in that they explicitly estimate the likelihood before using the Bayes' rule and calculating the discriminant. Discriminant-based methods directly estimate the discriminants, bypassing the estimation of class densities [19]. The reason for using this method is because it is easy to implement and it solves the classification problem using completely different approach from the SVM and Bayes' theorem, which gives useful insight for methods efficiency comparison.

3 Experiments and Results

We retrieved colorectal microarray data from the ArrayExpress biological database [2]. To obtain realistic modeling of the specific genes gene expression probability distribution, we performed a series of analyzes according the methodology presented in 2.2 that leaded to these results.

As far as we normalized gene expression levels (Table 1), we continued with genes reduction methods. Starting from the initial condition of 24,526 genes for 52 tissues we implemented a few statistical tests to separate the significant genes suitable for classification modeling, i.e., the data preprocessing. At first we removed the genes with low variability in their expression values using the low entropy filter (Table 2). Assuming the whole-genome distribution follows a normal distribution and most of the genes are not differentially expressed, we performed t-test statistics to find marker genes that discriminate carcinogenic from healthy tissue. The number of genes significantly reduced to approximately 3500 for up expression and 2900 for down expression. To remove the false positives, we used the FDR method which eliminated nearly 400 genes at both up and down expression. Eventually, using the volcano plot visual tool, we cut off the genes considering both statistical and biological significance, which resulted in a set of nearly 200 genes, most of them down expressed. The results are given in Table 3.

Table 1. Normalization results for the gene expression levels

Tissue	Statistics	Before	QN	RSN
tumor tissue	Sample min.	6,3517	6,3884	6,5971
	1st Quartile	6,9229	6,9719	7,1066
	Median	7,6698	7,7381	7,7613
	2nd Quartile	9,4721	9,5295	9,3357
	Sample max.	13,2958	13,3551	12,6789
	Outliers	425	430	659
normal tissue	Sample min.	6,3624	6,4057	6,6123
	1st Quartile	6,9220	6,9618	7,0968
	Median	7,6770	7,7213	7,7439
	2nd Quartile	9,4879	9,5542	9,3498
	Sample max.	13,3289	13,4410	12,7262
	Outliers	460	417	676

Table 2. Removing homogenous gene expressions

Filter	QN	RSN
Low entropy	22073	22073

Table 3. Finding significant marker genes

Norm. / Methods	T-test		FDR		Volcano Plot		
	up	down	up	down	up	down	**sum**
QN	3515	2865	3108	2598	50	165	**215**
RSN	3729	2968	3410	2736	41	151	**192**

Once we discovered marker genes that discriminate carcinogenic from healthy tissue, we used them to make a model according to which we can diagnose the patients' health condition. Since we have the a priori knowledge such as the gene expression levels and the two possible health conditions, we used few supervised learning methods in order to choose the one with best performance.

First, we used generative approach - modeling the prior distributions by ourselves. We modeled the prior distributions (Fig.1 and Fig.2) and used them in the Bayes' theorem to calculate the posterior probability. Thus, we maintained very high correct rate, especially for the carcinogenic samples, which is very important in the diagnosing process (Table 4).

Table 5 represents the results obtained from the SVM classification. When training the classifier we used three types of kernels. We used hold-out cross-validation technique which involved ⅔ of the samples in the training set and ⅓ in the testing set. In addition, we used bootstrapping method, but it gave very poor results. The SVM method produced good results, but they vary in every subsequent trial depending on the chosen training set.

Furthermore, we used Binary Decision Trees because of their simplicity and the different approach of discriminant calculation. The results in Table 6 show that it correctly classifies healthy tissues.

According to the overall results, the Bayes' theorem is the most accurate classification method in the problem of classifying colorectal carcinogenic tissue.

Table 4. Bayesian posterior probability classification

Bayes' theorem	**Cancer**		**Healthy**		**Total**	
	all	test	all	test	all	test
QN	100%	100%	92.30%	83.33%	96.15%	91.67%
RSN	96.15%	100%	96.15%	100%	96.15%	100%

Table 5. Support Vector Machines classification

SVM results	**Linear kernel**		**Quadratic kernel**		**GRB**	
	cancer	healthy	cancer	healthy	cancer	healthy
QN	100%	87.5%	75%	87.5%	0%	100%
RSN	87.5%	100%	87.5%	100%	100%	25%
	total		total		total	
QN	93.75%		81.25%		50%	
RSN	93.75%		93.75%		62.5%	
Bootstrapping	cancer	healthy	cancer	healthy	cancer	healthy
QN	20%	35%	10%	90%	70%	30%
RSN	20%	45%	45%	70%	45%	35%

Table 6. Binary Decision Tree classification

BDT results	Cancer	Healthy	Total
QN	75%	100%	87.5%
RSN	87.5%	100%	93.75%

The ability of the test to correctly classify positive and negative samples is measured as sensitivity and specificity respectively. Sensitivity refers to the true positive rate; whereas specificity takes into consideration the true negative rate. This analysis also indicates that the Bayes' theorem is the most suitable classifier in this case.

Table 7. Classifiers' Sensitivity and Specificity

Results	Bayes' theorem	Linear kernel	Quadratic kernel	GRB	BDT
Sensitivity	1	1	0.8750	1	0.75
Specificity	0.9231	0.8750	0.8750	0	1

4 Summary and Conclusions

As we are well introduced with the incidence and mortality caused by the colorectal cancer worldwide, we used DNA microarray data to observe its gene expression behavior. We assumed that the responsibility for its occurrence lies in the disrupted gene expression levels, and therefore, we performed different statistical tests to unveil those genes. Those tests discovered approximately 200 marker genes that discriminate carcinogenic from healthy tissue, which we used to build an accurate diagnostic system. Histogram representation confirmed different gene expression pattern at carcinogenic and healthy tissues distinctively. Subsequently, we used few different classification methods in order to choose the most accurate one. The best results were achieved using the Bayes' theorem - we obtained over 90% correctness when classifying the tissues. We can conclude that the reason the Bayes learning model was most accurate for this problem is in the realistic modeling of the a priori probability.

The results from this paper can be used for future research in upgrading the model in order to obtain even more accurate diagnostic system. Furthermore, the unveiled significant marker genes can be used in building ontology which can be very useful in developing new pharmaceutical molecules.

References

1. GLOBOCAN 2008 (2008), `http://globocan.iarc.fr/factsheets/cancers/colorectal.asp/`
2. Weisenberger, D.J., Van Den Berg, D., Laird, P.W., Hinoue, T.: Gene Expression Analysis of Colorectal Tumors and Matched Adjacent Non-Tumor Colorectal Tissues. In: EMBL-EBI, ArrayExpress, Experiment: E-GEOD-25070 (2011)

3. LaPointe, L.C.: Gene Expressions Biomarkers for Colorectal Neoplasia. Flinders University of South Australia, School of Medicine, Dept. of Medicine (2008), http://theses.flinders.edu.au/public/adt-SFU20091011.090028/index.html
4. Hinoue, T., Weisenberger, D.J., Lange, C.P.E., Shen, H., Byun, H.M., Van Den Berg, D., Malik, S., Pan, F., Noushmehr, H., Van Dijk, C.M., Tollenaar, R.A.E.M., Laird, P.W.: Genome-scale Analysis of Aberrant DNA Methylation in Colorectal Cancer. Genome Res., pp. 271–282 (2011) (February 22, 2012)
5. Li, B.Q., Huang, T., Liu, L., Cai, Y.D., Chou, K.C.: Identification of Colorectal Cancer Related Genes with mRMR and Shortest Path in Protein-Protein Interaction Network. PLoS ONE 7, e33393 (2012), doi:10.1371/journal.pone.0033393
6. Muro, S., Takemasa, I., Oba, S., Matoba, R., Ueno, N., Maruyama, C., Yamashita, R., Sekimoto, M., Yamamoto, H., Nakamori, S., Monden, M., Ishii, S., Kato, K.: Identification of Expressed Genes Linked to Malignancy of Human Colorectal Carcinoma by Parametric Clustering of Quantitative Expression Data. Genome Biol. 4, R21 (2003)
7. Wu, Z., Aryee, M.J.: Subset Quantile Normalization Using Negative Control Features. Journal of Computational Biology 17(10), 1385–1395 (2010)
8. Du, P., Feng, G., Kibbe, W.A., Lin, S.: Using Lumi, a Package Processing Illumina Microarray (2012)
9. Du, P., Lin, S.: Towards an Optimized Illumina Microarray Data Analysis Pipeline. In: Midwest Symposium on Computational Biology & Bioinformatics (2007)
10. Kohane, I.S., Kho, A.T., Butte, A.J.: Microarrays for an Integrative Genomics. MIT (2003)
11. Butte, A.J., Kohane, I.S.: Mutual Information Relevance Networks: Functional Genomic Clustering Using Pairwise Entropy Measurements. In: Pacific Symposium on Biocomputing, vol. 5, pp. 415–426 (2000)
12. Needham, C.J., Manfield, I.W., Bulpitt, A.J., Gilmartin, P.M., Westhead, D.R.: From Gene Expression to Gene Regulatory Networks in Arabidopsis Thaliana. BMC Systems Biology 3, 85 (2009)
13. Yu, H., Tu, K., Xie, L., Li, Y.Y.: Digout: Viewing Differential Expression Genes as Outliers. Journal of Bioinform. and Comput. Biol. 8(suppl. 1), 161–175 (2010)
14. Storey, J.D., Tibshirani, R.: Statistical Significance for Genomewide Studies. Proceedings of the National Academy of Sciences of the United States of America 100(16), 9440–9445 (2003)
15. Tarca, A.L., Romero, R., Draghici, S.: Analysis of Microarray Experiments of Gene Expression Profiling. American Journal of Obstetrics and Gynecology 195(2), 373–388 (2006)
16. Bishop, C.M.: Pattern Recognition and Machine Learning. Springer (2006)
17. Brown, M.P.S., Grundy, W.N., Lin, D., Cristianini, N., Sugnet, C.W., Furey, T.S., Ares Jr., M., Haussler, D.: Knowledge-Based Analysis of Microarray Gene Expression Data by Using Support Vector Machines. Proceedings of the National Academy of Sciences of the United States of America 97(1), 262–267 (2000)
18. Refaeilzadeh, P., Tang, L., Liu, H.: Cross-Validation. In: Enc. of Database Systems (2009)
19. Alpaydin, E.: Introduction to Machine Learning. MIT Press, Cambridge (2010)
20. Bogdanova, A.M., Ackovska, N.: New Support Vector Machines-Based Approach over DNA Chip Data. In: Innovations in Information Technology, December 16-18, pp. 16–19. IEEE, Al Ain (2008) 978-1-4244-3397-1/08
21. Bogdanova, A.M.: DNA Chips in Bioinformatics. In: Computational Intelligence and Information Technologies, CIIT 2007, Molika, Macedonia, January 21-25 (2007)
22. Bogdanova, A.M., Ackovska, N.: Data Driven Intelligent Systems. In: Proceedings of the ICT Innovations (2010) ISSN 1857-7288

Mobile Users ECG Signal Processing

Emil Plesnik and Matej Zajc

University of Ljubljana, Faculty of Electrical Engineering, Trzaska cesta 25, 1000 Ljubljana, Slovenia
{emil.plesnik,matej.zajc}@fe.uni-lj.si

Abstract. In recent years we have witnessed the growth of a number of multimedia, health, cognitive learning, gaming user applications which include monitoring and processing of the users' physiological signals, also termed biosignals or vital signs. The acquisition of the biosignals should be non-invasive and should not affect the activities and arousal of the user to achieve relevant results. Developments in mobile devices, e.g. smartphones, tablet PCs, etc. and electrode design have enabled unobtrusive acquisition and processing of biosignals, especially for mobile and non-clinical applications. The paper reviews recently developed non-clinical applications that exploit biosignal information. The paper analyses challenges for digital signal processing that arise from data acquisition from the mobile user are presented with focus on the electrocardiogram (ECG). Influence of the analysed challenges is demonstrated on a selected QRS detection algorithm by using signals from the MIT-BIH Noise stress test database (*nstdb*). Results confirm that algorithms for processing signals of mobile users need a more thorough preprocessing procedure as opposed to simple band-pass filtering.

Keywords: Digital signal processing, biosignal, electrocardiogram, mobile user, application.

1 Introduction

Mobile, wireless, ubiquitous computing and communication are changing the medical treatment process. With the use of wireless sensing, advanced signal processing algorithms and communication protocols the patient - medical staff interaction is more convenient and more time- and cost-effective. Additionally, the clinical aspects of monitoring and processing physiological signals (e.g. electrocardiogram – ECG, electroencephalogram – EEG, electromyogram – EMG, blood pressure, etc.), also termed biosignals or vital signs, are penetrating the field of non-clinical applications. This paper is focused on the ECG.

Among biosignals, the electrocardiogram (ECG) has an important role. It is an indicator of heart-beating and heart-related medical conditions. Most of the relevant information in the ECG signal can be derived from the analysis of the QRS complex and the P-Q-R-S-T characteristic points of the ECG cycle, e.g. the R-R interval, heart rate variability, heart rate, etc. Automated detection of these characteristic points has been addressed before by using different methods, such as signal derivatives and

S. Markovski and M. Gusev (Eds.): *ICT Innovations 2012*, AISC 207, pp. 315–324.
DOI: 10.1007/978-3-642-37169-1_31

filtering, wavelet transformation, neural networks, hidden Markov models, Hilbert transformation, phase portrait analysis, etc. In this paper we selected the phase portrait analysis method described in [1] to continue and upgrade previous research.

Reliable detection of characteristic complexes and points is essential for determining cardiac abnormalities in medicine. Additionally, analysis of the ECG characteristic points is also useful for numerous non-medical mobile applications such as ambient intelligence [2], driver monitoring [3, 4], biometrics [5, 6], sport training [7, 8], emotion recognition [9, 10], mobile applications of health-monitoring [11, 12]. These applications acquire and process the signals in a non-clinical environment with small, portable devices enabling local data storage and wireless communication. However, they are designed to monitor only 1 – 3 ECG leads at once and do not support monitoring of the standard clinical 12-lead ECG.

On the other hand, modern health systems are undergoing radical changes due to ageing population and increased incidence of chronic diseases. Together with medical care they are moving from reactive towards preventive care and from hospital care to at home care. Substantial efforts are oriented to research and support of this paradigm shift that will reduce costs of overburdened health systems. For example, monitoring and analysis of biosignals could be used to achieve a closed-loop disease management solution for serving heart failure and coronary heart disease patients similar to the EU FP 7 Heartcycle project [13]. The system design should contain two interlaced closed loops: a patient loop and a professional loop. The professional loop should involve medical professionals and hospital information systems to ensure optimal and personalized patient care. The patient loop should interact directly with the patient, support the daily treatment and increase patient motivation to improve and maintain health. To enable monitoring patients at their homes, small, mobile, powerful devices supporting data storage and processing would be needed. Although operating in a non-clinical environment at home, the devices would have to satisfy certain medical standards because of their use for clinical purposes.

Clearly there is a need for devices for acquiring and especially algorithms for analysing biosignals of mobile users in non-clinical environments for medical and non-medical purposes. However, due to low ECG signal amplitudes on skin surface and consequently high sensitivity for different noise sources, there are some additional requirements for the mobile system to be able to address the challenges arising in the mobile users' world. The paper analyses these challenges for digital signal processing of the ECG in mobile applications and demonstrates their influence on a selected QRS detection algorithm by using signals from the MIT-BIH Noise stress test database (*nstdb*). Results confirm that algorithms for processing signals of mobile users need a more thorough preprocessing procedure as opposed to clinically acquired signals.

The rest of the paper is organized as follows: in Section 2 a description of the mobile user ECG monitoring device and its basic requirements is given, in Section 3 the main signal processing challenges in mobile applications are analysed, in Section 4 the results of selected algorithm evaluation on the *nstdb* are given, and Sections 5 and 6 give the discussion and the conclusion, respectively.

2 Mobile User

Environment of the mobile user is different from the clinical environment in medical facilities, although the architecture of a mobile ECG monitoring device (Fig. 1) does not imply that on the first glance. Being able to move around freely outside and in the buildings as opposed to lying still on a bed is the fundamental difference between the two surroundings. The mobile user's world also has some other distinct properties which lead to specific system requirements. For example, due to user mobility the design of the monitoring device has to be small, lightweight, thin, flexible, one-piece and wireless [14]. Additionally, to further ensure unobtrusiveness, the device must not interfere with body movement, has to be easy to place and remove, has to have low difficulty and frequency of maintenance and has also to comply with social and fashion concerns [15]. To enable mobility and flexibility of the user, the device has to support secure and reliable wireless communications that ensure user's privacy and confidentiality [16–18]. For the monitoring device to remain unobtrusive, minimal interaction with the user is required [18]. However, when the interaction does take place, it has to run over a user-friendly interface which has to be intuitive, simple, self-explanatory and easy to use [18, 19]. Since portable, mobile devices run on batteries with limited energy supply, the applications that run on them have to be optimized for low energy consumption [12, 15]. This results in striving to minimal memory usage, minimal user interface, shifting heavy processing to the network, running simple feature extraction algorithms, enabling low-power running modes, supporting local and cloud-based storage, selecting the appropriate communication protocol, etc. [15, 18, 19].

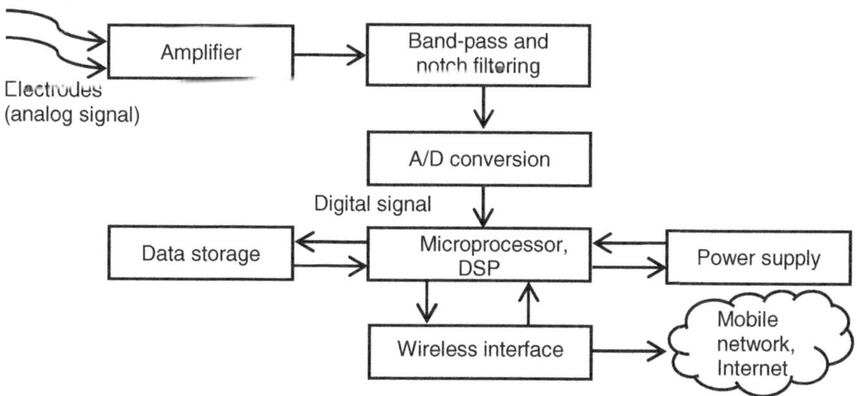

Fig. 1. Architecture of a mobile device for ECG (biosignal) monitoring

It is practically impossible to optimally satisfy all the requirements at once. Especially, because they are often eliminating each other, e.g. higher communication bandwidth requires higher energy consumption which has to be as low as possible. Therefore, to build an optimal mobile monitoring device is often a trade-off between all requirements which is an application specific task and cannot be determined universally.

3 Signal Processing Challenges in Mobile ECG Applications

For mobile ECG applications, reliable and robust noise removal from acquired signals has high priority. In general there are different kinds of noise entering at different points in a mobile acquisition system (Fig. 2) that can be present in the ECG: power line interference, instrumentation noise of electronic components, artifacts due to electrode displacement and movement or due to physical activity and breathing of the user. These disturbances are even more pronounced in a mobile environment with constant motion. In this section possible denoising solutions are discussed.

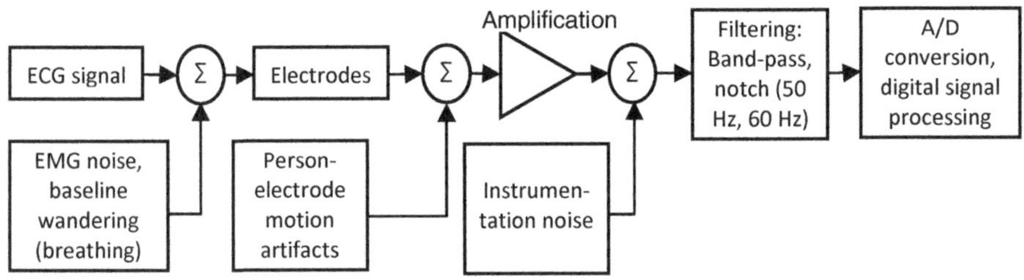

Fig. 2. Main noise entry points in a mobile ECG acquisition system

3.1 Person-Electrode Motion Artifacts

Person-electrode motion artefact reduces the reliability of the signal measurements and consequently the analysis results, with automatized computer assisted analysis. These motion artifacts can cause false alarms during monitoring and therefore reduce confidence in the equipment, alarms, etc.

Person-electrode motion artifacts are either caused by changes in the impedance of the electrode-skin contact or changes in the skin potential due to skin stretch. The first source of artifact has been significantly reduced with the use of wet adhesive (Ag-AgCl) electrodes, but has risen again with the development of capacitive ECG acquisition. The skin stretch remains a serious problem. Standard methods for reducing the effect of skin stretch require skin abrasion which causes discomfort and prevents the use outside clinical environment.

The artifacts resemble the characteristics of baseline wander, but are more problematic to remove since their spectral content (1 – 10 Hz) overlaps that of the ECG cycle considerably. In the ECG these artifacts can be manifested as large amplitude waveforms that can be mistaken for QRS complexes and cause falsely detected heartbeats. Different approaches are used to remove person-electrode motion artifacts from the ECG signal, e.g. adaptive filtering [20], wavelet denoising [21], empirical mode decomposition [22], independent component analysis [23], etc.

Severe example of these artifacts is shown in Fig. 3, where the monitored person is performing intensive exercise which caused loss of electrode-skin contact at certain moments (indicated by interrupted signal). The signal in Fig. 3 was acquired with Ag-AgCl gel electrodes. Although the selected QRS complex detector [1] was designed to overcome a certain level of noise, it did not manage to completely eliminate all the artifacts on the interval between sample numbers 24.000 and 25.500 in Fig. 3.

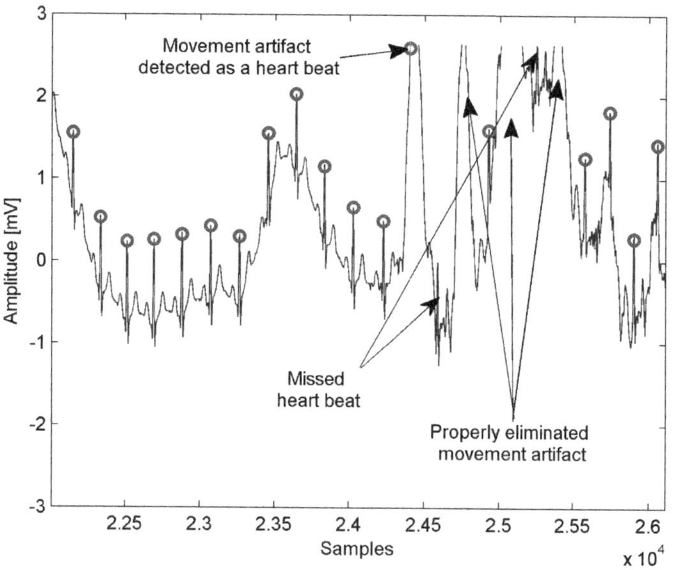

Fig. 3. Person-electrode movement artifacts (between sample numbers 24.000 and 25.500) in an ECG signal acquired from a mobile user with Ag/AgCl adhesive electrodes

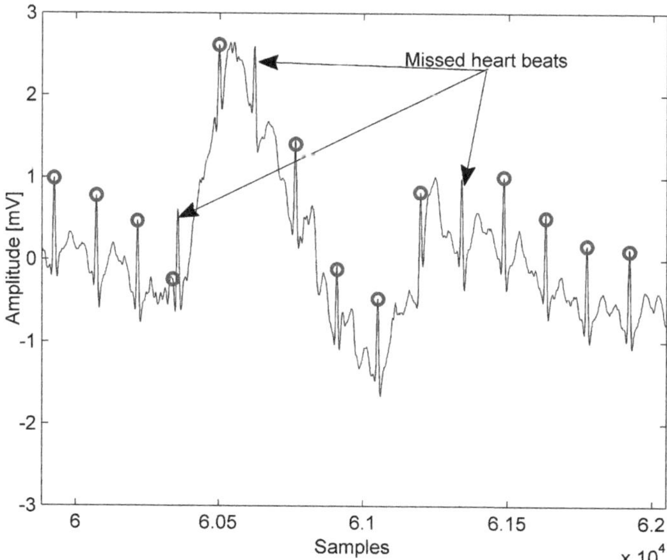

Fig. 4. Baseline wandering in an ECG signal segment acquired with Ag-AgCl gel electrodes from a user doing squat exercises

3.2 Baseline Wandering

Baseline wandering is an exterior, low-frequency activity in the ECG (Fig. 4), caused by respiratory activity, body motion, etc. It may interfere with the signal analysis and cause inaccurate interpretation due to its spectral content overlapping with that of the signal. The frequency spectrum of the baseline wandering usually includes frequencies under 1 Hz [24], although higher frequency components are possible especially during more intensive movements, e.g. during exercising. For example, ECG analysis using reference to the isoelectric line, e.g. ST segment analysis, detection of onset and offset of characteristic points, cannot be performed in the presence of baseline wandering which prevents defining the isoelectric line. According to [24], two major techniques are employed for the removal of baseline wander from the ECG, namely linear filtering and polynomial fitting. However, other methods are also known, such as wavelets [25], empirical mode decomposition [26], morphological filtering [27], etc.

For example, in Fig. 4 there is an ECG signal segment contaminated with baseline wander. The signal was acquired with Ag-AgCl gel electrodes during squat exercise of the monitored person. The selected QRS detector [1] missed three heart beats in a seven-second long interval.

3.3 Electromiographic (EMG) Noise

The presence of EMG noise is caused by muscle contractions which result in change of electric potential similarly as contractions of the heart muscle. The EMG noise is a direct consequence of body movement and enters the monitoring system at its source together with the wanted ECG signal (Fig. 2). Unfortunately, the frequency spectrum of the EMG noise (between dc and 10 kHz) considerably overlaps that of the ECG cycle and represents a very difficult filtering problem as opposed to e.g. baseline wandering. Additionally, muscle noise often has higher amplitude values than the ECG signal due to more powerful contractions of body muscles, e.g. during exercising, running. However, the EMG noise is not repetitive as the ECG signal, and it usually occurs in short term bursts. The EMG noise causes QRS like artifacts in the ECG signal resulting in false beat detections and disabling ECG delineation. Known solutions to EMG noise removal are, wavelet transform [21], time-varying low-pass filtering [24], SVD filtering [28], genetic particle filtering [29], etc.

In Fig. 5, an ECG signal segment also contaminated with EMG noise is shown. The signal was acquired with Ag-AgCl gel electrodes during fast walking of the monitored person. The selected QRS detector [1] had much difficulties with QRS detection. On the interval there are many missed and also false detected heart beats.

Although the individual missed or falsely detected heart beats do not present a significant issue for the results, this can quickly turn around with longer, strenuous activity of the mobile user. For example, if the user doesn't properly adjust the electrodes before the use of the monitoring device, heavy noise contamination can occur in the signal, causing unnecessary false alarms and interventions or can prevent proper treatment advances and supervision. That is why the users should be provided with reliable equipment using efficient preprocessing and processing techniques.

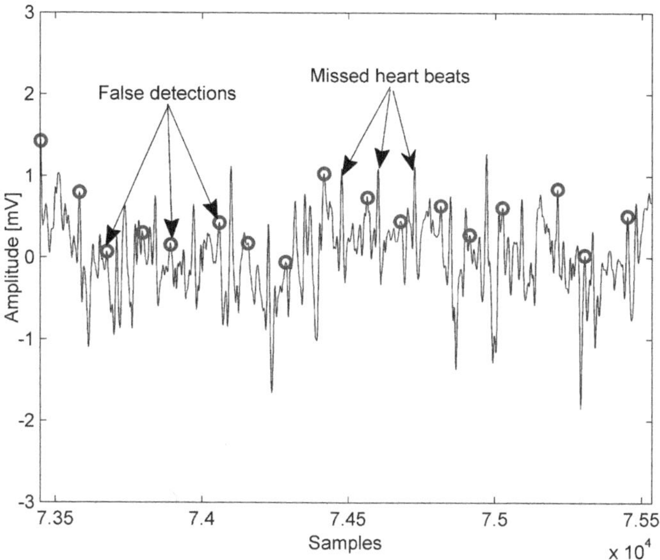

Fig. 5. EMG noise in an ECG signal segment acquired with Ag-AgCl gel electrodes from a user during fast walking

4 Results

The performance of the selected method [1] with the presence of noise was evaluated for the QRS detection with the *nstdb* [30]. The signals in the database are gradually contaminated with a mixture of electrode movement artifacts, baseline wander and EMG noise, simulating the mobile user signal acquisition system in Fig. 2. The selected algorithm constructs the ECG signals in a 2D phase space using the delay method, where the amplitudes of the signal and its delayed version are presented on the x and y axis, respectively. Delay value of one sample was applied. Adaptive thresholding was used in the phase portrait for the QRS complex detection. The performance of the selected algorithm, which used a simple Butterworth 4th order band-pass filtering in the frequency range between 0.5 and 100 Hz was assessed by using the sensitivity (Se) (1) and the positive predictivity (PPV) (2)

$$Se = TP/(TP+FN) \tag{1}$$

$$PPV = TP/(TP+FP), \tag{2}$$

where TP, FN and FP represent the numbers of true positive, false negative and false positive QRS detections, respectively. The results of the noise stress test are presented in Table 1. They show a good success rate for a SNR of more than or equal to 12 dB, where both the Se and the PPV values are above 97 %. For a SNR of less than 12 dB the Se value dropped below 91 % and the PPV value dropped below 93 %.

Table 1. Noise stress test using the selected method

Signal	No. of annotated QRS complexes	No. of detected QRS complexes	TP	FP	FN	Se [%]	PPV [%]
118e24	2278	2272	2272	0	6	99.74	100.00
118e18	2278	2272	2272	0	6	99.74	100.00
118e12	2278	2254	2246	8	32	98.60	99.65
118e06	2278	2139	1988	151	290	87.27	92.94
118e00	2278	2119	1692	427	586	74.28	79.85
118e_6	2278	1825	1343	782	935	58.96	73.59
119e24	1987	1987	1985	2	2	99.90	99.90
119e18	1987	1987	1985	2	2	99.90	99.90
119e12	1987	1994	1944	50	43	97.84	97.49
119e06	1987	2138	1804	334	183	90.79	84.38
119e00	1987	2241	1556	685	431	78.31	69.43
119e_6	1987	1889	1089	800	898	54.81	57.65

5 Discussion

The *nstdb* was used to evaluate the noise sensitivity of the QRS detection for the selected algorithm [1]. The results of the noise stress test presented in Table 1 indicate that the algorithm is noise-sensitive for a SNR below 12 dB for signals in the *nstdb*. For SNRs above 12 dB, all the parameters are above 97 %. The ANSI EC13 standard requires reliable detection QRS complexes with amplitudes ranging from 0.5 to 5 mV in the presence of 0.1-Hz, 4 mV peak-to-peak noise, which is basically baseline wandering noise. This means that the detector should be able to cope with SNR as low as – 18 dB. There is yet no standardized minimum required performance in the presence of other types of noise. Similarly, there is no clear definition of QRS detection performance requirement in the available literature. The generally accepted performance indicators are the Se and the PPV, which are desired to be as high as possible (in most cases in the available literature around 99 %). Therefore, according to results in Table 1, the preprocessing of the selected algorithm should be improved to achieve better results with more noisy recordings with SNR below 12 dB.

6 Conclusion

The paper analyses the challenges resulting from digital signal processing of the mobile users ECG. Their influence is demonstrated by signals from the MIT-BIH Noise stress test database (*nstdb*) [30] used for evaluation of the selected QRS detection algorithm combined with band-pass filtering for preprocessing. Evaluation results indicate that the algorithm is effective (Se and PPV parameters above 97 %) for SNRs above 12 dB. This confirms that the band-pass filtering is not sufficient for effective QRS detection in signals contaminated with higher noise amplitudes which cause lower SNRs and often occur in mobile user applications.

References

1. Plesnik, E., Malgina, O., Tasič, J.F., Zajc, M.: Detection of the electrocardiogram fiducial points in the phase space using the euclidian distance measure. Medical Engineering & Physics 34, 524–529 (2012)
2. Fayn, J., Rubel, P.: Toward a Personal Health Society in Cardiology. IEEE Transactions on Information Technology in Biomedicine 14, 401–409 (2010)
3. Wartzek, T., Eilebrecht, B., Lem, J., Lindner, H.-J., Leonhardt, S., Walter, M.: ECG on the road: robust and unobtrusive estimation of heart rate. IEEE Transactions on Biomedical Engineering 58, 3112–3120 (2011)
4. Jung, S.-J., Shin, H.-S., Chung, W.-Y.: Highly Sensitive Driver Health Condition Monitoring System using Nonintrusive Active Electrodes. Sensors and Actuators B: Chemical 171-172, 691–698 (2012)
5. Agrafioti, F., Gao, J., Hatzinakos, D.: Heart Biometrics: Theory, Methods and Applications. In: Yang, J. (ed.) Biometrics. InTech (2011)
6. Singh, Y.N., Singh, S.K., Gupta, P.: Fusion of electrocardiogram with unobtrusive biometrics: An efficient individual authentication system. Pattern Recognition Letters 33, 1932–1941 (2012)
7. Yong, X., Tingting, B., Chunhua, B., Qianli, M.: Design of the Athlete's Electrocardiogram Monitoring and Evaluation System Based on Wireless Sensor Network. In: 2011 First International Workshop on Complexity and Data Mining (IWCDM), pp. 60–63 (2011)
8. Xu, L., Zhong, Y., Yin, S., Zhang, Y., Shen, Y., Hao, D., Hu, Y., Zhang, R.: ECG and pulse variability analysis for exercise evaluation. In: 2011 IEEE International Conference on Automation and Logistics (ICAL), pp. 52–57 (2011)
9. Xu, Y., Liu, G., Hao, M., Wen, W., Huang, X.: Analysis of affective ECG signals toward emotion recognition. Journal of Electronics (China) 27, 8–14 (2010)
10. Agrafioti, F., Hatzinakos, D., Anderson, A.K.: ECG Pattern Analysis for Emotion Detection. IEEE Transactions on Affective Computing 3, 102–115 (2012)
11. Morak, J., Kumpusch, H., Hayn, D., Modre-Osprian, R., Schreier, G.: Design and Evaluation of a Telemonitoring Concept Based on NFC-Enabled Mobile Phones and Sensor Devices. IEEE Transactions on Information Technology in Biomedicine 16, 17–23 (2012)
12. Klingeberg, T., Schilling, M.: Mobile wearable device for long term monitoring of vital signs. Computer Methods and Programs in Biomedicine 106, 89–96 (2012)
13. HeartCycle Project, http://www.heartcycle.eu/
14. Chuo, Y., Marzencki, M., Hung, B., Jaggernauth, C., Tavakolian, K., Lin, P., Kaminska, B.: Mechanically Flexible Wireless Multisensor Platform for Human Physical Activity and Vitals Monitoring. IEEE Transactions on Biomedical Circuits and Systems 4, 281–294 (2010)
15. Bonfiglio, A., Rossi, D.D. (eds.): Wearable Monitoring Systems. Springer, NY (2010)
16. Zvikhachevskaya, A., Markarian, G., Mihaylova, L.: Quality of Service Consideration for the Wireless Telemedicine and E-Health Services. In: IEEE Wireless Communications and Networking Conference, WCNC 2009, pp. 1–6 (2009)
17. Varshney, U.: Pervasive Healthcare Computing: EMR/EHR, Wireless and Health Monitoring. Springer, NY (2009)
18. Kailas, A., Chong, C.-C., Watanabe, F.: From Mobile Phones to Personal Wellness Dashboards. IEEE Pulse 1, 57–63 (2010)

19. Olla, P.: Mobile Health Solutions for Biomedical Applications. Medical Information Science Reference, NY (2009)
20. Eilebrecht, B., Wartzek, T., Willkomm, J., Schommartz, A., Walter, M., Leonhardt, S.: Motion Artifact Removal from Capacitive ECG Measurements by Means of Adaptive Filtering. In: Jobbágy, Á. (ed.) 5th European Conference of the International Federation for Medical and Biological Engineering. IFMBE Proceedings, vol. 37, pp. 902–905. Springer, Heidelberg (2011)
21. Mithun, P., Pandey, P.C., Sebastian, T., Mishra, P., Pandey, V.K.: A wavelet based technique for suppression of EMG noise and motion artifact in ambulatory ECG. In: 2011 Annual International Conference of the IEEE, Engineering in Medicine and Biology Society, EMBC, pp. 7087–7090 (2011)
22. Lee, J., McManus, D.D., Merchant, S., Chon, K.H.: Automatic Motion and Noise Artifact Detection in Holter ECG Data Using Empirical Mode Decomposition and Statistical Approaches. IEEE Transactions on Biomedical Engineering 59, 1499–1506 (2012)
23. Milanesi, M., Martini, N., Vanello, N., Positano, V., Santarelli, M., Landini, L.: Independent component analysis applied to the removal of motion artifacts from electrocardiographic signals. Medical and Biological Engineering and Computing 46, 251–261 (2008)
24. Sörnmo, L., Laguna, P.: Bioelectrical Signal Processing in Cardiac and Neurological Applications. Academic Press, San Diego (2005)
25. Sayadi, O., Shamsollahi, M.B.: Multiadaptive Bionic Wavelet Transform: Application to ECG Denoising and Baseline Wandering Reduction. EURASIP Journal on Advances in Signal Processing, 41274 (2007)
26. Pal, S., Mitra, M.: Empirical mode decomposition based ECG enhancement and QRS detection. Computers in Biology and Medicine 42, 83–92 (2012)
27. Zhang, F., Lian, Y.: QRS Detection Based on Morphological Filter and Energy Envelope for Applications in Body Sensor Networks. Journal of Signal Processing Systems 64, 187–194 (2009)
28. Paul, J.S., Reddy, M.R., Kumar, V.J.: A transform domain SVD filter for suppression of muscle noise artefacts in exercise ECG's. IEEE Transactions on Biomedical Engineering 47, 654–663 (2000)
29. Li, G., Zeng, X., Lin, J., Zhou, X.: Genetic particle filtering for denoising of ECG corrupted by muscle artifacts. In: 2012 Eighth International Conference on Natural Computation (ICNC), pp. 562–565 (2012)
30. The MIT-BIH Noise Stress Test Database, `http://www.physionet.org/physiobank/database/nstdb/`

Top-Down Approach for Protein Binding Sites Prediction Based on Fuzzy Pattern Trees

Georgina Mirceva and Andrea Kulakov

Ss. Cyril and Methodius University in Skopje,
Faculty of Computer Science and Computer Engineering, Skopje, Macedonia
{georgina.mirceva,andrea.kulakov}@finki.ukim.mk

Abstract. The understanding of the relation between the protein structure and protein functions is one of the main research topics in bioinformatics nowadays. Due to the complexity of the methods for determining protein functions, there are many proteins with unknown functions. Hence, many researchers investigate various computational methods for determining protein functions. We focus on investigating methods for predicting the protein binding sites, and afterwards their characteristics could be used for annotating protein structures. In order to overcome the problem of sensitivity on data changes, we already introduced the fuzzy theory for protein biding sites prediction. In this paper we introduce an approach for detecting protein binding sites using a top-down induction of fuzzy pattern trees. This approach outperforms the existing bottom-up approach for inducing fuzzy pattern trees, and also most of the examined approaches which are based on classical classification algorithms.

Keywords: Protein function, Protein binding sites, Fuzzy pattern trees (FPTs), Top-down, BIND database.

1 Introduction

Proteins are involved in various processes in the living organisms, so the knowledge of their functions is crucial for designing new drugs. Experimental methods for investigating the protein functions exist, but these methods are very expensive and complex. Due to this reason, there are numerous protein structures with unknown functions. Hence, there is a reasonable necessity for development of methods for investigating the protein functions in an automated manner.

There are various state of the art methods for determining protein functions. First group of methods [1] investigates the protein functions by analyzing protein-protein interaction networks. Nevertheless, the acquisition of the knowledge of the interacting proteins with some given protein structures is expensive. Second group of methods [2] inspects the homology in the protein structures and/or sequences. However, this group of methods is able to determine only global similarity between the structures, while common functions could be shared even between protein structures that have local similarities although they do not have global similarity. Third group of methods [3] analyses the conservation of the protein structures and/or sequences. Fourth group

S. Markovski and M. Gusev (Eds.): *ICT Innovations 2012*, AISC 207, pp. 325–334.
DOI: 10.1007/978-3-642-37169-1_32

of methods determinates the protein functions by analyzing the characteristics of the protein binding sites [4]. The protein binding sites are the amino acids where protein structures interact. In this work we focus on investigating automated method for predicting the protein binding sites.

Many researches considered different characteristics of the amino acid residues of the protein structures for predicting the protein binding sites. The Accessible Surface Area (ASA) [5], depth index (DPX) [6], protrusion index (CX) [7] and hydrophobicity [8] are among the most commonly used characteristics for protein binding sites prediction. However, any of these characteristics does not make a distinction between the binding and non-binding sites. Hence, most commonly several characteristics are used together in order to provide better protein binding sites prediction.

In the literature, there are many methods for protein binding sites prediction [9, 10, 11, 12, 13]. Nevertheless, during evolution the characteristics of the protein amino acid residues slightly change, and these methods are very sensitive to these changes. To overcome this problem, we have already introduced the fuzzy theory for protein binding sites prediction. First, the fuzzy decision trees (FDTs) are introduced. Several methods for inducing FDTs are presented in [14] and [15], while [16] and [17] give some optimizations of FDTs. The authors of [18] point out the advantages and disadvantages of the fuzzy decision trees versus classical decision trees. In [19] we have already introduced a method for protein binding sites prediction based on FDTs. However, FDTs can use only a single fuzzy aggregation operator, while fuzzy pattern trees (FPTs) [20] are able to combine different fuzzy aggregation operators in the process of induction of the model. In our recent publication [21], we have introduced an approach for predicting the binding sites of the protein structures by inducing FPTs. In [22], the authors proposed several improvements of the bottom-up approach for inducing fuzzy pattern trees [20]. The first improvement is regarding the direction of the induction of the model. Namely, in [20] the FPT is induced in bottom-up direction, while in [22] the induction is in top-down direction. In this way the current model tree is modified slightly, instead of merging two trees in a completely new tree as in the bottom-up approach. With the second improvement, a better termination criterion is introduced. In this way the termination of the induction of the model is adapted according to the complexity of the problem.

In this paper, we introduce a fuzzy pattern tree based approach for protein binding sites prediction where the model is induced in top-down direction. In the training phase, several characteristics of the amino acid residues are extracted, and then using the top-down approach for building fuzzy pattern trees [22] the models are induced. In the test phase, the query amino acid residues are classified as binding or non-binding sites. We present some experimental results regarding the prediction power of the approach. Also the approach is compared with the bottom-up approach for inducing FPTs and with several classical classification algorithms that can be used for this purpose.

In section 2, we present a top-down approach for protein binding sites prediction based on fuzzy pattern trees. In section 3 some experimental results of the evaluation of the approach are presented, while in section 4 we conclude the paper and give directions for additional improvements.

2 Top-Down Approach for Protein Binding Sites Prediction Based on Fuzzy Pattern Trees

In this research paper we introduce a top-down approach for protein binding sites prediction based on fuzzy pattern trees (FPTs). In the training phase, several characteristics of the amino acid residues are extracted, and then using the top-down approach for building fuzzy pattern trees [22] the models are induced. We have already introduced a similar approach [21] for protein binding sites prediction based on fuzzy pattern trees where the induction of the models is in bottom-up direction. In the test phase, the query amino acid residues are classified as binding or non-binding sites.

2.1 Extraction of the Characteristics of the Amino Acid Residues

In this research we consider the following characteristics of the amino acid residues: accessible surface area (ASA), depth index (DPX), protrusion index (CX) and hydrophobicity. In the state of the art literature, these characteristics are most commonly used for protein binding sites prediction. As it was already mentioned, any of these characteristics does not make a perfect distinction of the binding and non-binding sites. Therefore, usually these characteristics are combined with the intention to provide more accurate protein binding sites prediction.

The accessible surface area (ASA) is first described in 1971 by Lee and Richards [23], and it is among the most important characteristics regarding the preferences of the amino acid residues to be involved in protein-protein interactions. The ASA of an amino acid residue is usually calculated using the rolling ball algorithm [5] where a rolling sphere (probe) with some radius is rolled around the protein structure. In this research we use the most commonly used value for the radius of the rolling probe, i.e. 1.4 Å. In this way, the ASA of each atom is calculated. Since an amino acid residue is constituted by a number of atoms, its ASA is calculated as a sum of the accessible surface areas of the atoms that constitute the given amino acid residue.

Some of the amino acid residues are deeply in the protein interior, so they cannot be involved in protein-protein interactions. Due to this, in order to avoid unnecessary predictions concerning these amino acid residues, we make predictions only about the surface amino acids. For that purpose, we make an assessment about the surface amino acid residues based on their ASA values. In accordance with [24], we consider a given amino acid residue as a surface amino acid if at least 5% of its surface is accessible by the rolling sphere. Later, when the models for protein binding sites prediction will be induced, only the surface amino acids will be taken into account.

Besides ASA, other commonly used characteristic of the amino acid residues is the depth index (DPX) [6]. The depth index of an atom is the Euclidean distance from that atom to its nearest atom whose ASA is greater than zero, while the depth index of an amino acid residue is calculated as an average of the depth indices of the atoms that constitute the given amino acid.

The third amino acid residues' characteristic that we consider in this research is the protrusion index (CX) [7]. For each non-hydrogen atom, the number of heavy atoms within a given radius r is calculated. In this research we use a radius of 10 Å [7]. The volume occupied in the sphere V_{int} is calculated by multiplying the number of heavy atoms within the sphere and the average volume of atoms. The protrusion index CX is calculated as ratio of the remaining volume V_{ext} and the occupied volume V_{int}, where the reaming volume is calculated as $4r^3\pi/3 - V_{int}$. The non-hydrogen atoms surrounded by many heavy atoms would have low values for CX, while the non-hydrogen atoms surrounded by few heavy atoms would have large values for CX. The protrusion index of an amino acid residue is calculated as an average of the protrusion indices of the non-hydrogen atoms that constitute that amino acid.

The last amino acid residues' characteristic that we consider is hydrophobicity, which is related with the hydrophobic effect. To be precise, in the protein interior more commonly the hydrophobic amino acids are found, while hydrophilic amino acids are more commonly positioned towards the protein surface. In this research, we use the scale introduced by Kyte and Doolittle [8] that is the most widely used scale.

2.2 Top-Down Induction of Fuzzy Pattern Trees

After extraction of the amino acid residues' characteristics, we induce fuzzy pattern trees (FPTs) [20] for protein binding sites prediction using the top-down induction method introduced in [22]. In the training phase the model is induced and it is constituted from two fuzzy pattern trees, one for each class (binding and non-binding sites). In the test phase, the query amino acid residue is presented to both trees, and it is classified in the class for which the highest similarity is obtained. Since the top-down approach [22] for inducing fuzzy pattern trees is an improvement of the bottom-up approach introduced by Huang et al. [20], we will first describe the bottom-up approach. Afterward, we will present the changes that are introduced in the top-down approach.

The induction of fuzzy pattern trees starts with fuzzification of the data set. In this research we use the straight-line fuzzy membership functions (FMFs) introduced by Zadeh [25]. Nevertheless, the convex FMFs can significantly increase the prediction accuracy of the models [25]. Consequently, in this research we also use the Gaussian fuzzy membership function. During fuzzification, the data set is labeled with fuzzy terms. In our research we consider four amino acid residues' characteristics and if we set the number of fuzzy membership functions per attribute (N) to 5, then in the bottom-up induction process we will obtain 20 primitive trees. Next, using the Root mean squared error (RMSE) similarity measure we calculate the similarity between the membership values of the corresponding fuzzy term for an amino acid residues' characteristic and the membership values of a given fuzzy term for the class attribute. Later, the primitive tree with highest similarity is selected.

The primitive trees are too simple and could not achieve acceptable prediction accuracy, so these trees are additionally aggregated using fuzzy aggregation operators. In the literature, several types of fuzzy aggregation operators can be found. In this research we use only the basic operators, i.e. AND, OR, MAX and MIN. As we already said, in the induction of fuzzy pattern trees [20] different types of fuzzy aggregation operators could be used, which is not a case with the fuzzy decision trees. Due to this reason, with fuzzy pattern trees more accurate models can be induced.

As it was mentioned, the primitive trees, which are trees at level 0, could not gain high prediction power. Therefore, the primitive tree with highest similarity is aggregated with the other primitive trees, thus obtaining 19 trees at level 1. Next, the trees at level 2 are formed by aggregation of the tree at level 1 with highest similarity and the trees from level 0 and level 1 that do not have highest similarity at their levels. The same iterative procedure is repeated until some termination criterion is satisfied. In this way the fuzzy pattern tree is induced in bottom-up direction.

In the aggregation, the tree with highest similarity at the current level could be aggregated with primitive trees, or even with more complex trees. Based on this, we can obtain two types of models: simple models (SMs) where the current tree with highest similarity is aggregated only with primitive trees excluding the primitive tree with highest similarity, and general models (GMs) where in the aggregation besides the primitive trees at level 0 also the trees from the other levels are considered.

In the bottom-up approach for inducing fuzzy pattern trees there are two termination criteria. The first criterion is satisfied if none of the candidate trees at the current level do not improves the similarity. With the second criterion the model tree is terminated when some predefined depth of the tree is reached. In this way also the model complexity is controlled. However, this termination criterion is not well adapted to the complexity of the problem that we are solving.

In [22], the authors propose several improvements of the bottom-up approach for inducing fuzzy pattern trees [20]. First, the algorithm reverses the direction of the tree construction from bottom-up to top-down. Second, a novel termination criterion is introduced taking in consideration the complexity of the problem.

Regarding the first improvement, the idea is to modify the current tree slightly, instead of merging two trees into completely different tree as in the bottom-up approach. This is done by replacing the leaf nodes by trees where two characteristics are aggregated by some fuzzy aggregation operator. With this top-down induction, the lower operators in the model tree have lower influence in the decision than the operators found at higher levels in the model tree.

Regarding the termination criterion, the authors in [22] performed analysis which of the termination criteria used in [20] are more frequently satisfied. In [22], they used forty datasets and in 90% of the cases the maximum-depth criterion is satisfied, while only in 10% of the cases the no-improvement criterion is satisfied. Some classes can be represented by very simple models, while some classes require more complex models with higher depths. Hence, it is obvious that the maximum-depth criterion is not an appropriate termination criterion since it could stop the induction too early or too late because it does not consider the complexity of the problem. In order to provide an adaptive termination criterion, the authors in [22] analyze the relative improvement of the models in two consecutive iterations. Namely, the induction process terminates when $sim_{max}(t) < (1 + \epsilon)\, sim_{max}(t - 1)$, where ϵ is a user-defined value in the interval [0,1], while $sim_{max}(t)$ and $sim_{max}(t - 1)$ correspond to the highest similarities at the t - th and $(t - 1)$ - th iteration. In this research we have set ϵ to 0,25% as suggested in [22].

Besides these two major improvements, the authors in [22] additionally introduce changes in the fuzzification of the data set where no specification of the type of fuzzy membership function is required. Also, the authors propose a way for handling missing values.

3 Evaluation

Next, we evaluate the top-down approach for protein binding sites prediction based on fuzzy pattern trees. We use a part of the BIND database [26] that is a database with knowledge obtained in experimental manner. Using the selection criterion given in [27], we take a representative data set from the BIND database thus each pair of protein chains has less than forty percents sequence similarity. Using the same selection criterion, from this representative data set the test set is formed so each pair of test protein chains has less than twenty percents sequence similarity, while the other protein chains are considered in the training data set. The obtained training data set contains 1062 protein chains, while the test set contains 1858 protein chains. Next, we perform filtering of the surface amino acid residues, which are the amino acids with at least 5% accessible surface area. In the problem that we are solving, the class of non-binding sites is dominant over the class of binding sites. Due to this, we balance the training data set to avoid inducing models which are biased towards the dominant class. In this way we obtain a training data set with 128877 binding and 128398 non-binding sites, and a test set with 47501 binding and 437136 non-binding sites. It is worth to mention that we make balancing only on the training data set, so the test data set is still not balanced. Therefore, we must choose an evaluation measure which is suitable for unbalanced data sets. In this research we use the Area under ROC curve (AUC-ROC) measure as a measure of the prediction power of the models. This measure is among the most appropriate measures when the classes have extremely different number of samples, as in our case. The Area under ROC curve is calculated as TPR*TNR+TPR*(1-TNR)/2+TNR*(1-TPR)/2=(TPR+TNR)/2, where TPR is the True Positive Rate and TNR is the True Negative Rate. TPR is calculated as TP/(TP+FN), while TNR is calculated as TN/(TN+FP), where TP is the number of true positives (correctly predicted positive amino acid residues), TN is the number of true negatives (correctly predicted negative residues), FP is the number of false positives (non-binding sites residues predicted as binding sites) and FN is the number of false negatives (binding sites residues predicted as non-binding sites). The AUC-ROC has a value in the interval [0,1], where 0 corresponds to inverse prediction and 1 corresponds to perfect prediction.

First, we will evaluate the results obtained by the bottom-up approach for inducing fuzzy pattern trees. In Table 1, the results of the bottom-up approach for protein binding sites prediction based on FPTs are presented, where the influence of the number of fuzzy membership functions (FMFs) per characterictic (N) and the type of FMFs is inspected. In this analysis the RMSE similarity measure is used, and from the fuzzy aggregation operators only AND and OR operators are considered. In this analysis we induce only simple models (SMs). From Table 1 we can see that the most powerful model is obtained using 5 trapezoidal FMFs per characteristic. For the other types of FMFs (triangular and Gaussian), it is better to take 4 FMFs per characteristic. However, in this analysis we used the most basic types of FMFs. Also other convex FMFs (like bell, log-normal etc.) can be used. Regarding similarity measure, in this analysis we used RMSE. We can also apply other more sophisticated similarity measures in order to improve the approach. Regarding fuzzy aggregation operators, in this analysis we used only the algebraic AND and OR operators. Further, we can incorporate other fuzzy aggregation operators.

Table 1. The results for AUC-ROC obtained by the bottom-up approach using different number (N) of fuzzy membership functions (FMFs) per characteristic

	Triangular FMF	Trapezoidal FMF	Gaussian FMF
N=3	0,5564	0,5168	0,5565
N=4	**0.5643**	0,5568	**0,5642**
N=5	0,5503	**0,5648**	0,5435

Next, in Table 2 we present the results obtained by the top-down approach for protein binding sites prediction based on fuzzy pattern trees. In this analysis we investigate the influence of the types of fuzzy aggregation operators that are considered in the aggregations. In this analysis we have set ϵ to 0,25% as suggested in [22]. Regarding fuzzy aggregation operators, we considered the MIN, MAX, AND and OR operators.

Table 2. The results obtained by the top-down approach using different fuzzy aggregation operators

Fuzzy aggregation operators	TPR	TNR	AUC-ROC
MIN, MAX	0,5649	0,5573	0,5611
AND,OR	0,6265	0,5362	**0,5813**
MIN, MAX, AND, OR	0,6193	0,5326	0,5760

Based on the results we can conclude that with the top-down approach better results are obtained. Regarding the fuzzy aggregation operators, when only MIN and MAX operators are used, the model obtains comparable value for AUC-ROC with the best models obtained by the bottom-up approach. However, when the AND and OR fuzzy aggregation operators are used, the prediction power is increased, thus obtaining 0,5813 value for AUC-ROC. It is interesting that when all four operators are combined together (MIN, MAX, AND and OR), lower AUC-ROC is obtained than the case when only AND and OR operators are used. This is due to the fact that the top-down approach is greedy approach.

Besides comparison of the bottom-up and top-down approaches for protein binding sites prediction based on inducing fuzzy pattern trees, in this research we also make comparison of the top-down approach with several approaches based on classical classification algorithms. In this comparison we used the following algorithms: C4.5 Tree [28], REPTree (a variant of C4.5 where reduced-error pruning is used), ADTree [29] and its multi class version LADTree [30], NBTree [31], Random Forest [32], Naïve Bayes [33], Bayes Net [34], Multilayer perceptron [35] and RBF network [36]. From the results given in Table 3, it can be seen that C4.5, REPTree and Random Forest algorithms favor the non-active class and they obtain lower AUC-ROC. Naïve Bayes is better than C4.5, REPTree and Random Forest, but because the characteristics of the amino acid residues are dependent, this algorithm is not the best choice. Bayes Net takes into account the dependences between the characteristics, and thus obtains higher AUC-ROC. NBTree, which is a mixture of Naïve Bayes and Decision Trees showed as an excellent method for protein binding sites prediction. Also the examined Multilayer perceptron and RBF network have comparable prediction power with the other methods. The top-down approach outperforms all the methods considered in this research excluding the ADTree and LADTree algorithms. However, it still has comparable prediction power with these algorithms.

Table 3. The results obtained using different algorithms

Algorithm	TPR	TNR	AUC-ROC
Bottom-up approach for FPTs	0,5566	0,5730	**0,5648**
Top-down approach for FPTs	0,6265	0,5362	**0,5813**
C4.5 Tree	0,4353	0,6642	0,5498
REPTree	0,3818	0,6840	0,5329
ADTree / LADTree	0,5298	0,6443	**0,5871**
NBTree	0,6203	0,5390	0,5796
RandomForest	0,1824	0,8730	0,5277
Naïve Bayes	0,4772	0,6620	0,5696
Bayes Net	0,6203	0,5390	0,5796
Multilayer perceptron	0,7651	0,3922	0,5787
RBF network	0,4959	0,6467	0,5713

4 Conclusion

In this paper, we introduced a top-down approach for protein binding sites prediction based on fuzzy pattern trees. This top-down approach is an improvement of the bottom-up approach where two major changes are introduced. With the first improvement, the direction of the induction is reversed, while with the second improvement the termination criterion is adapted based on the complexity of the learning problem.

As it was expected, the top-down approach outperformed the existing bottom-up approach for protein binding sites prediction based on fuzzy pattern trees. In this research we also made comparison of the top-down approach with several approaches based on classical classification algorithms. The top-down approach outperformed all the methods considered in this analysis excluding the ADTree and LADTree algorithms. Nevertheless, it still has comparable prediction power with them.

In future, we plan to improve the approach by choosing the most appropriate similarity measure for this learning problem, and also by applying more appropriate fuzzufication of the data set. Also, we can incorporate other fuzzy aggregation operators.

References

1. Kirac, M., Ozsoyoglul, G., Yang, J.: Annotating proteins by mining protein interaction networks. Bioinformatics 22(14), e260–e270 (2006)
2. Todd, A.E., Orengo, C.A., Thornton, J.M.: Evolution of function in protein superfamilies, from a structural perspective. J. Mol. Biol. 307(4), 1113–1143 (2001)
3. Panchenko, A.R., Kondrashov, F., Bryant, S.: Prediction of functional sites by analysis of sequence and structure conservation. Protein Science 13(4), 884–892 (2004)
4. Tuncbag, N., Kar, G., Keskin, O., Gursoy, A., Nussinov, R.: A survey of available tools and web servers for analysis of protein-protein interactions and interfaces. Briefings in Bioinformatics 10(3), 217–232 (2009)
5. Shrake, A., Rupley, J.A.: Environment and exposure to solvent of protein atoms. Lysozyme and insulin. J. Mol. Biol. 79(2), 351–371 (1973)
6. Pintar, A., Carugo, O., Pongor, S.: DPX: for the analysis of the protein core. Bioinformatics 19(2), 313–314 (2003)
7. Pintar, A., Carugo, O., Pongor, S.: CX, an algorithm that identifies protruding atoms in proteins. Bioinformatics 18(7), 980–984 (2002)
8. Kyte, J., Doolittle, R.F.: A simple method for displaying the hydropathic character of a protein. J. Mol. Biol. 157(1), 105–132 (1982)
9. Aytuna, A.S., Gursoy, A., Keskin, O.: Prediction of protein-protein interactions by combining structure and sequence conservation in protein interfaces. Bioinformatics 21(12), 2850–2855 (2005)
10. Neuvirth, H., Raz, R., Schreiber, G.: ProMate: a structure based prediction program to identify the location of protein-protein binding sites. J. Mol. Biol. 338(1), 181–199 (2004)
11. Bradford, J.R., Westhead, D.R.: Improved prediction of protein-protein binding sites using a support vector machines approach. Bioinformatics 21(8), 1487–1494 (2005)
12. Ogmen, U., Keskin, O., Aytuna, A.S., Nussinov, R., Gursoy, A.: PRISM: protein interactions by structural matching. Nucleic Acids Res. 33(2), W331–W336 (2005)
13. Jones, S., Thornton, J.M.: Prediction of protein-protein interaction sites using patch analysis. J. Mol. Biol. 272(1), 133–143 (1997)
14. Janikow, C.Z.: Fuzzy decision trees: issues and methods. IEEE Transactions on Systems, Man, and Cybernetics 28(1), 1–14 (1998)
15. Olaru, C., Wehenkel, L.: A complete fuzzy decision tree technique. Fuzzy Sets and Systems 138(2), 221–254 (2003)
16. Suárez, A., Lutsko, J.F.: Globally optimal fuzzy decision trees for classification and regression. IEEE Transactions on Pattern Analysis and Machine Intelligence 21(12), 1297–1311 (1999)
17. Wang, X., Chen, B., Olan, G., Ye, F.: On the optimization of fuzzy decision trees. Fuzzy Sets and Systems 112(1), 117–125 (2000)
18. Chen, Y.-L., Wang, T., Wang, B.-S., Li, Z.–J.: A Survey of Fuzzy Decision Tree Classifier. Fuzzy Information and Engineering 1(2), 149–159 (2009)
19. Mirceva, G., Naumoski, A., Stojkovik, V., Temelkovski, D., Davcev, D.: Method for Protein Active Sites Detection Based on Fuzzy Decision Trees. In: Kim, T.-H., Adeli, H., Cuzzocrea, A., Arslan, T., Zhang, Y., Ma, J., Chung, K.-I., Mariyam, S., Song, X. (eds.) DTA/BSBT 2011. CCIS, vol. 258, pp. 143–150. Springer, Heidelberg (2011)
20. Huang, Z.H., Gedeon, T.D., Nikravesh, M.: Pattern trees induction: a new machine learning method. IEEE Transaction on Fuzzy Systems 16(3), 958–970 (2008)
21. Mirceva, G., Kulakov, A.: Fuzzy pattern trees for predicting the protein binding sites. In: The 9th Conference for Informatics and Information Technology, CIIT 2012 (2012)

22. Senge, R., Hüllermeier, E.: Top-Down Induction of Fuzzy Pattern Trees. IEEE Transactions on Fuzzy Systems 19(2), 241–252 (2011)
23. Lee, B., Richards, F.M.: The interpretation of protein structures: Estimation of static accessibility. J. Mol. Biol. 55(3), 379–400 (1971)
24. Chothia, C.: The Nature of the Accessible and Buried Surfaces in Proteins. J. Mol. Biol. 105(1), 1–12 (1976)
25. Zadeh, L.A.: Fuzzy sets. Information and Control 8, 338–353 (1965)
26. Bader, G.D., Donaldson, I., Wolting, C., Ouellette, B.F., Pawson, T., Hogue, C.W.: BIND: the Biomolecular Interaction Network Database. Nucleic Acids Res. 29(1), 242–245 (2001)
27. Chandonia, J.–M., Hon, G., Walker, N.S., Conte, L.L., Koehl, P., Levitt, M., Brenner, S.E.: The ASTRAL Compendium in 2004. Nucleic Acids Res. 32, D189–D192 (2004)
28. Quinlan, R.: C4.5: Programs for Machine Learning. Morgan Kaufmann Publishers, San Mateo (1993)
29. Freund, Y., Mason, L.: The alternating decision tree learning algorithm. In: Sixteenth International Conference on Machine Learning, pp. 124–133 (1999)
30. Holmes, G., Pfahringer, B., Kirkby, R., Frank, E., Hall, M.: Multiclass alternating decision trees. In: Elomaa, T., Mannila, H., Toivonen, H. (eds.) ECML 2002. LNCS (LNAI), vol. 2430, p. 161. Springer, Heidelberg (2002)
31. Kohavi, R.: Scaling Up the Accuracy of Naive-Bayes Classifiers: A Decision-Tree Hybrid. In: Second International Conference on Knowledge Discovery and Data Mining, pp. 202–207 (1996)
32. Breiman, L.: Random Forests. Machine Learning 45(1), 5–32 (2001)
33. John, G.H., Langley, P.: Estimating Continuous Distributions in Bayesian Classifiers. In: Eleventh Conference on Uncertainty in Artificial Intelligence, pp. 338–345 (1995)
34. Neapolitan, R.E.: Learning Bayesian Networks. Prentice Hall, Upper Saddle River (2004)
35. Rosenblatt, F.: Principles of Neurodynamics: Perceptrons and the Theory of Brain Mechanisms. Spartan Books, Washington DC (1961)
36. Chen, S., Cowan, C.F., Grant, P.M.: Orthogonal least squares learning algorithms for radial basis function networks. IEEE Transactions on Neural Networks 2, 302–309 (1991)

Component-Based Development: A Unified Model of Reusability Metrics

Bojana Koteska and Goran Velinov

Faculty of Computer Science and Engineering,
Rugjer Boshkovikj 16, P.O. Box 393 1000 Skopje, Macedonia
bojana.koteska@gmail.com, goran.velinov@finki.ukim.mk

Abstract. Inability to use standard software reusability metrics when measuring component reusability makes the choice of reusability metric a challenging problem in software engineering. In this paper, we give a critical review on the existing component reusability metrics and we suggest new attributes to be included as additional conditions when evaluating component reusability. Due to the incompleteness of already proposed metrics and the lack of a universally accepted and transparent model for measuring reusability, we define a unified model that could be adapted to different reusability requirements and various component solutions. In order to improve the process of measuring component reusability we create a prototype for modeling and combining metrics where reusability can be calculated using the existing or newly composed formulas. This prototype will facilitate the process of testing the component reusability and it will allow users easily to select the right component to be integrated in their system.

Keywords: Component-Based Development, Reusability Model, Reusability Metrics.

1 Introduction

One of the most important issues associated with component-based development approach are component reusability and its measurement. Since there are many components available, some metrics for identifying and determining the component reuse effectively must be defined [8]. The process of finding the component reusability in practice is difficult. The lack of unified reusability measurement model makes measuring and testing the component reusability challenging problems in software engineering.

In this paper we reference the existing component reusability metrics and we give a critical review on the existing ones. The review we have made gives us a conclusion that the already proposed reusability metrics are incomplete and with several deficiencies. Our goal is to enhance those metrics and to add new attributes that can be considered as important factors in reusability measurement process.

The open issue about reusability metrics is: Why still does not exist a unified model for calculating component reusability? Our paper shows that there

S. Markovski and M. Gusev (Eds.): *ICT Innovations 2012*, AISC 207, pp. 335–344.
DOI: 10.1007/978-3-642-37169-1_33

are many proposed reusability metrics, but no model is considered as globally accepted. Making a unified model for measuring component reusability requires formalized approach, relevent choice of quality factors and including additional factors. As it is described in the section 3, our paper contributes to this issue in two ways. Our first contribution is choosing the attributes important for measuring component reusability and the second is including some additional factors.

We propose a unified model also that could be adapted to different component solutions and a prototype which could help users to measure reusability effectively. Using this prototype users will be able to define their own metrics and express them mathematically with some parameters that reusability depends on.

The paper is organized as follows. In Section 2 we give an overview of the background work and critical review on software reusability metrics. In Section 3 we present some new criteria that have to be added as parameters to existing metrics, our unified model and prototype for evaluating component reusability. In Section 4 we give conclusion and explain our future work.

2 Survey of Component Reusability Metrics

Since components are black-box entities and they provide their functionality through a set of interfaces, they can be reused as a black-box entities without knowing their internal parts [9]. However, sometimes component needs modifications to fit successfully in the software project. In that case, the parts that require changes have to be identified and after the changes have been made testing should be performed to check if the component satisfies the intended requirements. This process is called white-box reusing and it requires knowledge of implementation details [12].

2.1 Black-Box Component Reusability Metrics

This subsection lists the reusability metrics for black-box components. In [10] table of reusability metrics is presented including three columns: *Metric Name*, *Description* and *Contributor*. We express metrics by specifying: *Contributor*, *Characteristic that affect reusability* and *Formula*. Additionally, we include several different metrics.

In [7], *V. P. Venkatesan* and *M. Krishnamoorthy* proposed several metrics for measuring reusability. First, metrics below measure understandability because it is important to measure the usability as characteristic that affect reusability:

- Quality of Manuals (QM). There are several formulas expressing QM:

$$QM_i = \frac{NO_i}{Total\, No.of\, Functional\, Elements}, \quad i = 1,..,4 \ . \tag{1}$$

  where NO_1 is number of functional elements in the manual, NO_2 is number of tables, NO_3 is number of figures and NO_4 is number of design diagrams.

- Coverage of Demos (CD). Formula that express CD:

$$CD = \frac{NO. of\, Functional\, Elements\, shown\, in\, Demo}{Total\, No. of\, Functional\, Elements} \quad . \tag{2}$$

- Size of Manuals (SM). Formula that express SM is:

$$SM = \frac{NO. of\, Functional\, Elements\, in\, the\, Manuals}{Total\, No. of\, Pages} \quad . \tag{3}$$

Next, metrics important to measure complexity as characteristic that affect reusability are:

- Component Coupling (COC). Formula for COC is:

$$COC = \frac{NO. of\, other\, components\, sharing\, attribute\, or\, method}{Total\, No. of\, possible\, sharing\, pairs\, in\, the\, comp. - based\, app.} \quad . \tag{4}$$

- Constraint Complexity (CTC). Formula that express CTC is:

$$CTC = \frac{NO. of\, constraints}{No. of\, properties\, and\, operations\, in\, an\, interface} \quad . \tag{5}$$

- Configuration Complexity (CFC). CFC can be expressed as:

$$CFC = \frac{NO. of\, configurations}{No. of\, context\, of\, use\, of\, the\, component} \quad . \tag{6}$$

Authors said that another factor that directly affects reusability is portability. It can be expressed with External Dependency metric (ED) as follows:

$$ED = \frac{NO. of\, Methods\, with\, parameters\, passed\, and\, return\, values}{Total\, No. of\, Methods\, Excluding\, Read/Write\, Methods} \quad . \tag{7}$$

The last factor that should be measured and that directly affects reusability is confidence. Authors presented confidence with following metrics:

- Maturity (Mat). Mat can be expressed as:

$$Mat = DF + CR \quad . \tag{8}$$

  where DF is the number of faults detected and CR is the number of change requests.

In [13], *H. Washizaki*, *H. Yamamoto* and *Y. Fukazawa* presented several reusability metrics. All the measurement values in the following metrics should be normalized to a number between 0 and 1. As metrics that measure understandability they proposed:

- Existence of meta-information (*EMI*). Its confidence interval is between 0.5 and 1.

$$EMI = \begin{cases} 1 & \text{(If BeanInfo class exists)} \\ 0, & \text{(Otherwise)} \end{cases} \quad . \tag{9}$$

- Rate of Component Observability (*RCO*). Confidence interval is between 0.17 and 0.42.

$$RCO(c) = \begin{cases} \frac{P_r(c)}{A(c)}, & A(c) > 0 \\ 0, & \text{(Otherwise)} \end{cases} \quad . \tag{10}$$

where $P_{\mathrm{r}}(c)$ is the number of readable properties in component c and $A(c)$ is the number of fiels in the component c's Facade class.

- Rate of Component Customizability (*RCC*). Authors considered adaptability as important factor that depends on customizability. Its confidence interval is between 0.17 and 0.34:

$$RCC(c) = \begin{cases} \frac{P_w(c)}{A(c)}, & A(c) > 0 \\ 1, & \text{(Otherwise)} \end{cases} \quad . \tag{11}$$

where $P_{\mathrm{w}}(c)$ is the number of readable properties in component c.

The last quality factor in this paper that affects reusability is portability. Portability can be expressed with Self-Completeness of Components Return Value metric which confidence interval is between 0.61 and 1 and Self-Completeness of Components Parameter metric which confidence interval is between 0.42 and 0.77.

- Self-Completeness of Components Return Value ($SCC_r(c)$). $SCC_r(c)$ can be expressed as:

$$SCC_r(c) = \begin{cases} \frac{B_v(c)}{B(c)}, & (B(c) > 0) \\ 1, & \text{(Otherwise)} \end{cases} \quad . \tag{12}$$

where $B_{\mathrm{v}}(c)$ is the number of business methods without return value in component c and $B(c)$ is the number of business methods in component c.

- Self-Completeness of Components Parameter ($SCC_p(c)$). $SCC_p(c)$ can be expressed as:

$$SCC_p(c) = \begin{cases} \frac{B_p(c)}{B(c)}, & (B(c) > 0) \\ 1, & \text{(Otherwise)} \end{cases} \quad . \tag{13}$$

where $B_{\mathrm{p}}(c)$ is the number of business methods without return parameters in component c.

In [14], *L. Yingmei, S. Jingbo, X. Weining* considered reusability as a factor that depends on functionality, reliability, utilizability, maintainability and portability. For measuring the component reusability, the Reusability measure Value metric (*RMV*) is proposed:

$$RMV = W_1 * F + W_2 * R + W_3 * U + W_4 * M + W_5 * P \quad . \tag{14}$$

where W_{i}(i=1,...,5) denotes weights, F the functionality, R reliability, U utilizability, M maintainability and P portability.

In [11], *A. Sharma*, *R. Kumar* and *P.S. Grover* surveyed some component reusability metrics. One of them is Component Reusability metric (CR). It can be expressed as:

$$CR = \frac{\sum IM}{TIM} \quad . \tag{15}$$

where IM is interface method providing commonality functions in a domain and TIM is the sum of total interface methods.

2.2 White-Box Component Reusability Metrics

As we mentioned before, component can be reused also as a white-box entity. It is important when some source code changes have to be done before reusing the component. In this subsection white-box component metrics for measuring reusability are presented.

In [1], *S. Bi*, *X. Dong* and *S. Xue* proposed two formalized metrics:

- Metric based on the amount for reuse and new code:

$$R = \frac{L_r}{L_r + L_n} \quad . \tag{16}$$

  where R denotes the component reusability, L_{r} denotes lines of source code and Ln denotes lines of new code.

- Metric based on relationship of component reusability and software complexity. It can be expressed as:

$$R = \frac{K}{C} \quad . \tag{17}$$

  where R denotes some certain component reusability, C denotes component complexity and K denotes reuse coefficient in domain. K depends on universality of the total components in the field. There are to ways to calculate C. It can be calculated according to McCabe or Halstead method. According to McCabe method, component complexity can be expressed as:

$$V(G) = M - N + P \quad . \tag{18}$$

  where $V(G)$ is the number of direction cycles in connected program graph, M is the number of arcs in the program graph, N is the number of nodes in the program graph and P is the number of separate components in the program graph (P is always 1 because the program graph is connected.) According to Halstead method, formula about complexity is:

$$N = N1 + N2 \quad . \tag{19}$$

  where N is the length of component code, $N1$ is the total number which component has operator and $N2$ is the total number which component has operand.

2.3 Critical Review

Software metrics are quantitative indicators of component attributes. They can help in component comparison, fault prediction and cost estimation. Defining metrics is a difficult process because components usually have more interfaces and interface methods. Their size is often unknown also [11].

To summarize the metrics that have been presented in this section and attributes they measure, we created Fig. 1 that shows the relation between metrics and attributes when component is viewed as a black-box entity. Table helped us to confirm that only few attributes are used in the reusability measurement process and that motivated us to add new attributes that we considered as important in the component reusability measurement process.

The proposed reusability metrics, summarized by authors, are somewhat limited because the authors include only a certain number of attributes, i.e. those attributes that they considered to be important for the process of reusability measurement. However, some of the attributes are expressed only by one metric, so the relevance of the calculations becomes questionable. For example, the authors cover only five attributes in [7], four in [13], five in [14], etc.

Attribute/ Metric Name	understandability	usability	complexity	portability	adaptability	confidence	customizability	reliability	utilizability	maintainability	functionality
Quality of Manuals	✓	✓									
Coverage of Demos	✓	✓									
Size of Manuals	✓	✓									
Component Coupling			✓								
Constraint Complexity			✓								
Configuration Complexity			✓								
External Dependency				✓							
Maturity						✓					
Existence of meta-information	✓										
Rate of Component Observability	✓										
Rate of Component Customizability					✓		✓				
Self-Completeness of Component's Return Value				✓							
Self-Completeness of Component's Parameter				✓							
Reusability Measure Value				✓				✓	✓	✓	✓

Fig. 1. Matrix of correlation between attributes and reusability metrics

The fact that black-box component evaluation is still real problem due to the incompleteness and lack of maturity of proposed reusability metrics [8] and the absence of generally accepted model also give us idea to create a unified model for measuring component reusability where any attribute considered as relevant can be added in the process of measuring component reusability.

3 A Unified Model for Measuring Component Reusability

3.1 Adding New Reusability Factors

We propose component security and installability to be added as reusability factors that can be defined as attributes in our measurement model. These attributes have not been formalized yet as factors that component reusability depends on, but according to [15] and [16] they are significant for software reusability.

Component security is very important aspect in process of component selection. The real problem is that if component is secure in one environment, it does not mean that it will be secure in other. That is the reason why many companies want to test components security before the decision to implement the component in a new system. We suggest adding security as a relevant factor in the component reusability model. The idea is to measure the security of a component as a black-box entity by applying metrics to components security services such as classes and functions.

According to the Common Criteria standard namely ISO/IEX 15408, security functional requirements consist of eleven classes for security requirements: security audit, communication, cryptographic support, user data protection, identification and authentication, security management, privacy, protection of system security functions, resource utilisation, system access and trusted path and channels [6]. To measure the security, weight factor is assigned to each security objective and each class, relative to the importance. The final score of the security class of a component can be calculated using these equations:

$$Fi, j = \text{MIN}(Di, j, k), k \geq 1 \ . \tag{20}$$

$$Oi = \text{MIN}(Fi, j), j \geq 1 \ . \tag{21}$$

$$C = \sum_{i=1}^{N} W_i O_i \ . \tag{22}$$

where C is a class, O is a security objective, F is a security function and W is the percentage weight of an objective. D_i, j, k represents the scores of the dependencies k of the security function j of security objective i.

Next attribute we propose is installability. We considered it as a crucial factor because every component has to be installed in the development environment successfully before it is being used. ISO 9126-3 [4] defines Installation flexibility metric as follows:

$$X = \frac{A}{B} \ . \tag{23}$$

where A is number of implemented customizable installation operation as specified and confirmed in review and B is number of customizable installation operation required.

3.2 A Generic Formula

In order to define a generic formula, we use the recursive definition of a formula in predicate logic, given in [3] where open formula is formed by combining atomic formulas. We adjust this defininion to formula inspired by equation 14 and we propose a unified reusability measurement model that could be easily adapted to different component solutions and reusability requirements. Regard to our model, component reusability can be formally expressed as:

$$CR = \sum_{i=1}^{N} W_i M_i \ . \tag{24}$$

where N is the number of attributes that will be considered as important in the measurement process, W_i is weight factor for the i-th attribute and M_i is the metric that measure the i-th attribute.

Any of the proposed attributes shown in Fig. 1, security and installability also could be used as relevant, but additional attributes could be added too depending on the users' needs. Weight factor is a value that indicates the impact of the attribute in a set of attributes included in the component reusability measurement formula.

We define metric as a formula that express a certain attribute. The question here is what if that formula contains attribute that could be also expressed with any other formula. In that case, we could say that metric is an open formula consisted of atomic formulas. We formalize the term metric as follows:

$$M = X(\circ X)^* \ . \tag{25}$$

where X is atomic formula or term that is represented by a real number, $\circ$ is any binary arithmetic operation and * means that expression in brackets can be repeated 0 or more times. In our case, we use the word term to present any parameter (variable), constant or n-ary function of several terms.

3.3 A Reusability Metrics Modeling Prototype

To present practically the process of determining the component reusability metric we implement a prototype that enables modeling of metrics, creating metrics or using the existing ones. The prototype is implemented in C Sharp and it uses the CodeDom Calculator [2] as a parser. This calculator allows writing C Sharp algorithms. The prototype has a simple user interface shown in Fig.2. This prototype enables users to make a choice of the component which reusability should be calculated and the attribute that has to be included as relevant in the measurement process. When the attribute is selected, a list of metrics that measure that attribute becomes available for choice. After the metric is chosen, the metric's formula appears in the programming window and declarations of parameters included in that formula are written also. The formula could be modified and new parameters from the list of parameters on the left side could be added. Declarations of the added parameters will be written automatically. User could also add new variables and blocks of C Sharp commands. When the modifications are made, user is able to save the formula by inserting metric name and attribute that the new metric will measure.

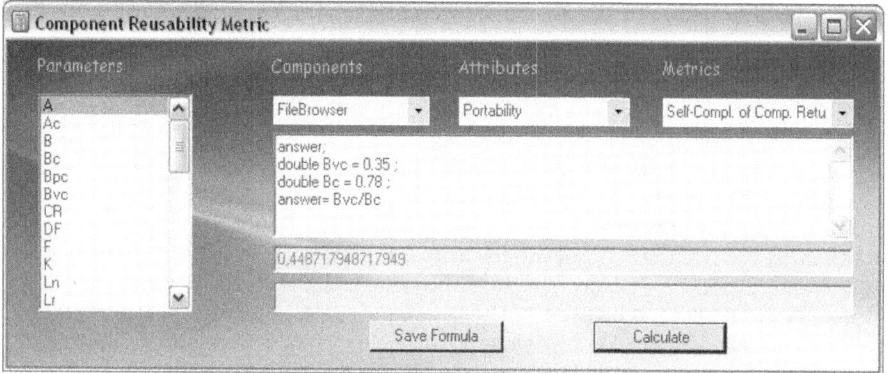

Fig. 2. Prototype for measuring component reusability

To get the result of measurement, user should enter the values for every parameter used in the formula.

4 Conclusion and Future Work

Our work presented in this paper solves the problem of finding a unified method for measuring component reusability. Many component reusability metrics were proposed, but none of them is considered as a standard. The unified model we defined enables making combination of all existing metrics and it could be easily adapted to any future definitions of component reusability metric.

Another way in which this paper contribute is by improving the existing metrics with the addition of security and installability, which in our opinion, are significiant in terms of raising the criteria for selection of the component when reusability is considered as crucial. Our aim was also to implement a prototype that will summarize all component reusability metrics that we found as relevant and will give users an incentive to expand the existing metrics and to define new ones.

This is our first version of the component reusability measurement model and it needs to be validated by measuring the reusability of some real components and applying appropriate verification strategies [5]. The practical implementation of the measurement process will be discussed in our next research paper. It will be interesting to analyze the final results obtained by changing the weight factors and to determine when and which of the attributes that reusability depends on should be considered as most important in the process of measuring component reusability.

References

1. Bi, S., Dong, X., Xue, S.: A Measurement Model of Reusability for Evaluating Component. In: Proceedings of the 2009 First IEEE International Conference on Information Science and Engineering, pp. 20–22. IEEE Computer Society, Washington, DC (2009)
2. CodeDom Calculator, http://www.c-sharpcorner.com/uploadfile/mgold/codedomcalculator08082005003253am/codedomcalculator.aspx
3. Davis, S., Gillon, B.S.: Semantics: A Reader. Oxford University Press, New York (2004)
4. ISO 9126-3 (2003), http://www.iso.org
5. Kaloyanova, K., Ignatova, P.: Software Testing Automation. In: Proceedings of the Second International Scientific Conference Computer Science 2005, Chalkidiki, Greece, pp. 220–225 (2005)
6. Khan, K.M., Han, J.: Assessing Security Properties of Software Components: A Software Engineers Perspective. In: Proceedings of the Australian Software Engineering Conference, pp. 199–210. IEEE Computer Society, Washington, DC (2006)
7. Prasanna Venkatesan, V., Krishnamoorthy, M.: A Metrics Suite for Measuring Software Components. JCIT: Journal of Convergence Information Technology 4, 138–153 (2009)
8. Sarbjeet, S., Manjit, T., Sukhvinder, S., Gurpreet, S.: Software Engineering - Survey of Reusability Based on Software Component. International Journal of Computer Applications 8, 39–42 (2010)
9. Sametinger, J.: Software Engineering with Reusable Components. Springer-Verlag New York, Inc., New York (1997)
10. Shanmugasundaram, G., Prasanna Venkatesan, V., Punitha Devi, C.: Reusability metrics - An Evolution based Study on Object Oriented System, Component based System and Service Oriented System. Journal of Computing 3, 30–38 (2011)
11. Sharma, A., Rajesh, K., Grover, P.S.: A Critical Survey of Reusability Aspects for Component-Based Systems. In: Proc. World Adacemy of Science Engineering and Technology, pp. 411–415. Citeseer, Philadelphia (2007)
12. Sharma, A., Rajesh, K., Grover, P.S.: Managing Component-Based Systems With Reusable Components. International Journal of Computer Science and Security 1, 52–57 (2007)
13. Washizaki, H., Yamamoto, H., Fukazawa, Y.: A Metrics Suite for Measuring Reusability of Software Components. In: Proceedings of the 9th International Symposium on Software Metrics, pp. 211–223. IEEE Computer Society, Washington, DC (2003)
14. Yingmei, L., Jingbo, S., Weining, X.: On Reusability Metric Model for Software Component. In: Wu, Y. (ed.) Software Engineering and Knowledge Engineering. AISC, vol. 114, pp. 865–870. Springer, Heidelberg (2012)
15. Kath, O., Schreiner, R., Favaro, J.: Safety, Security, and Software Reuse: A Model-Based Approach. In: Fourth International Workshop in Software Reuse and Safety (RESAFE 2009), Washington, DC, USA (2009)
16. Sharma, V., Baliyan, P.: Maintainability Analysis of Component Based Systems. International Journal of Software Engineering and Its Applications 5, 107–118 (2011)

Scalability of Gravity Inversion with OpenMP and MPI in Parallel Processing

Neki Frasheri and Betim Çiço

Polytechnic University of Tirana, Faculty of Information Technology, Tirana, Albania
nfrasheri@fti.edu.al, betim.cico@gmail.com

Abstract. The geophysical inversion of gravity anomalies is experimented using the application GMI based on the algorithm CLEAR, run in parallel systems. Parallelization is done using both OpenMP and MPI. The scalability in time domain of the iterative process of inversion is analyzed, comparing previously reported results based in OpenMP with recent data from tests with MPI. The runtime for small models was not improved with the increase of the number of processing cores. The increase of user runtime due to the size of the model resulted faster for MPI compared with OpenMP and for big models the latter would offer better runtime. Walltime scalability in multi-user systems did not improved with the increase of processing cores as result of time sharing. Results confirm the scalability of the runtime at the order of $O(N^8)$ relative to the linear size N of 3D models, while the impact of increasing the number of involved cores remains disputable when walltime is considered. Walltime upper limit for modest resolution 3D models with 41*41*21 nodes was 10^5 seconds, suggesting the need of using MPI in multi-cluster systems and of GPUs for better resolution. The results are in framework of FP7 Infrastructure project HP-SEE.

Keywords: High performance computing, gravity inversion, OpenMP, MPI.

1 Introduction

In this paper we analyze a case of parallelization of gravity anomaly inversion using MPI. Geophysical inversion represents a difficult problem, requesting the evaluation of physical parameters of a 3D geosection from values measured in a 2D surface array [1]. The geophysical inversion belongs to the category of problems considered as “ill posed” by Hadamard [2]. The process of inversion represents a kind of extrapolation from a 2D array of data (the surface distribution of the measured physical field) into a 3D spatial array of physical proprieties. Using different algorithms or even with minor changes of initial parameters, different authors have obtained alternate solutions with differences between local and global optimal points ([3], [4], [5], [6]). Despite decades of work, solutions remain “uncertain” [7].

The complexity of the problem is increased by the volume of data necessary to represent 3D spatial geological structures and their physical properties. Geological sections may extend for kilometers while key geological features may have thickness

S. Markovski and M. Gusev (Eds.): *ICT Innovations 2012*, AISC 207, pp. 345–352.
DOI: 10.1007/978-3-642-37169-1_34

measures in meters, which imply the need for huge 3D arrays to represent geosections [8]. Complexity of geological structures and lack of complete information about their spatial extension and physical properties implies the need for multiple calculation alternatives and interactive software.

The huge volume of data requires considerable processing time and in recent years the interest is shown for the application of high performance computing platforms in geophysical inversion ([11], [12], [13], [14]).

During the first stage of our work the parallelization with OpenMP was achieved and tested in the HPCG system of the Center at the Institute of Information and Communication Technologies (IICT-BAS) in Sofia, Bulgaria, and in the SGE system of the NIIFI Supercomputing Center at University of Pécs, Hungary; with results presented in [8] and [10]. The results from OpenMP experiments with complex geological models are presented in [9]. The actual work is focused in parallelization with MPI tested in both SGE and HPCG sites.

2 Methodology of Study

The geophysical gravity problem in geophysics is treated in details in classical literature as [15]. We used a simple iterative inversion algorithm for our application "GIM" (*Gravity Inversion Modeling*) based on the logic of the method "CLEAN" developed by Högbom [16].

We represented the 3D geosection with a 3D prismatic array of nodes, within which different anomalous bodies are situated. The 2D rectangular array of points is situated over this 3D array, representing the ground surface where field measurements are taken (Fig. 1) [8].

Experiments were carried out using a 3D geosection of 4000m*4000m*2000m, discretized with spatial steps 400m, 200m, and 100m with respectively 11*11*6, 21*21*11, 41*41*21 nodes, using a multiplication spatial factor of 2. The ground surface 2D array of points has similar spatial extension and discretization.

The anomalous body was modeled with a prismatic structure with dimensions 400m*400m*1800m and density 5 G/cm^3 compared with the density zero of the environment 0 G/cm^3. The density steps 1.0 G/cm^3 and 0.1 G/cm^3 were used for a maximal density per node 5 G/cm^3. The number of iterations (and related runtime) was expected inverted proportional with the density step; but the use of low values for the density step would be necessary to separate bodies with small difference of respective mass densities.

The algorithm searches iteratively the best 3D node in the geosection's array, which surface anomalous effect offers the least squares error. The mass density of the best node is increased with the predefined step. The effect is subtracted from the surface anomaly and the iteration repeated. Iterations continue until the normalized least squares error becomes smaller than a predefined level [8].

The complexity of the algorithm was evaluated taking into consideration as "elementary calculation" the volume of calculation when the effect of one node is calculated for one point. If the number of linear nodes in one edge of the array is N, the volume of elementary calculations for one iteration is of the order O(N^5).

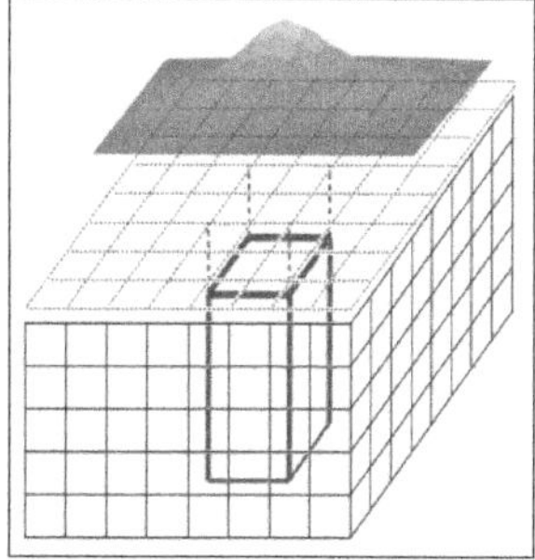

Fig. 1. – The discretized model of 3D geosection array of nodes (—) and the 2D ground array of points (▪ ▪ ▪), with an anomalous body at the center of geosection (▬)

Increase of the number of linear nodes N for the same model implies the decrease of the spatial discretization step and of the contribution of elementary calculations per iteration, leading to an order $O(N^3)$ of extra calculations. The overall order of the volume of elementary calculations per iteration is expected $O(N^8)$.

The MPI version of the software was run in 1, 2, 4, 8, 16, 32, 64, 128 and 256 cores. The software was executed through the command "/usr/bin/time", which evaluates the execution time parameters: *user*, *system*, *elapsed*, and the percentage of the usage of CPU. The scalability of the algorithm was evaluated with the net user runtime for 100% of CPU. The waiting time the end-user would expect, when running the application in a production system, was measured with the elapsed time.

At the end, we cross compared both the runtime and walltime of MPI and OpenMI versions of the software in both HPCG and SGE systems, in order to support a better planning methodology in the selection of the platform for the specific software and related models.

3 Computational Results

In this paper we focused only on the scalability of the MPI version of the software tested in SGE and HPCG sites. For the same models the number of iterations and least squares error using the OpenMP version of the software, tested in SGE and HPCG, are analyzed in detail in [8] and [10].

The runtime decreased inverted proportionally with the number of cores, except for small models (Fig. 2). For small models (the case of spatial step 400m) the scalability did not followed the expected trend due to the higher time overhead related with inter-process communication compared with the computational time. For big models, the decrease rate of the runtime slowed down when switching from 8 cores to 16 and 32 cores in HPCG – this system is composed by computer nodes with 8 hyper-thread cores, interconnected in a 20 GB switch, and the phenomenon is related with delays in communication between processes in different computer nodes.

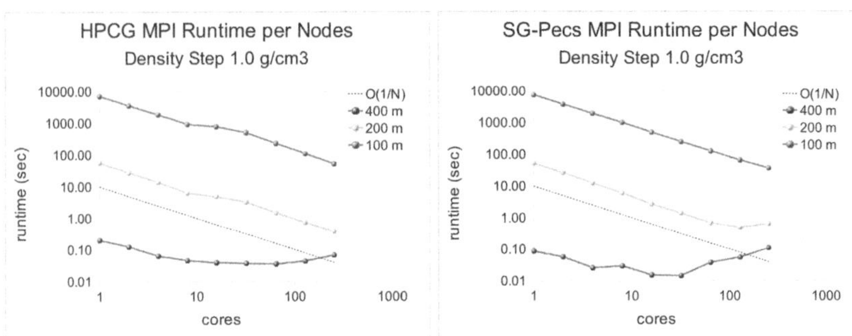

Fig. 2. Runtime as function of the number of cores for three models with discretization step 400m, 200m and 100m

Seen from another perspective, the runtime of MPI increases with the size of the model as it is shown in Fig. 3.

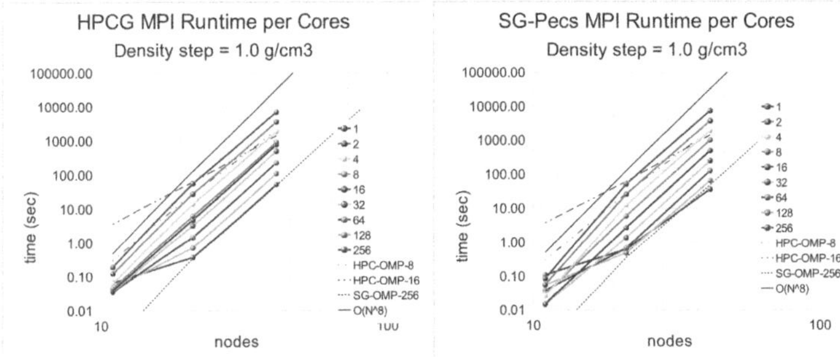

Fig. 3. Runtime as function of linear nodes of models (inverted discretization step) for different number of cores. OpenMP references for the runtime order are included.

Increase of the MPI runtime per model size was of the order of $O(N^8)$, similar with OpenMP results obtained in both systems. When using OpenMP in HPCG, the increase rate of the runtime resulted of lower order. HPCG nodes have 8 hyper-threading cores each, and OpenMP may be used with up to 16 hyper-threads, which leads to lower order of the increased rate of runtime. Compared with OpenMP, the absolute runtime with MPI resulted with lower values.

The scalability was disturbed when the MPI walltime time was used (Fig. 4), instead of the net runtime (Fig. 3), as result of time sharing between different users (the case of HPCG). MPI data were compared with respective data obtained with OpenMP. OpenMP in both HPCG and SGE gave lower and similar increase rate of the walltime. The rate of the walltime resulted of the order O(N^7) lower than the rate of the runtime, indicating that the couple algorithm+system works better for big models in many cores.

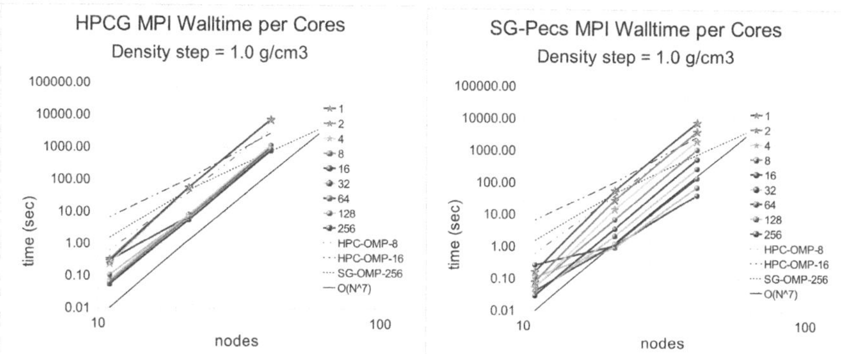

Fig. 4. Walltime as function of linear nodes of models (inverted discretization step) for different number of cores. OpenMP references for the walltime order are included.

We compared the relative to each other performance of systems HPCG and SGE, when running MPI and OpenMP versions of the software, using respective runtime and walltime for different model sizes.

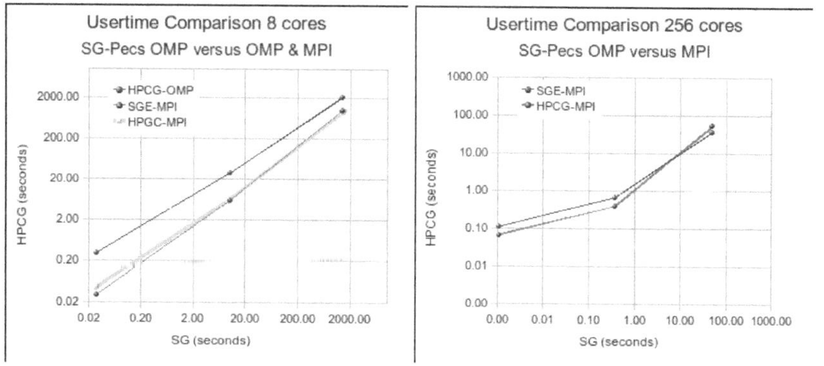

Fig. 5. Cross-comparing the usertime obtained with OpenMP and MPI in HPCG and SGE for three model sizes, using 8 and 256 cores

Comparison for OpenMP in HPCG is done only for 8 cores because in that system the maximal number of OpenMP parallel threads is limited in a single computer node with 8 cores permitting a maximum of 8 OpenMP parallel simple threads (not hyper-threads). The cross-comparison of runtime is presented in Fig. 5.

For 8 cores the OpenMP in SGE gave similar runtime compared with MPI in both SGE and HPCG, while OpenMP in HPCG performed worse (for this application and tested models).

For 256 cores the OpenMP in SGE gave similar runtime compared with MPI in both systems but only for medium and big models. For small models there is the increase of the runtime due to the interprocess communication overhead.

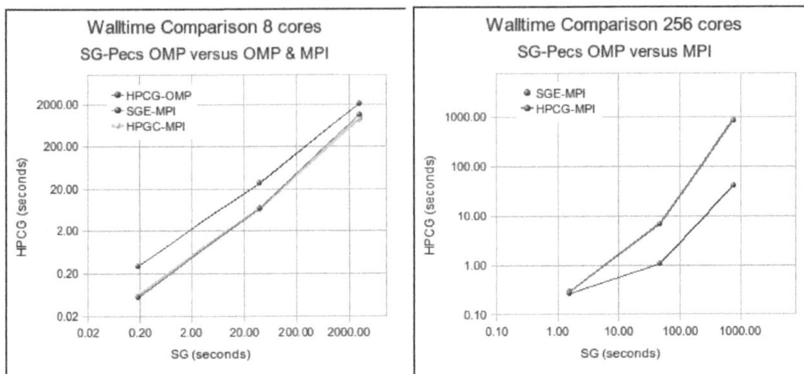

Fig. 6. Cross-comparing the walltime obtained with OpenMP and MPI in HPCG and SGE for three model sizes, using 8 and 256 cores

Cross-comparison of walltime is given in Fig. 6. For 8 cores the OpenMP walltime was similar in both systems. The MPI wallime was similar in both systems with improved values (for this application and tested models) compared with OpenMP.

For 256 cores the MPI walltime in both systems resulted with lower values when compared with the OpenMP in SGE. The degradation of the walltime for small models resulted less visible compared with the respective cases of runtime. Differences between MPI walltime in both systems are due to the multi-user daily workload during tests.

4 Conclusions

The use of MPI for parallelization of 3D gravity inversion algorithm GIM resulted similar with OpenMP, but with the advantage that most of available parallel systems in the region offer full capacities for running parallel processes while limited capacities for parallel threads.

The increase order of the runtime with the linear size N of the model resulted as expected at $O(N^8)$, the decrease order of the runtime per cores remained at $O(N^{-1})$. In our tests for models with 41*41*21 nodes and spatial discretization step of 100m and density step 0.1 G/cm^3 the runtime reached values of 100,000 seconds (28 hours). All this makes difficult the proposed inversion algorithm for detailed geomodels, necessary for engineering works, but it remains useful for regional studies.

The phenomenon of degradation of the runtime scalability for small models run in high number of computing cores is visible in both OpenMP and MPI solutions. As expected, for the MPI solution the degradation of runtime resulted worse (see also [10]).

Alternatives for the increase of parallelization of 3D inversion algorithms include GPUs and distributed grid platforms (multi-clusters). GPUs support up to thousand parallel threads in a single desktop, which is a capacity similar with most of MPI platforms in the region. Multi-cluster solutions theoretically may permit exploitation of many thousand CPUs in a distributed grid environment such as pan-European Grid Infrastructure, but a careful analysis is necessary to evaluate the impact of inter-process communication delays between different clusters, a phenomenon identified already in our results.

Tests with multiple bodies and real field data reconfirmed the need of using initial solutions defined on basis of other geological factors, and the need for careful interpretations of results [9]. In this context the consideration of programming alternatives and simplification of models may help in reducing the runtime and improve inversion solutions.

Acknowledgments. This work makes use of results produced by the High-Performance Computing Infrastructure for South East Europe's Research Communities (HP-SEE), a project co-funded by the European Commission (under contract number 261499) through the Seventh Framework Programme. HP-SEE involves and addresses specific needs of a number of new multi-disciplinary international scientific communities (computational physics, computational chemistry, life sciences, etc.) and thus stimulates the use and expansion of the emerging new regional HPC infrastructure and its services. The work is supported by the HP Cluster Platform Express 7000 operated by the Institute of Information and Communication Technologies, Bulgarian Academy of Sciences in Sofia, Bulgaria and the SGE system of the NIIFI Supercomputing Center at University of Pécs, Hungary,. Full information is available at http://www.hp-see.eu/.

References

1. Sen, M., Stoffa, P.: Global Optimization Methods in Geophysical Inversion. Elsevier Science B.V. (1995)
2. Hadamard, J.: Sur les prolemes aux derivees partielles et leur signification physique, vol. 13, pp. 1–20. Bull Princeton Univ. (1902)
3. Shamsipour, P., Chouteau, M., Marcotte, D., Keating, P.: 3D stochastic inversion of borehole and surface gravity data using Geostatistics. In: EGM 2010 International Workshop, Adding New Value to Electromagnetic, Gravity and Magnetic Methods for Exploration Capri, Italy, April 11-14 (2010)
4. Silva, J., Medeiros, W.E., Barbosa, V.C.F.: Gravity inversion using convexity constraint. Geophysics 65(1), 102–112 (2000)
5. Zhou, X.: 3D vector gravity potential and line integrals for the gravity anomaly of a rectangular prism with 3D variable density contrast. Geophysics 74(6), 143–153 (2009)
6. Zhou, X.: Analytic solution of the gravity anomaly of irregular 2D masses with density contrast varying as a 2D polynomial function. Geophysics 75(2), 11–19 (2010)
7. Wellmann, F.J., Horowitz, F.G., Schill, E., Regenauer-Lieb, K.: Towards incorporating uncertainty of structural data in 3D geological inversion. Elsevier Tectonophysics TECTO-124902 (2010), `http://www.elsevier.com/locate/tecto` (accessed September 07, 2010)
8. Frasheri, N., Cico, B.: Analysis of the Convergence of Iterative Geophysical Inversion in Parallel Systems. In: Kocarev, L. (ed.) ICT Innovations 2011. AISC, vol. 150, pp. 219–226. Springer, Heidelberg (2012)

9. Frasheri, N., Bushati, S.: An Algorithm for Gravity Anomaly Inversion in HPC. In: Frasheri, N., Bushati, S. (eds.) SYNASC 2011 - 13th International Symposium on Symbolic and Numeric Algorithms for Scientific Computing, Timisoara, Romania, September 26-29 (2011)
10. Frasheri, N., Cico, B.: Convergence of Gravity Inversion using OpenMP. In: Information Technologies IT 2012, Zabljak Montenegro, February 27-March 02 (2012)
11. Rickwood, P., Sambridge, M.: Efficient parallel inversion using the Neighborhood Algorithm. Geochemistry Geophysics Geosystems Electronic Journal of the Earth Sciences 7(11) (November 1, 2006)
12. Loke, M.H., Wilkinson, P.: Rapid Parallel Computation of Optimized Arrays for Electrical Imaging Surveys. In: Loke, M.H., Wilkinson, P. (eds.) Near Surface 2009 – 15th European Meeting of Environmental and Engineering Geophysics, Dublin, Ireland, September 7-9 (2009)
13. Hu, Z., He, Z., Wang, Y., Sun, W.: Constrained inversion of magnetotelluric data using parallel simulated annealing algorithm and its application. In: SEG Denver Annual Meeting 2010 – SEG Expanded Abstracts. EM P4 Modeling and Inversion, vol. 29 (2010)
14. Wilson, G., Čuma, M., Zhdanov, M.S.: Massively parallel 3D inversion of gravity and gravity gradiometry data. PREVIEW - The Magazine of the Australian Society of Exploration Geophysicists (June 2011)
15. Lowrie, W.: Fundamentals of Geophysics. Cambridge University Press (2007)
16. Högbom, J.A.: Aperture Synthesis with a Non-Regular Distribution of Interferometer Baselines. Astr. Astrophys. Suppl. 15, 417 (1974)

Getting Beyond Empathy

A Tougher Approach to Emotional Intelligence

Andrew James Miller, Shushma Patel, and George Ubakanma

London South Bank University, London, UK
{millea10,shushma,ubakang}@lsbu.ac.uk

Abstract. Emotional intelligence (EI) is considered to be an essential part of a project manager's skill set, often meaning the difference between project success or failure. Project managers with naturally high EI have a distinct advantage, employing their interpersonal sensitivity and tempered communication style to convince and persuade even the most difficult team members or stakeholders.

However, when negotiating over resources, budgets and schedules that are critical for the project's success, an overemphasis on openness, trust and willingness to compromise can leave high EI negotiators vulnerable to exploitation, particularly when dealing with more ruthless lower EI counterparts.

This paper examines emotional resilience and tough tactics in negotiation as important elements of EI in project management that paradoxically can be especially difficult for naturally high EI individuals. It assesses current theory and suggests areas for further research.

Keywords: Emotional Intelligence, Empathy, Project Management, Conflict, Negotiation, Machiavellianism.

1 Introduction

Emotional Intelligence is considered to be an essential requirement for managerial and leadership effectiveness, and according to some it is twice as important as cognitive abilities [1]. When applied to project management, the focus is often placed on the importance of "soft skills" such as communicating, listening, sensitivity, influencing, and motivating [2] and how they achieve optimum results by prioritizing people over process.

These skills are firmly tested during the difficult negotiations and conflicts that arise during a project. Yet if we go by Goleman's four domains of Emotional Intelligence, while a significant amount of research and best practice exists for the Social Awareness element of EI that focuses on the above-mentioned soft skills, rather less attention has been paid to its equally important counterpart, Self Awareness [3]. That is to say, while project managers are trained to look outwards, observing and reacting to the emotions of others, they are not necessarily trained to look inwards to first understand and manage their own emotions.

This has led to some curious contradictions in research findings on Emotional Intelligence in negotiation contexts, such as the discovery that while high EI negotiators

S. Markovski and M. Gusev (Eds.): *ICT Innovations 2012*, AISC 207, pp. 353–361.
DOI: 10.1007/978-3-642-37169-1_35

are better at resolving conflicts, they often resolve them at their own cost [4] and that high EI negotiators consistently allow themselves to be manipulated by their lower EI counterparts [5-6]. A number of studies suggest that a high degree of Emotional Intelligence can negatively impact negotiation performance [7-8].

Equally concerning for proponents of applied EI is the growing research suggesting overt displays of negative emotions such as anger can be extremely effective in negotiations [9], even though this goes against the core principles of Emotional Intelligence in negotiation and conflict settings [10-11].

Such bullying tactics should not work on the emotionally intelligent negotiator, yet in many instances they do. This paper looks at existing attitudes and possible misconceptions surrounding Emotional Intelligence in a business and project management context, and suggests that EI is often erroneously equated to having a natural inclination towards empathizing and accommodating, and prioritizing the emotional wellbeing of others over the achievement of business objectives.

Furthermore, it attempts to elaborate a more complete picture of Emotional Intelligence, focusing on how the four key skills should be applied to tough negotiations and conflict situations, and how the emotionally intelligent project manager needs to be particularly mindful of the Self Awareness and Self Management dimensions of applied EI. Finally, it looks at wider applications for EI in a project management environment, and the extent to which it can be utilized to counterattack and overcome deceptive and aggressive opponents during conflict resolution. The ethical considerations of adopting such a strategy are also discussed.

2 Emotional Intelligence and Conflict

For many, the word “conflict” has a particular emotive resonance, often highly negative. So much so that “negotiation”, the process of constructively dealing with conflict, is equally emotionally charged to the point of making all but the most seasoned of project managers “inherently uncomfortable” [12]. So what is it about conflict and negotiation that makes project managers so uneasy?

Fisher & Ury [10] suggest that in every negotiation the parties involved have two primary concerns, “Every negotiator wants to reach an agreement that satisfies his substantive interests. That is why one negotiates. Beyond that, a negotiator also has an interest in the relationship with the other side.” The balancing of these concerns is reflected in the five approaches to dealing with conflict (Accommodate, Collaborate, Compromise, Withdraw, Defeat), also known as the negotiating styles model [13-14].

Recent research indicates high EI negotiators are more likely to be pushed into accommodating in win / lose scenarios. Despite their hypothesis that high EI negotiators are more likely to manipulate their lower EI counterparts using their Emotional Intelligence as a strategic advantage, Der Foo et al. [4] discovered that while high EI negotiators create more value in negotiations, they are often unable to claim it, concluding that high EI negotiators show too much sympathy to their low EI opponents, leaving themselves vulnerable to exploitation. Kong & Bottom [15] offer a different explanation, arguing that high EI negotiators strive to stay true to their “idealistic self”, valuing symbolic rewards such as trust, rapport and relationships over material gain.

However, if one goes back to Goleman's four key skills model neither of these explanations feels particularly satisfactory, as they seem to accept the high EI negotiators' preoccupation with empathy and relationship management as archetypal, without questioning the absence of Self Awareness and Self Management skills. This feeds into the distorted perception of Emotional Intelligence as simply the art of accommodating others, a tenet that would doom any project manager to failure.

3 Can Emotional Intelligence Be a Weakness?

Can putting too much stock in emotions become a distraction, and even undermine rational decision-making? In a negotiation setting, Van Kleef et al. [16] observed that negotiators who carefully scrutinize the information within their opponents emotions instinctively follow the cues and back down to angry opponents, whereas negotiators who do not engage in such information processing are unaffected by their opponents' emotions and take a more rational stance.

A study by Hjertø & Schei [8] found that the self emotional appraisal dimension of Emotional Intelligence was negatively related to group performance in negotiations. This suggests that high EI negotiators may over-analyze their performance to the point of undermining their ability to take firm decisions.

Emotions such as worry and disappointment that provoke strong empathetic responses can be equally effective in gaining larger concessions from opponents [6]. Contrarily, emotions driven by empathy such as guilt and regret appear to motivate opponents to concede less [6]. This could play doubly against high EI negotiators, as their susceptibility to their own empathetic responses and heightened self emotional appraisal would lead them to concede more, feeling in part responsible for their opponent's negative emotions. In turn, their opponents would pick up on these signals of guilt and regret and therefore concede less.

Such concerns about EI are not limited to a negotiation setting. Assessing EI in terms of leadership effectiveness, Antonakis [7] states high self-monitors are less likely to adopt firm strategic positions or communicate a consistent vision on key issues, and that individuals with high need for affiliation are submissive, inconsistent and troubled by contentious issues. Antonakis argues that being immune to detecting emotional nuances in others may actually be useful for leaders as they would not be "derailed" by negative emotions, pandering to individuals or the need to be agreeable [7].

This is an interesting argument, as it assumes a leader cannot not have any control over how the emotions of others affect him/her or his/her actions. Again, we are getting a selective interpretation of Emotional Intelligence. This time, while the Self Awareness dimension of EI is recognized, Self Management is completely ignored.

The reality, as Jordan & Sheehan [17] explain, is that managers low in empathy and with poor emotional regulation are more likely to abuse staff and use bullying tactics to get what they want.. Apart from the obvious ethical concerns, research shows that such behavior actually works against achieving organizational objectives, and can be hugely costly for the organization long term [18].

Nevertheless, Antonakis does highlight the potential strengths of low EI individuals, and despite the vaunted claims of EI's indispensability for business success, evidence suggests that high EI individuals' heightened sensitivity to emotional stimulus can leave them open to exploitation, especially in the resource-scarce, time-pressured and heavily task-driven environments frequently encountered in project management. In contrast, low EI individuals who are uninhibited by how their actions affect others and unfazed by conflict, are able to simply seize the advantage wherever they see it.

And it may be worse still for the high EI project manager. Curiously, rather than being passive beneficiaries of this phenomenon due to their emotional obtuseness, research suggests that low EI individuals may deliberately manipulate the sensibilities of high EI individuals to gain a strategic advantage.

4 Emotional Intelligence and Self Management

Thus far, a potential weak-point has been identified for emotionally intelligent individuals: Self Management. Paradoxically, high EI individuals find themselves at a particular disadvantage when dealing with hard, unscrupulous low EI opponents who are able to capitalize on this vulnerability and actively exploit it in a calculated way to get what they want. How can this be?

The causality of the phenomenon may be complex. A study by Austin, Farrelly, Black, & Moore [5] found that high Mach (Machiavellian) individuals frequently utilize emotional manipulation to get what they want, despite lacking emotion-related skills. The fact that Machs are low in EI [5] suggests that in order to be taken in, EI individuals, albeit not consciously, are complicit in their own manipulation.

This creates a complicated problem for proponents of EI. Graham [19] describes low Machs as "chronic losers in interpersonal encounters" with their more wily co-workers and business associates, while Fulmer, Barry, & Long [9] assert that negotiators who use deceptive tactics often achieve better outcomes in negotiations. This can be an extremely delicate issue for the PM driven by the belief that successful project managers must place people before process, who is all too readily exploited by ruthless low EI individuals who have the luxury of a singularity of focus, devoid of any concern for the feelings of others.

At first this seems to bring into question the validity of the EI-based principled negotiation and leadership styles advocated by the widely accepted best practice [10], [20]. However, on closer inspection it just brings us back to the core skills of Emotional Intelligence, and above all the importance of Self Management.

The manipulative tactics of high Mach negotiators are not difficult to neutralize. A high EI negotiator just needs to stay attentive for deceptive negotiation tactics and structure the situation so that such tactics can be detected [21]. As Yunus & Derus [22] point out, high Machs are highly moderated by situational factors, and only are able to dominate when they interact face-to-face, the situation is uncontrolled with a minimum number of rules and regulations, and when an emotional involvement with details irrelevant to winning distracts their high EI opponents. If such situational factors are not in place, or are controlled by their opponent, the high Mach negotiator's advantage is lost.

Research by Grams & Rogers [23] shows that if deceptive negotiators meet resistance to their manipulative tactics they quickly change to a rational, reason-based strategy. The emotionally intelligent negotiator with an understanding of deceptive tactics and strong Self Management can easily diffuse the threat and turn the situation to their advantage. The key is good preparation going into the negotiation, defining a strategy beforehand based on what you know about the opponent and what is at stake, and identifying and calling out deliberately manipulative tactics if you think they are being used [24-25]. This can be applied to all conflict situations, and is particularly useful for project managers who often have to negotiate and re-negotiate with the same people.

It is also important for high EI individuals to resist the instinctive urge to shy from conflict through avoidance or appeasement. If handled appropriately conflict can often improve project collaboration [26], and as Shell [20] points out, not confronting issues of tension or unfair treatment from others only leads to exploitation and the ultimate collapse of the relationship.

When applied in a business context, the focus of Emotional Intelligence should always be the achievement of business objectives. Within their principled negotiating style framework Fisher & Ury [10] acknowledge that "in positional bargaining, a hard game dominates a soft one" and that being too friendly and sympathetic with a tough opponent can make you unnecessarily vulnerable. A number of authorities on ethics in negotiation even recognize deceptive behavior such as bluffing as a standard component of bargaining [9], [27].

While EI may be most strongly identified with its facet of empathy and taking into consideration the emotions of others, that is certainly not the whole story.

5 A Tougher Approach to Emotional Intelligence

Taking into account what we have seen in the previous sections, particularly concerning the potential weaknesses associated with high EI individuals, how can we get the most practical benefits from applied EI in negotiations and conflict settings?

Looking at the evidence, one might conclude that being emotionally sensitive can be a liability, and that to shield oneself from manipulative opponents an emotionally intelligent negotiator should attempt to suppress their emotional instincts. This is not necessarily the case.

As Fromm [24] emphasizes, the key is to stay aware of your emotions as they arise and to know when to distance yourself from destructive emotional responses while using your emotional intuition to good effect. As Fisher & Shapiro [11] point out, we cannot stop ourselves having emotions any more than we can stop having thoughts, and in many circumstances attempting to suppress our emotions can actually be detrimental [28]. The challenge then, is to stimulate helpful emotions within ourselves and those with whom we negotiate [11].

While using Emotional Intelligence to stimulate positive emotions to foster greater trust and collaboration is well known and widely practiced, when a tougher stance is needed with difficult opponents who refuse to cooperate, can EI still get results?

Nicholson [29] debunks the idea that an empathy-based approach to dealing with difficult people means making concessions. He argues that empathizing with someone who is causing you problems allows you find their locus of energy and "leverage it to achieve your ends". If you are able to establish what will be possible to gain from the individual, you can reframe your expectations accordingly and still channel their energies into meeting your goals. If you prompt them to release their negativity in a controlled way and get to the root cause behind it, it is much easier to identify issues that need to be resolved and subsequently drive the individual to arrive at a solution that will be beneficial for all.

Jordan & Troth [30] emphasize that "flexibility and response" are the hallmarks of EI, and a key part of Self Management and Relationship Management is being able to judge which emotions will be most strategically useful to you at any given time.

In the Thomas-Kilmann model of dealing with conflicts, no style is inherently better than the other. Taking a defeating stance (defined by Goleman in a leadership context as the "commanding" style), while not to be used excessively, in certain circumstances can be extremely effective [31], [13].

For example, when counterattacking difficult negotiators Wade [25] suggests that selective screaming, employing aggressive "bad cops on very short leashes", insulting offers and walk-outs are all viable strategies. In a wider project management context, Schyns & Hansbrough [32] argue that displays of anger can be crucially important for leader effectiveness in decisive and defining moments of the project, as do Lindebaum & Fielden [33].

There are obvious dangers to employing aggressive approaches to leadership and problem-solving, the biggest of which is becoming over-reliant on using dominant behavior, which if overused very quickly loses its effectiveness and burns out relationships. As always, the key is to find the right balance.

When looking at the idea of flexibility and response, LI & Roloff [34] outline that a positive mood leads to heuristic and flexible thinking, whereas a negative mood leads to more systemic and rational decision-making. They argue that individuals high in strategic EI are more likely to use emotions strategically to leverage this to achieve specific goals at particular times during a negotiation. They emphasize the effectiveness of the "dual display" phenomenon, where studies suggest that a good cop / bad cop routine outperforms a good cop / good cop or bad cop / bad cop routine, even when employed by a single person. Transitioning coherently from one mood to the other, applying just the right amount of pressure at the right time to direct the tone of the negotiation, is an arena in which emotionally intelligent individuals should excel.

The effectiveness of combining contrasting negotiation strategies is also evidenced by Falbe & Yukl [35] whose research concluded that hard / soft combinations are usually more successful than single tactics, though careful selection and employment is required. Similarly, Kopelman et al. [36] analyze three seemingly contradictory approaches to negotiations, "rational" (emotionless), "positive" and "irrational" (negative) and conclude that while each strategy has its strengths and weaknesses, a successful blending of all three approaches could be extremely effective, because a negotiator could get the concessions they wanted from their opponent while at the same time smoothing over the relationship.

It seems that the most pragmatic approach to applying Emotional Intelligence to conflict and negotiation settings is to stay aware of and manage one's own emotions internally, while strategically employing emotional behavior according to the needs of the situation. This can of course be as subtle as being empathetic to your counterpart's needs to create a more positive, cooperative atmosphere, or standing up to an aggressive opponent by showing emotional firmness and a refusal to be manipulated. However, EI can also be employed to control the tonal shifts of the interpersonal elements of the negotiation, positively or negatively, in order to leverage the emotional context to obtain maximum benefit.

This is where the ethical dimension comes in, and questions have to be asked as to the moral acceptability of manipulating the emotions of others for one's own personal gain. One could argue that it depends on the opponent. Lovell [37] emphasizes that when dealing with deceptive (high Mach) individuals it is essential for a project manager to adopt a different strategy to when dealing with low Machs. Wade [25] suggests the use of harsh tactics only when dealing with difficult opponents as a form of self-defense.

From practical experience, the employment of EI in project management often results in the blurring of distinctions between 'persuading' and 'manipulating' and inevitably requires at least a degree of emotional theatre, as in emotionally-charged contexts even the apparent absence of emotion can either create or alleviate tension.

There is no clear line that separates the ethical and non-ethical use of Emotional Intelligence in such circumstances, particularly when dealing with unethical opponents where one could argue that the end ultimately justifies the means. This is not to say that we should not endeavor to establish that line, however, as consistently taking the moral high ground has been shown to reap its own long-term rewards [38], and appealing to moral relativism in business ethics can be a very slippery slope.

6 Conclusion

In recent years there has been a backlash against applied Emotional Intelligence in business, as it has gained a reputation for being soft and easily taken advantage of in today's increasingly tough business environment. We would argue that this is based on a misconception of the fundamental principles of EI, particularly with regards to the Self Awareness and Self Management skills and how they should be applied in tough negotiation settings.

We would question research suggesting that "high EI" negotiators are more likely to be emotionally exploited and manipulated by lower EI opponents, as vulnerability to such exploitation is a clear indicator of a lack of Emotional Intelligence. We would argue that more research is needed into how people are assessed on the Emotional Intelligence scale, particularly with regards to Self Management.

Finally, we suggest that more research is needed into the alleged "dark side" of EI [5] and to what extent Emotional Intelligence can be legitimately utilized strategically to obtain material gains through manipulating others. The delicate balance of ethics and pragmatism is extremely difficult in any aspect of business, and the employment of Emotional Intelligence is no different. There is an important ethical dimension here that needs to be further explored.

Acknowledgments. We would like to thank the Technology Strategy Board for the support in the Knowledge Transfer Partnership programme that funded this work with Taylor & Francis.

References

1. Reynolds, A.: EIand Negotiation. Tommo Press, Hants (2003)
2. Alexander, T., Caldwell, J., Gonzalez, M., Harvey, D., Nye, B., Rodgers, C., Washer, A.: The role of EI in PM over the next 5 years. Organizational Acceleration (2010)
3. Goleman, D.: Emotional Intelligence. Bantam Books, New York (1995)
4. Der Foo, M., Elfenbein, H.A., Tan, H.H., Aik, V.C.: Emotional Intelligence and Negotiation: The Tension Between Creating and Claiming Value. Int. J. of Conflict Management 15(4), 411–429 (2004)
5. Austin, E.J., Farrelly, D., Black, C., Moore, H.: Emotional intelligence, Machiavellianism and emotional manipulation: Does EI have a dark side? Personality and Individual Differences 43(1), 179–189 (2007)
6. Van Kleef, G.A., De Dreu, C.K.W., Manstead, A.: Supplication and Appeasement in Conflict and Negotiation: The Interpersonal Effects of Disappointment, Worry, Guilt, and Regret. J. of Personality and Social Psyc. 91(1), 124–142 (2006)
7. Antonakis, J.: Why "Emotional Intelligence" does not Predict Leadership Effectivness: A Comment on Prati, Douglas, Ferris, Ammeter, and Buckley. International Journal of Organizational Analysis 11(4), 355–361 (2003)
8. Hjertø, K.B., Schei, V.: When Understanding Your Feelings Hurts Performance: Emotional Intelligence in Negotiating Groups. In: IACM 23rd Annual Conference Paper (2010), SSRN: `http://www.dx.doi.org/10.2139/ssrn.1612510`
9. Barry, B., Fulmer, I.S., Long, D.A.: Lying and Smiling: Informational and Emotional Deception in Negotiation. J. of Bus. Ethics 88, 691–709 (2008)
10. Fisher, R., Ury, W., Patton, B.: Getting to Yes: Negotiating an Agreement Without Giving In, 2nd edn. Penguin Books, New York (1991)
11. Fisher, R., Shapiro, D.: Beyond Reason: Using Emotions as You Negotiate. Penguin Books, New York (2005)
12. Pinto, J.K.: Understanding the Role of Politics in Successful Project Management. International Journal of Project Management 18(2), 85–91 (2000)
13. Thomas, K.W., Kilmann, R.H.: Thomas-Kilmann Conflict Mode Instrument. Xicom, Mountain View (1974)
14. HRDQ: Insight Newsletter (March 2011), `http://www.hrdqstore.com/Insight-Newsletter-March-2011_df_117.html`
15. Kong, D.T., Bottom, W.P.: Emotional Intelligence, Negotiation Outcome, & Negotiation Behaviour. Academy of Management Annual Meeting (2010)
16. Van Kleef, G.A., De Dreu, C.K.W., Manstead, A.: The Interpersonal Effects of Emotions in Negotiations: A Motivated Information Processing Approach. Journal of Personality and Social Psychology 87(4), 510–528 (2004)

17. Jordan, P.J., Sheehan, M.: Stress and managerial bullying: Affective antecedents and consequences. In: Transcending Boundaries: Integrating People, Processes and Systems, pp. 201–207. The School of Management, Griffith University, Brisbane (2000)
18. Sheehan, M.: Workplace Bullying: Responding with Some Emotional Intelligence. International Journal of Manpower 20, 57–69 (1999)
19. Graham, J.H.: Machiavellian Project Managers: Do They Perform Better? International Journal of Project Management 14(2), 67–74 (1996)
20. Shell, G.R.: Bargaining for Advantage: Negotiation Strategies for Reasonable People. Penguin Books, New York (2006)
21. Stuhlmacher, A.F., Adair, C.K.: Personality & Negotiation. In: Benoliel, M. (ed.) Negotiation Excellence. World Scientific, Singapore (2011)
22. Yunus, O.M., Derus, M.M.: Must Entrepreneurs be Machiavellians? In: 2nd International Conference on Entrepreneurship (2010)
23. Grams, W.C., Rogers, R.W.: Power and Personality: Effects of Machiavellianism, Need for Approval, and Motivation on Use of Influence Tactics. The Journal of General Psychology 117(1), 71–82 (1990)
24. Fromm, D.: Mind Games: Power, Personality and Emotion in Business Negotiation (2006), http://www.negotiatormagazine.com
25. Wade, J.: Negotiating with Difficult People, Queensland (2009)
26. Vaaland, T.I.: Improving Project Collaboration: Start with the Conflicts. International Journal of Project Management 22, 447–454 (2004)
27. Carson, T.L.: Second Thoughts About Bluffing. Business Ethics Quarterly 3, 317–341 (1993)
28. Lindebaum, D.: Rhetoric or remedy?: A Critique on Developing Emotional Intelligence. Ac. of Management Learning & Ed. 8(2), 225–237 (2009)
29. Nicholson, N.: How to Motivate Your Problem People. Harvard Business Review (January 2003) (Reprint R030D1)
30. Jordan, P.J., Troth, A.C.: Managing Emotions During Team Problem Solving: Emotional Intelligence and Conflict Resolution. Human Performance 17(2), 195–218 (2004)
31. Goleman, D.: Leadership That Gets Results. Harvard Business Review (March-April 2000) (Reprint R00204)
32. Schyns, B., Hansbrough, T.: When Leadership Goes Wrong: Destructive Leadership, Mistakes, and Ethical Failures. IAP, Charlotte (2010)
33. Lindebaum, D., Fielden, S.: 'It's Good to be Angry': Enacting anger in construction project management to achieve perceived leader effectiveness. Human Relations 64(3), 437–458 (2011)
34. Li, S., Roloff, M.E.: Strategic Emotion in Negotiation: Cognition, Emotion, and Culture (2006), http://www.neurovr.org/emerging/book7/9_8_Li.pdf
35. Fable, C., Yukl, G.: Consequences for Managers of Using Single Influence Tactics and Combinations of Tactics. Ac. of Mgmt. J. 35(3), 638–652 (1992)
36. Kopelman, S., Rosette, A.S., Thompson, L.: The Three Faces of Eve: Strategic Displays of Positive, Negative, and Neutral Emotions in Negotiations. Organizational Behavior& Human Decision Processes 99(1), 81–101 (2006)
37. Lovell, R.J.: Power and the Project Manager. International Journal of Project Management 11(2), 73–78 (1993)
38. Spiller, R.: Ethical Business and Investment: A Model for Business and Society. Journal of Business Ethics 27(1-2), 149–160 (2000)

Implementation in FPGA of 3D Discrete Wavelet Transform for Imaging Noise Removal

Ina Papadhopulli and Betim Çiço

Polytechnic University of Tirana, Faculty of Information Technology, Tirana, Albania
{ipapadhopulli,betim.cico}@gmail.com

Abstract. Discrete Wavelet Transform (DWT) is one of the most commonly used signal transformations. This transformation uses wavelets as filters, resulting in a frequency-time-amplitude dependence of the signal. Analyzing the obtained high frequency coefficients gives the possibility to remove the noise from the signal.

In this paper to achieve noise removal from the images we perform 3D-DWT (1D-DWT in three directions) using the biorthogonal Daubechies (9/7) filters. Due to their symmetry this filters are more suitable for image transformation. To reduce the number of multiply and accumulate resulting during signal filtering the Distributed Arithmetic (DA) technique is used taking advantage of the symmetry of the 9/7 filters. VHDL is used as the hardware description language and some modules are programmed in MATLAB. The implementation is mapped in FPGA Virtex-5 XUPV5-LX110T platform.

Keywords: 3D-DWT, wavelet, Daubechies 9/7 filter, Distributed Arithmetic.

1 Introduction

DWT is one of the most important types of signal transformation algorithms [7]. It is especially used in image processing. Comparing to other transformations such as Fast Fourie Transform (FFT), DWT gives a frequency-time-amplitude dependence. It allows us to remove frequency components at specific times in the data, giving the capability to throw out the "bad" and keep the "good" part of the data. This is what is needed for denoising and compression of images [4]. As filters, DWT uses wavelets, which are waveforms with limited duration and average value of zero. In contrast to the sinusoids used in FT wavelts are extremely adaptable due to their ability to stretch and shift and they can match an event unusually well.

In contrast to Continuous Wavelet Transform (CWT), DWT is a pure discrete transformation. The inverse transformation (IDWT) is computed easily and if the filter bank is chosen properly DWT ensures perfect reconstruction. The heart of DWT is the process of filtering which is the operation of convolution between the signal data and the filter coefficients. This is a high compute intensive task because of the large amount of data to be processed. FPGAs are appropriate devices to implement DWT because they are low-cost, fully support Digital Signal Processing (DSP) algorithms, and are dynamically reprogrammable. The multiply-and-accumulate (MAC) function during filtering can be implemented using Distributed Arithmetic (DA) [9] which uses

S. Markovski and M. Gusev (Eds.): *ICT Innovations 2012*, AISC 207, pp. 363–372.
DOI: 10.1007/978-3-642-37169-1_36

primarily look-up tables (LUTs). Xilinx FPGA architectures are based on LUTs [3] [11], which are part of their slices. MAC is replaced with shifts and look-up operations, resulting in more effective hardware utilization and low power consumption.

This paper is organized as follows. Section 2 gives a brief description of 1D-DWT, 3D-DWT and the 9/7 filters. Section 3 presents the logic of DA and the filtering architecture. Section 4 discusses how can be noise removed from images using DWT. Section 5 gives the implementation in FPGA Virtex-5 XUPV5-LX110T platform and finally brief conclusions are given in Section 6.

2 Discrete Wavelet Transform

2.1 1D-DWT

DWT gives a multiscale representation of a signal. DWT has two phases: the analysis phase (the decomposition of the signal) and the synthesis phase (the reconstruction of the signal). During all the analysis phase there are used two filters, the low-pass filter (L) and the high-pass filter (H). In DWT the signal is shrinked dyadically (power of 2) and is compared to the unchanged wavelet filters [4]. After filtering, the signal has half of its frequency band and based on Nyquist's theorem we can downsample by 2 without losing information. To achieve better noise removal DWT is performed several times and the final output is obtained by recursively applying lowpass/highpass frequency decomposition to the lowpass output of the previous stage. The number of levels depends on the type of noise to be removed and the size of the image.

Synthesis (IDWT) is the phase when the results obtained from DWT are upsampled by two and filtered with the high-pass (H') and low-pass (L') reconstruction filters respectively. H and H' combine to produce a highpass halfband filter, whereas L and L' produce a lowpass halfband filter. These two halfband filters produce an all-pass filter. H, L, H', L' are called a filter bank. If the right filter bank is chosen perfect reconstruction of the original signal can be achieved.

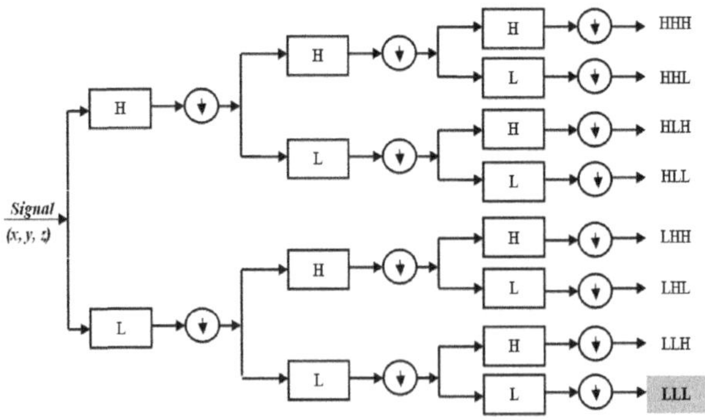

Fig. 1. One-level 3D-DWT

2.2 3D-DWT

3D-DWT is the combination of 1D-DWT in x, y, z directions [5] (see figure 1). Filtering during analysis and synthesis is done using the same filter bank for all the directions.

If another level of 3D-DWT will be performed only the data in the LLL subband will be further decomposed to the next level because these coefficients contain most of the image information.

In Figure 2 is shown one level of 3D-IDWT, where the decomposed coefficients (after have been processed for denoising) are upsampled by two and filtered with the synthesis high-pass and low-pass filters. As an output the reconstructed signal is taken. With perfect reconstruction the output is the original signal (input of 3D-DWT).

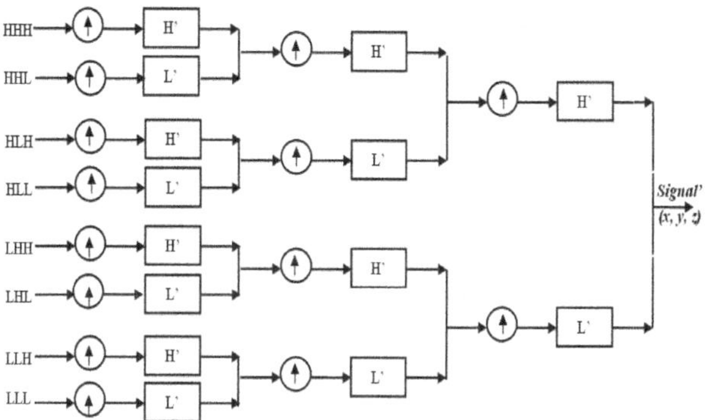

Fig. 2. One-level 3D-IDWT

2.3 9/7 Filters

There are a lot of different kinds of wavelets which can be used as filters in DWT [4]. One group of them is the orthogonal filters. These have the capability of perfect reconstruction and allias cancellation. Another group is the biorthogonal filters, which except the capabilities of the orthogonal filters allow perfect symmetry and vanishing moments.

Table 1. Daubechies 9/7 coefficients

K	Low-Pass	High-Pass
0	0.60294901	1.11508705
+/-1	0.26686411	-0.59127176
+/-2	-0.07822326	-0.05754352
+/-3	-0.01686411	0.09127176
+/-4	0.02674874	

In this work to implement DWT for image processing the biorthogonal Daubechies 9/7 (Db 9/7) filters are used. These filters are widely used, e.g. JPEG 2000 standard apply DWT with Db 9/7 for compression. In Db (9/7) the length of the low-pass filter is 9 and the length of the high-pass filter is 7 [6]. In Table 1 K is the coefficient number, Low-Pass is the column of the low-pass filter coefficients, High-Pass is the column of the high-pass filter coefficients.

These filters have perfect symmetry in the time domain and have linear phase in the frequency domain (see Figure 3) [4]. Symmetry and linear phase are desirable in Image Processing applications because human vision is more tolerant to symmetric errors [12].

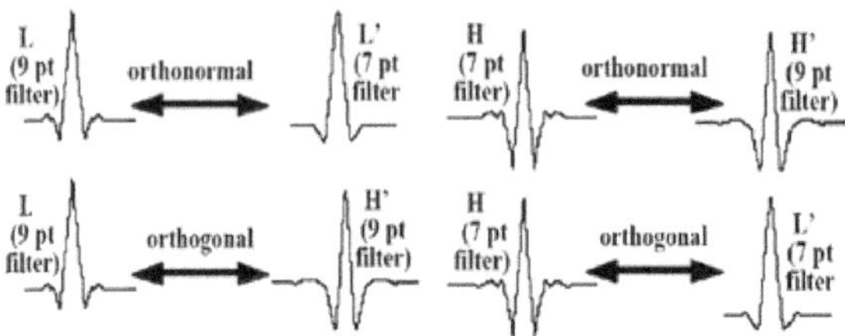

Fig. 3. Daubechies 9/7 filter bank

3 Distributed Arithmetic

During DWT the signal is filtered several times. Filtering is the operation of convolution of the signal with the filter coefficients. This calculation in hardware needs the use of many multipliers and adders. In this project the Distributed Arithmetic technique is used to reduce the MAC operations. As is explained below this approach employs no explicit multipliers in the design, only look-up tables (LUTs), shift registers and a scaling accumulator [8]. The basic operations required are a sequence of table look-ups, additions, subtractions and shifts of the input data sequence. All of these functions efficiently map to FPGAs [10].

DA is basically a bit-level rearrangement of the multiply and accumulate operation [9]. To illustrate the idea of DA we consider the equation 1 which has MAC operations.

$$y = \sum_{k=1}^{K} A_k x_k \tag{1}$$

where: A = $[a_1, a_2,\ldots,a_k]$ are the filter coefficients, x = $[x_1, x_2,\ldots, x_k]$ is a matrix of input data and x_k : $\{b_{k0}, b_{k1}, b_{k2}\ldots\ldots, b_{k(N-1)}\}$.

Using DA we arrive at the following equation where the input data words have been represented by the 2's complement number presentation in order to bound number growth under multiplication. b_{kn} is the n-th bit of b_k word and b_{k0} is the sign bit.

$$y = -\sum_{k=1}^{K} A_k \bullet (b_{k0}) + \sum_{n=1}^{N-1}\left[\sum_{k=1}^{K} A_k \bullet b_{kn}\right]2^{-n} \tag{2}$$

In this design the input data samples are 8 bits (N), whereas K is 9 for the low-pass filtering and 7 for the high-pass filtering operation of the signal.

Since A_k is constant then equation (2) is a function of b_{1n}, $b_{2n}\ldots$ b_{kn}, and has 2K possible values so we have the following dependence:

$$\left[\sum_{k=1}^{K} A_k b_{kn}\right] = f_n(b_{1n} b_{2n} \cdots b_{Kn}) \tag{3}$$

This equation can be pre-calculated for all possible combinations of b_{1n}, $b_{2n}\ldots$ b_{kn}. We can store the results in a look-up table of 2^K words addressed by K-bits and there is no need to do the multiply and accumulate operations.

The sample throughput of the filter is decoupled from the filter length when using DA. In our case, when we are using the Daubechies 9/7 filter, the size of the look-up table must have 29 words for the low pass filter and 27 words for the high pass filter.

Input samples are presented to the input parallel-in, serial-out shift register (PISO) at the input signal sample rate [9]. The output of the PISO shift register is presented to a cascade of serial-in, serial-out shift registers (SISO) with 8 bits. We have the same number of SISO shift registers as filter coefficients. These registers store the input sample history in a bit-serial format and are used to form the required inner-product computation. The outputs of these registers are used as address inputs to the look-up table which stores all possible combinations. The output of the LUT is added with the previous partial sum which is shifted by one. This is done for the eight bit positions and then the final result is computed.

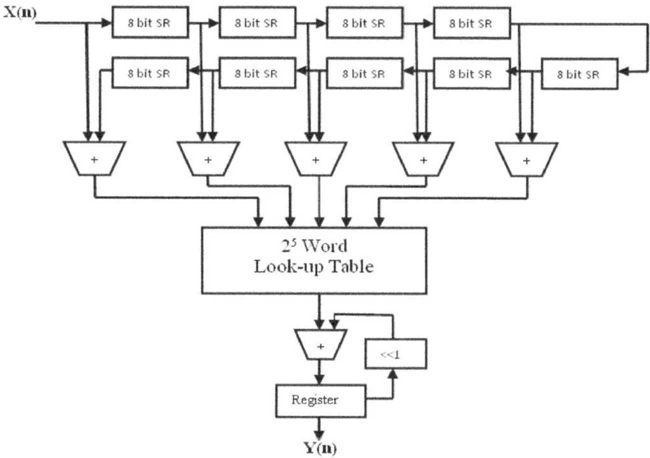

Fig. 4. Architecture for low-pass filtering

To compute inner calculations are required 8 clock cycles, because the input samples are 8 bit serial. To decrease the computation time, we can use parallel DA, where the input samples are partitioned. Fully parallelism is achieved when one input sample is 1 bit [1]. Increasing parallelism increases the hardware resources of the FPGA being used. Daubechies 9/7 filters have a symmetric impulse response. Based on this symmetry the size of the Look-Up Table can be decreased.

Due to symmetry, at the same time, symmetric bits from the output of the cascaded shift registers will be multiplied with the same filter coefficient, so we can add them together and as a consequence the input address of the LUT is halved (see Figure 4). For low-pass filtering the LUT has 25 words and for the high-pass filtering the LUT has 24 words.

The architecture for high filtering is the same as in Figure 4. In this case seven 8 bit shift registers, four adders, one 24 Word LUT and one scaling accumulator must be used.

4 Denoising

As is explained previously DWT gives a frequency–time dependence. Performing 1D-DWT with several levels (n) results in 1 group of approximation coefficients (from the low-pass filtering of the last level) and n groups of detail coefficients (from the high filtering of each level). The approximation coefficients contain most of the information of the image (a smoothed result of the image), whereas the detail coefficients contain high frequency events. Most of them generally equal zero. In Figure 5 are shown the amplitudes of two subband coefficients resulting from 4 levels of DWT. The subbands are LLLL (coefficients derived from four times low filtering) and LLLH (coefficients derived from four times low filtering). It is clearly noticed the difference in the amplitude of this two subband coefficients.

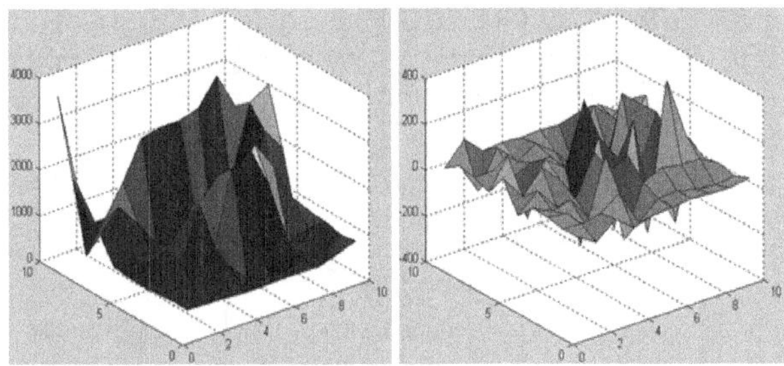

Fig. 5. Amplitude of coefficients resulting from 4 levels of DWT (a) LLLL subband (b) LLLH subband

Especially the detail coefficients derived from the high-pass filtering of the first levels of transformation contain little information about the signal. Figure 6 illustrates how 2D-DWT (with one level) decomposes an image.

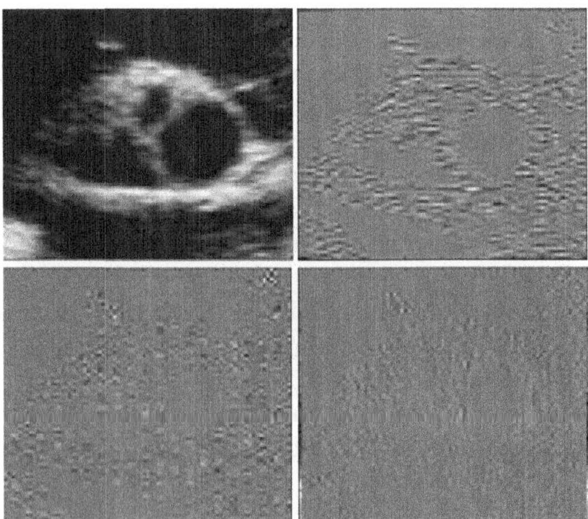

Fig. 6. LL, LH, HL, HH subbands after 2D DWT

Images usually suffer from noise which is generated by electronic devices or can be introduced from the communication channels used to transfer the image.

Unlike FFT in DWT we can discard some frequencies, which we think are part of the noise, but for certain time intervals [4]. Most common noise starts and stops in time.

This work is based on the white noise. This type of noise is very common and it appears at nearly all frequencies. Noise is typically considered as a linear addition to a useful information signal. The approach to remove noise after performing DWT is effective and simple. Since noise is distributed all over the image we assume that it is present in the detail coefficients in small amplitudes. Here the soft threshold method is used, where each coefficient which is smaller than the threshold (T) is set to zero, whereas to the other coefficients T is subtracted [8].According to [2] T is defined as:

$$T = \sigma\sqrt{2\log(N)} \tag{4}$$

where σ is the noise standard deviation and N is the number of noise coefficients. After decomposing the image, performing the soft threshold algorithm is simple, but previously it is necessary to find the optimal threshold (T) and the optimal number of DWT levels (L) in order to achieve the best noise removal for a given image.

According to Equation (4) T depends on the type of noise. The optimal number of DWT levels to perform depends on the number of image coefficients. Decomposing too much means changing the "good" portion of the coefficients because in higher levels of DWT the details come from the low frequencies.

To find T and L, we compute the difference between the coefficients of the image (size 512x512) achieved after adding random noise and removing it with the above technique and the coefficients of the original pure image is computed. Then the Root Mean Square (RMS) of the difference matrix is determined. This procedure is done for 1:5 levels of DWT and 0:40 values of threshold. The result is presented in the graphics in Figure 7. The graphics is derived using MATLAB R2010b.

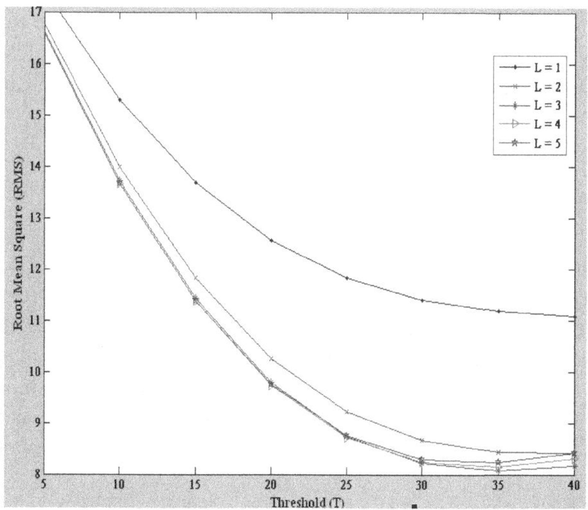

Fig. 7. RMS versus Threshold and DWT Levels

As we can see from Figure 7 we achieve better noise removal performing DWT with 3 levels and performing the soft-threshold algorithm with T = 35 for the given image, because this is the point where the RMS has the lowest value.

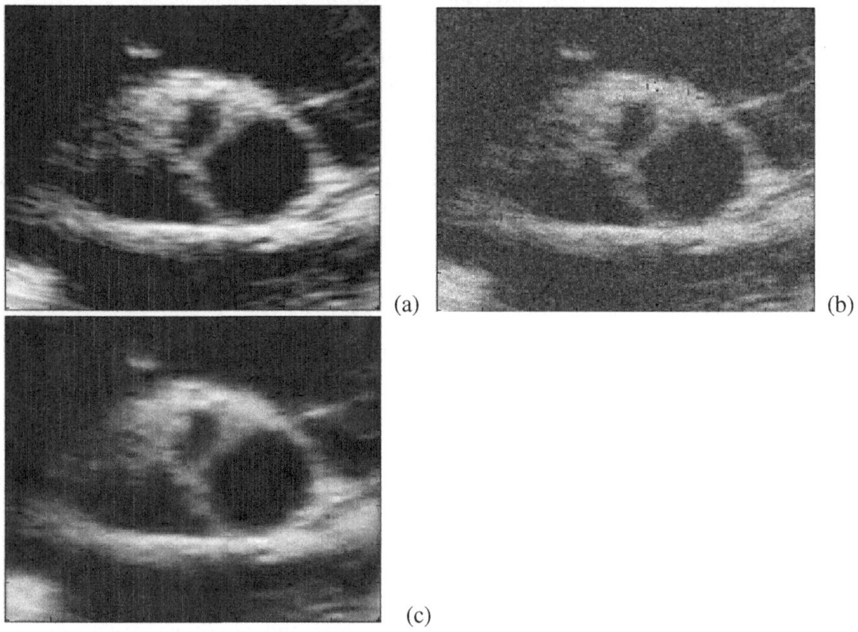

Fig. 8. (a) Original image (b) Image with noise (c) Image after 2D-DWT and noise removal

In Figure 8 is shown the original image, the image after adding random noise and the image after performing 2D- DWT with 3 levels and removing noise with the soft threshold algorithm with T = 35. Performing DWT does not decrease or increase the size of the original image. The image have the same resolution due to downsampling by two and upsampling by two during analysis and synthesis. The RMS between the original and the reconstructed image is 8.06. From Figure 8, it is clear that great part of the noise is removed.

5 Implementation in FPGA

As a software tool for synthesis and analysis is used Xilinx ISE 12.4. The 3D-DWT algorithm is implemented in VHDL as a hardware description language. Some of the modules are implemented and tested using MATLAB and translated into hardware implementation using AccelDSP 12.4, a Xilinx tool.

This design is mapped in Virtex-5 XUP VLX110T. Figure 9 shows the Register Transfer Level (RTL) schematic of the design. The dwt3d module is responsible for the analysis phase whereas the idwt3d is responsible for the synthesis phase.

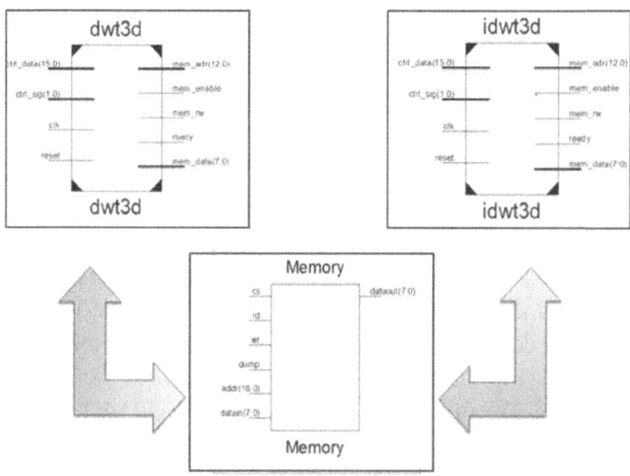

Fig. 9. RTL schematic of 3D DWT

After the process of mapping the device utilization summary is given in Table 2.

Table 2. Device utilization summary

Logic Utilization	Used
Number of Slice Registers	422
Number of Slice LUTs	907
Number of occupied Slices	293
Number of bonded IOBs	44

Virtex-5 XUP-VLX110T has the hardware resources to support this architecture of 3D-DWT.

According to [6], a DWT processor with RNS parallel structure uses 3335 slices and a DWT processor with parallel CSD uses 738 slices.

6 Conclusions

In this paper we presented the implementation of 3D-DWT in FPGA for imaging noise removal which is critical in certain areas. Daubechies 9/7 filters due to their symmetry give better results in noise removal. A RMS of 8.06 between the noiseless image and the denoised image can be achieved using 3D-DWT, if we choose the right threshold and number of DWT levels to perform. Virtex 5 slices include 6-input LUTs which can be configured to replace the MAC operations during filtering with DA technique. The symmetric coefficients of 9/7 filter and DA decrease the hardware resources to implement 3D-DWT and it can be fully supported by Xilinx Virtex-5 XUPV5-LX110T platform.

References

1. Al-Haj, A.M.: An FPGA-Based Parallel Distributed Arithmetic Implementation of the 1-D Discrete Wavelet Transform (February 2004)
2. Barsanti, R.J., Lehman, C.: Application of a Wavelet-Based Receiver for the Coherent Detection of FSK Signals (March 2008)
3. Clive Maxfield "FPGAs Instant Access"
4. Lee Fugal, D.: Conceptual Wavelets in Digital Signal Processing (An In-Depth Practical Approach for the Non-Mathematical) (2010)
5. Tripathy, M.R., Sachdeva, K., Talhi, R.: 3D Discrete Wavelet Transform VLSI Architecture for Image Processing (2009)
6. Jiang, R.M., Crookes, D.: FPGA Implementation of 3D DWT for Real Time Medical Imaging. In: ECCTD, pp. 519–522 (August 2007, 2008)
7. Polikar, R.: The Wavelet Tutorial (The Engineers Ultimate Guide to Wavelets Analysis), http://users.rowan.edu/~polikar/wavelets/wttutorial.html
8. Cai, S., Li, K., Selesnick, I.: MATLAB Implementation of Wavelet Transforms, http://eeweb.poly.edu/iselesni/WaveletSoftware/
9. Tutorial: Distributed Arithmetic (Implementations and Applications)
10. Xilinx: Distributed Arithmetic FIR Filter v9.0, http://www.xilinx.com/ipcenter/catalog/logicore/docs/da_fir
11. Xilinx: ML505/ML506/ML507 Evaluation Platform (October 2009), http://www.xilinx.com
12. Hong, Y.K., Lee, S.J., Kim, Y.-H.: Image Processor Using 3D-DWT as Part of Health Care Management System (August 2009)

IPv6 Transition Mechanisms and Deployment of IPv6 at Ss. Cyril and Methodius University in Skopje

Goce Gjorgjijoski[1] and Sonja Filiposka[2]

[1] University Computer Centre at Ss. Cyril and Methodius University in Skopje, blvd. Goce Delcev 9, 1000 Skopje, R. Macedonia
gocegj@ukim.edu.mk

[2] Faculty of Computer Science and Engineering, Ss. Cyril and Methodius University in Skopje, Rugjer Boshkovikj 16, 1000 Skopje, R. Macedonia
sonja.filiposka@finki.ukim.mk

Abstract. Having in mind the vast amount of IP-enabled devices that are activated each day, it is clear that the deployment of IPv6 is a must. Many companies have already implemented some sort of IPv6 networks, but implementation of IPv6 is not a small task. One cannot just switch from IPv4 to IPv6. There are several strategies and different deployment models involved in the process, and there is no "one size fits all" solution. However, there are some methods that are more recommended than others. In this paper, we give an overview of the IPv6 transition mechanisms and their deployment in a specific case study. Our aim is to investigate the pros and cons of the transition mechanisms and decide on the best solution to the given problem.

Keywords. IPv6, transition, dual-stack, tunneling, translation, 6rd, NAT64, DNS64.

1 Introduction

Internet Protocol version 4 (IPv4) has reached its limit. It can no longer support the growth of the Internet. In February 2011, the Internet Assigned Numbers Authority (IANA) allocated the last five blocks of /8 addresses to each of the five regional Internet registries (RIRs) [1]. There are no more IPv4 addresses left. The only way to ensure the future growth of the Internet is to implement the new version of the Internet Protocol, Internet Protocol version 6 (IPv6) [2]. Having in mind the vast amount of IP-enabled devices that are activated each day, it is clear that the deployment of IPv6 is a must.

IPv6 has much larger address space and better address management than IPv4. It also has a more efficient routing infrastructure, a new IP header format, stateless and stateful address configurations, enhanced security, and standardized QoS support. Overall, it is better and improved version of IPv4, that has the capabilities to meet the future requirements of the Internet.

S. Markovski and M. Gusev (Eds.): *ICT Innovations 2012*, AISC 207, pp. 373–382.
DOI: 10.1007/978-3-642-37169-1_37

Many companies have already implemented some sort of IPv6 networks, but implementation of IPv6 is not a small task, due to the fact that IPv6 is not backward compatible with IPv4. One cannot just switch from IPv4 to IPv6. There are several strategies and different deployment models involved in the process, and there is no a single best solution [3]. However, there are some methods that are more recommended than others.

In this paper, we first start with a survey of different models for IPv6 deployment, called transition mechanisms. These transition mechanisms are divided by Internet Engineering Task Force (IETF) into three main categories: dual stack, tunneling and translation [4]. We make a brief overview of each of these categories, when and where they can be used, and what are their main advantages and disadvantages.

Next, we discuss the implementation of a sample mechanism from all three categories on a case study model: the University Computer Network at Ss. Cyril and Methodius University in Skopje (UKIM). We start with the already implemented dual stack model, followed by scenarios of how the other two main transition mechanisms, 6rd for tunneling and NAT64/DNS64 for translation, could be implemented.

2 Transition Mechanisms

Since IPv6 is not backward compatible with IPv4, one cannot just switch from IPv4 to IPv6. On the other hand, the implementation of IPv6 is a complex process. There are several different strategies and different deployment models, called transition mechanisms, involved in this process. Each of those transition mechanisms is used for its specific purpose, and there is no "one size fits all" solution. However, there are some methods that are more recommended than others.

Internet Engineering Task Force (IETF) has divided these transition mechanisms into three main categories: dual stack, tunneling and translation. Table 1 presents a list of the most popular transition mechanisms, divided into these three categories.

Table 1. List of the most popular IPv6 transition mechanisms

Dual stack	Tunneling			Translation
▪ Dual stack model	▪ 6rd	▪ 6over4	▪ 4in6	▪ NAT64
	▪ 6to4	▪ ISATAP	▪ 6PE	▪ DNS64
	▪ 6in4	▪ Teredo	▪ Softwire	▪ SIIT

2.1 Dual Stack

The dual stack model [4] is the simplest method for IPv6 deployment, and the recommended starting point for full IPv6 implementation. In this model, both IPv4 and IPv6 are enabled on all devices on an existing network, so both protocols are running at the same time. In this situation, the operating systems and applications are those that choose which protocol to use. Most of the new operating systems and applications are designed to use IPv6 first, if available.

This deployment model has several advantages and it is recommended for most networks, service providers and network managers [3]. It can be deployed on hosts and routers on the same interfaces as IPv4, which allows them to continue to reach IPv4 resources, while also adding IPv6 functionality. It also deals with most address selection and DNS resolution issues, avoids problems associated with the configuration of tunnels and Maximum Transmission Unit (MTU) settings, and these kinds of networks are robust and reliable.

However, there are also several disadvantages associated with this model. Because both IPv4 and IPv6 are running at the same time, everything on the network needs to be done twice. For example, in the case of security, every rule that is applied or changed for IPv4, needs to be applied or changed for IPv6 also. In the case of routers, there are two routing tables, and two routing processes, which means additional CPU utilization and memory consummation.

Another disadvantage associated with this model is that not all applications deal gracefully with situations where IPv6 destination addresses work unreliably. In this kind of situations it may take a long time for some applications to switch over from IPv6 to IPv4.

2.2 Tunneling

Tunneling is used for interconnection of hosts that run the same version of IP, across different version of IP. This applies to both IPv6-to-IPv6 across IPv4, and IPv4-to-IPv4 across IPv6 communications.

Tunneling should be used when direct connectivity cannot be established, such as when crossing another administrative domain or having a networking device that cannot be easily reconfigured. For example, tunneling technologies can be used in a situation where there is no direct IPv6 connectivity between two IPv6 networks (called IPv6 islands), but there is an IPv4 connectivity (and vice versa). In this situation tunnels are created and used to transfer IPv6 packets across an IPv4 network.

There are many types of tunneling technologies. Tunnels can be configured manually [4], or automatically with technologies like 6in4 [4], 6to4 [5], 6over4 [6], 4in6 [7], Intra-Site Automatic Tunnel Addressing Protocol (ISATAP) [8], Teredo [9], running IPv6 over MPLS with IPv6 Provider Edge Routers (6PE) [10], IPv6 Rapid Deployment (6rd) [11], Softwire [12] [13], tunnel brokers [14], and others.

There are several advantages and disadvantages associated with both manual and automatic tunneling. Manually created tunnels are simple to deploy and available on most platforms. However, the downside of manual tunneling is the fact that tunnels must be configured manually and that they represent a single point of failure.

The advantage of automatically created tunnels is that they do not require to be manually configured. They are available on most platforms, and can use BOOTP, DHCP or RARP (as well as manual configuration) to obtain IPv4 addresses.

There are a number of problems associated with specific automatic tunneling technologies. In the case of 6over4, IPv6 multicast is implemented over IPv4 multicast, and this technology is not scalable. On the other hand, 6to4 scales well for sites, but not for individual hosts, and at times causes IPv6 connectivity to appear to be available on a network when in fact there is no connectivity. ISATAP is designed for intra-site use, not for inter-site connectivity, requires more setup than other technologies, and has same security issues. Overall, the disadvantages of automatic tunneling depend on the chosen technology.

IPv6 Rapid Deployment (6rd). IPv6 Rapid Deployment (6rd) [11] is the most recommended tunneling mechanism [3]. It is an improved version of 6to4, and it is used for transferring IPv6 packets across an IPv4 network. The main difference from 6to4 is that it uses an Service Provider's (SP's) own IPv6 address prefix, rather than a well-known prefix. 6rd uses 32 bits of IPv6 address to map the entire IPv4 address space.

The 6rd architecture has two main routers: 6rd Customer Edge (6rd CE) that is at the customer side, and 6rd Border Relay (6rd BR) that can be any dual stack enabled routing capable networking device on the Internet that supports 6rd. There can be multiple 6rd CEs and 6rd BRs in a single network. In this architecture, the subnet that is behind 6rd CE can be a dual stack or IPv6-only, the connection between 6rd CE and 6rd BR is IPv4-only, and the rest of the network beyond 6rd BR can be a dual stack or IPv6-only.

The main advantage of 6rd is that from the perspective of customer sites and the IPv6 Internet at large, the IPv6 service provided is equivalent to native IPv6. The main disadvantage of 6rd is that it is a subject to routing loop attack [15].

2.3 Translation

Translation is used for interconnection of hosts on IPv6-only network with hosts on IPv4-only network, and vice versa. Translation mechanisms solve the problem of incompatibility between IPv4 and IPv6. With the use of these mechanisms an IPv6-only network can be created that is capable of communicating with both IPv6 and IPv4 hosts on the Internet.

Translation can be implemented by using Application Layer Gateways [16], or by using network-layer IPv6/IPv4 translation, with technologies such as Stateless IP/ICMP Translation Algorithm (SIIT) [17], Stateful NAT64 [18], DNS64 [19], and others.

Translation can be used with all kinds of devices, and the advantage here is that there is no need to change the configuration of the hosts. However, translation mechanisms have a conceptual disadvantage: they break the end-to-end connectivity, which is considered a core concept of the Internet.

NAT64/DNS64. The most recommended deployment model for translation is a combination of Stateful NAT64 (NAT64 in the rest of this paper) [18] and DNS64 [19], also known as NAT64/DNS64 [20].

NAT64/DNS64 is a combination of mechanisms used for translating IPv6 packets to IPv4 packets, and vice versa. In this scenario, there are two servers: a NAT64 server, and a DNS64 server. Both servers need to have at least two interfaces: one connected to the IPv4 network, and the other connected to the IPv6 network. There is also a well-known IPv6 prefix (64:ff9b::/96) associated with this model, used for representation of IPv4 addresses in the IPv6 world. For example, the representation of the 10.1.10.1 IPv4 address in the IPv6 world is 64:ff9b::10.1.10.1.

The NAT64 server is the one that translates packets between the IPv6 network and the IPv4 network, and vice versa. The DNS64 server is the one that creates representations of IPv4 addresses in the IPv6 world (AAAA records) for hosts that have only IPv4 addresses (hosts that have only A records).

For example, if an IPv6-only host wants to communicate with an IPv4-only web server (for example, www.website.test), the IPv6-only host asks for an AAAA record for www.website.test. In this situation, the DNS64 server creates an AAAA record for www.website.test from the A record for that domain, and sends the resulting AAAA record to the IPv6-only host. In this way, the IPv6-only host receives an IPv6 address and sends its packets to this IPv6 address. Because this IPv6 address has the well-known prefix (64:ff9b::/96), the router sends these packets to the NAT64 server, and not to the Internet. When the NAT64 server receives these packets, he translates them from IPv6 to IPv4, and sends them across the IPv4 network.

The main advantage of this solution is the ability for interconnection between IPv6-only hosts and IPv4-only hosts.

The main disadvantage associated with this model is that it works automatically only when an IPv6 host is the initiator of a communication. For an IPv4 host to be able to reach an IPv6 host, a static translation needs to be configured manually. There are other disadvantages associated with DNS64, particularly in a situation when there are other DNS servers on the network that are not aware of the DNS64 function, and when DNSSEC is used.

3 Deployment of IPv6 at Ss. Cyril and Methodius University in Skopje

The Ss. Cyril and Methodius University in Skopje (UKIM) represents a functional community of 23 faculties, 5 research institutes and 11 accompanying members. Most of the UKIM institutions are grouped into six main campuses: social sciences, architecture and civil engineering, technical, medicine, biotechnical sciences, and natural sciences and mathematics. All of the UKIM faculties in these six campuses and several institutes (a total of 28 parties) are connected to a single computer network, called the University Computer Network (see Figure 1). The backbone of the University Computer Network, that connects these 28 institutions, is maintained by the University Computer Centre, located at the UKIM Rector's Office, which is part of the social sciences campus. Each of these 28 institutions maintains its own part of the University Computer Network.

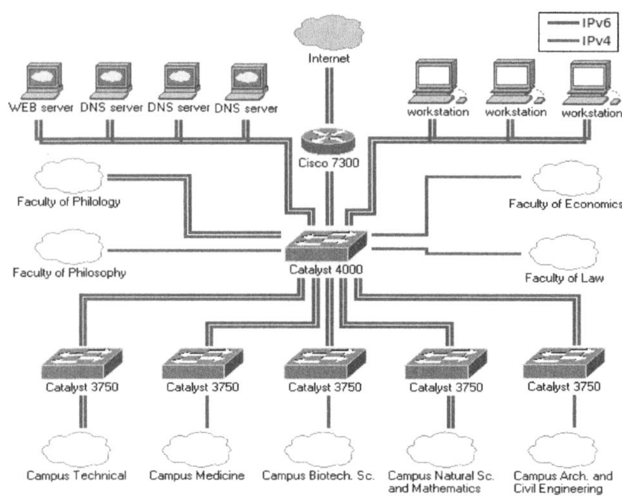

Fig. 1. Dual stack implementation at the University Computer Network

The University Computer Network is a large computer network that has a lot of managed and unmanaged networking devices, workstations, and servers that run on different kinds of operating systems. It offers the standard network services, like web, email, DNS, Active Directory, and many others.

Following the many universities and companies that have already implemented some sort of IPv6 networks, our goal is also to implement IPv6 at the University Computer Network, and finally to switch over to IPv6 only. We have considered the pros and cons of the three main strategies (dual stack, tunneling, and translation) for IPv6 deployment, and decided that the best method for deployment of IPv6 at the University Computer Network is the dual stack model.

In the following sections we present the implementation of the dual stack model at the University Computer Network, and two implementation scenarios for the other two strategies (tunneling and translation), that can be used for solving some of the challenges that we are facing after the dual stack implementation.

3.1 The Dual Stack Implementation

We have implemented the dual stack model (see Figure 1) at the backbone of the University Computer Network, by enabling the IPv6 protocol on every routing device, and assigning an IPv6 address to every interface that already had an IPv4 address. At this point, every UKIM faculty's main default gateway is a dual stack device. The University Computer Network has also a dual stack connection to the Internet.

There are three faculties that have partial implementation of the dual stack model: the Faculty of Computer Science and Engineering, the Faculty of Electrical Engineering and Information Technologies [21], and the Faculty of Philology. The other faculties haven't yet implemented the dual stack model, mostly because of lack of manpower and equipment support.

At the University Computer Centre, for production and testing purposes, we have configured the dual stack model on three workstations, one web server (the official UKIM website server), and three DNS servers (one of these is the primary DNS server for .mk ccTLD). We also plan to implement a full dual stack solution at the UKIM Rector's Office, and armed with this experience, to support full IPv6 implementation at UKIM faculties.

At this point we are facing two main challenges with the IPv6 implementation at the University Computer Network. First, those faculties that haven't yet implemented the dual stack model in their own backbone networks are not able to create an internal IPv6 subnet that is capable of reaching the IPv6 Internet. Second, if we create an IPv6-only subnet on the University Computer Network, the hosts on this subnet will not be able to reach the hosts on the IPv4 Internet.

One possible solution for the first problem is the implementation of some kind of tunneling technology, like 6rd, and for the second problem, the implementation of some kind of translation technology, like NAT64/DNS64. These solutions are presented in the following sections.

3.2 6rd Implementation Scenario

As it can be seen on Figure 1, there are several campuses and faculties that are only connected with IPv4 to the University Computer Network. One way for these faculties to implement some sort of IPv6 subnets is to use some tunneling technology, like 6rd (see Figure 2).

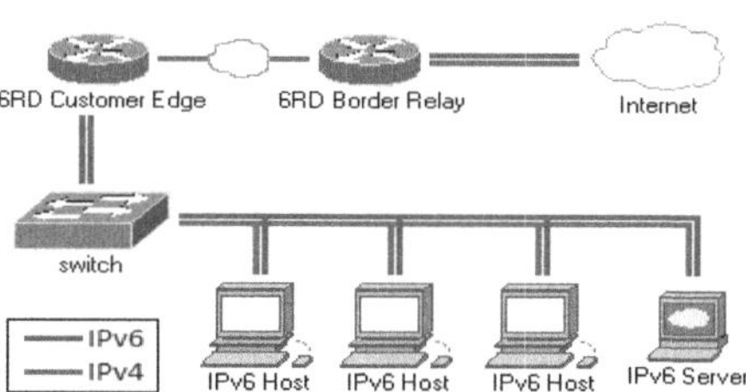

Fig. 2. 6rd implementation scenario

To be able to implement 6rd, faculties should create a subnet that contains several IPv6 enabled hosts and servers. This subnet needs to have a 6rd Costumer Edge (6rd CE) router. The advantage here is that there are a lot of cheap routers that can support 6rd, if their firmware is upgraded with the firmware downloaded from the DD-WRT website [22].

In this way, the 6rd CE router can be configured to create automatic tunnels with a 6rd Border Relay (6rd BR). Although, for 6rd BR any dual stack enabled routing capable networking device on the Internet that supports 6rd can be chosen, the best choice for 6rd BR at the University Computer Network is the Cisco 7300 router. In this way, it can be used as a 6rd BR for all UKIM faculties, without the drawback of additional routing.

With this 6rd scenario implemented, the IPv6 enabled hosts on the subnet can communicate with the IPv6 hosts on the Internet. This scenario can also be used for creating an IPv6-only subnet that will be capable of communicating only with the rest of the IPv6 world. For hosts on this subnet to be able to communicate with IPv4-only hosts, the dual stack model needs to be implemented, as shown on Figure 2.

3.3 NAT64/DNS64 Implementation Scenario

With the dual stack model implemented at the University Computer Network, and even with future implementation of 6rd, one fundamental problem still remains within the network: IPv6-only hosts cannot communicate with IPv4-only hosts. Dual stack hosts can communicate with directly connected IPv6 and IPv4 hosts. IPv6-only hosts can communicate only with directly connected IPv6 hosts, and with implementation of 6rd, they can communicate with IPv6-only hosts across IPv4 network, but they cannot communicate with IPv4-only hosts. For IPv6-only hosts to be able to communicate with IPv4-only hosts, a translation technology, like NAT64/DNS64, needs to be implemented.

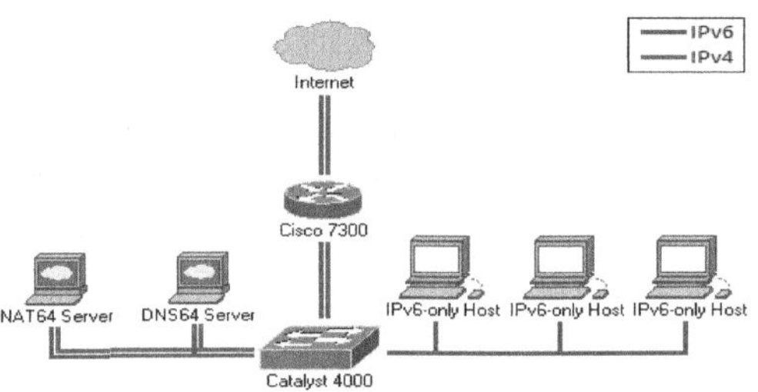

Fig. 3. NAT64/DNS64 implementation scenario

The scenario for the implementation of the NAT64/DNS64 translation technology at the University Computer Network is presented on Figure 3. As can be seen, in this scenario, the NAT64 server and the DNS64 server are connected to the Catalyst 4000 L3 switch, but they can be connected anywhere on the network, as long as they are connected to a dual stack enabled routing networking element that has access to the internal IPv6 network and the IPv4 Internet. The drawback of putting those servers anywhere else on the network is the additional routing that needs to be configured.

Another thing that should be kept in mind is that the IPv6-only hosts do not have to be connected directly to the Catalyst 4000 switch. It can be any IPv6-only host that has access to the internal IPv6 network.

Very important task in this scenario is to choose a router that will separate the NAT64 traffic from the rest of the IPv6 traffic on the network. NAT64 traffic needs to be sent to the NAT64 server, and the rest of the IPv6 traffic intended for the IPv6 Internet needs to be sent towards the Internet. For this role we will choose the Cisco 7300 router.

Although there are separate servers for NAT64 and DNS64 on Figure 3, it is important to keep in mind that those are logical functions, and can be carried by a single server or router. There are implementations of NAT64 functionality available for Linux and OpenBSD [23], and DNS64 functionality in the Bind [24] and Unbound [25] DNS servers.

4 Conclusion

IPv6 is the future of the Internet. There is no doubt about that. If the companies and institutions around the world want to be part of this future Internet, they must implement IPv6. It is only a question of time when the Internet will run on IPv6 only, and there is not much time left.

However, the implementation of IPv6 is not an easy task. As shown in this paper, there are different transition mechanisms that can be used for IPv6 deployment, and there is no a single best solution. Each of those transition mechanisms is used for its specific purpose. The dual stack model is used when there is need for both IPv4 and IPv6 to be running on the same network. Tunneling is used in a situation where there are separated IPv6 islands that need to be interconnected, and translation is used for interconnection of hosts that run a different version of the IP protocol.

It is very important for companies and institutions to choose the right transition mechanisms that can meet the requirements of their networks. For this purpose, we have presented a real life case study: the implementation of IPv6 at the Ss. Cyril and Methodius University in Skopje, the challenges that we are facing after the implementation, and the possible solutions to these challenges.

We have implemented the dual stack model at the backbone of the University Computer Network. After the implementation, we are facing two main challenges: there are faculties that are not able to create an internal IPv6 subnet that is capable of reaching the IPv6 Internet, and if we create an IPv6-only subnet, the hosts on this subnet will not be able to reach the hosts on the IPv4 Internet.

As a solution for these problems, we have proposed the following scenarios: the implementation of some kind of tunneling technology, like 6rd, for the first problem, and the implementation of some kind of translation technology, like NAT64/DNS64, for the second problem.

References

1. Goth, G.: The End of IPv4 is Nearly Here — Really. IEEE Internet Computing 16, 7–11 (2012)
2. Deering, S., Hinden, R.: Internet Protocol, Version 6 (IPv6) Specification, RFC 2460 (1998), `http://tools.ietf.org/html/rfc2460`
3. Arkko, J., Baker, F.: Guidelines for Using IPv6 Transition Mechanisms during IPv6 Deployment, RFC 6180 (2011), `http://tools.ietf.org/html/rfc6180`
4. Nordmark, E., Gilligan, R.: Basic Transition Mechanisms for IPv6 Hosts and Routers, RFC 4213 (2005), `http://tools.ietf.org/html/rfc4213`
5. Carpenter, B., Moore, K.: Connection of IPv6 Domains via IPv4 Clouds, RFC 3056 (2001), `http://tools.ietf.org/html/rfc3056`

6. Carpenter, B., Jung, C.: Transmission of IPv6 over IPv4 Domains without Explicit Tunnels, RFC 2529 (1999), http://tools.ietf.org/html/rfc2529
7. Conta, A., Deering, S.: Generic Packet Tunneling in IPv6 Specification, RFC 2473 (1998), http://tools.ietf.org/html/rfc2473
8. Templin, F., Gleeson, T., Thaler, D.: Intra-Site Automatic Tunnel Addressing Protocol (ISATAP), RFC 5214 (2008), http://tools.ietf.org/html/rfc5214
9. Huitema, C.: Teredo: Tunneling IPv6 over UDP through Network Address Translations (NATs), RFC 4380 (2006), http://tools.ietf.org/html/rfc4380
10. De Clercq, J., Ooms, D., Prevost, S., Le Faucheur, F.: Connecting IPv6 Islands over IPv4 MPLS Using IPv6 Provider Edge Routers (6PE), RFC 4798 (2007), http://tools.ietf.org/html/rfc4798
11. Townsley, W., Troan, O.: IPv6 Rapid Deployment on IPv4 Infrastructures (6rd) – Protocol Specification, RFC 5969 (2010), http://tools.ietf.org/html/rfc5969
12. Storer, B., Pignataro, C., Dos Santos, M., Stevant, B., Toutain, L., Tremblay, J.: Softwire Hub and Spoke Deployment Framework with Layer Two Tunneling Protocol Version 2 (L2TPv2), RFC 5571 (2009), http://tools.ietf.org/html/rfc5571
13. Wu, J., Cui, Y., Metz, C., Rosen, E.: Softwire Mesh Framework, RFC 5565 (2009), http://tools.ietf.org/html/rfc5565
14. Durand, A., Fasano, P., Guardini, I., Lento, D.: IPv6 Tunnel Broker, RFC 3053 (2001), http://tools.ietf.org/html/rfc3053
15. Nakibly, G., Templin, F.: Routing Loop Attack Using IPv6 Automatic Tunnels: Problem Statement and Proposed Mitigations, RFC 6324 (2011), http://tools.ietf.org/html/rfc6324
16. Baker, F., Li, X., Bao, C., Yin, K.: Framework for IPv4/IPv6 Translation, RFC 6144 (2011), http://tools.ietf.org/html/rfc6144
17. Li, X., Bao, C., Baker, F.: IP/ICMP Translation Algorithm, RFC 6145 (2011), http://tools.ietf.org/html/rfc6145
18. Bagnulo, M., Matthews, P., van Beijnum, I.: Stateful NAT64: Network Address and Protocol Translation from IPv6 Clients to IPv4 Servers, RFC 6146 (2011), http://tools.ietf.org/html/rfc6146
19. Bagnulo, M., Sullivan, A., Matthews, P., van Beijnum, I.: DNS64: DNS Extensions for Network Address Translation from IPv6 Clients to IPv4 Servers, RFC 6147 (2011), http://tools.ietf.org/html/rfc6147
20. Skoberne, N., Ciglaric, M.: Practical Evaluation of Stateful NAT64/DNS64 Translation. Advances in Electrical and Computer Engineering 11, 49–54 (2011)
21. Minov, S., Bojchev, D., Tentov, A.: IPv6 Implementation in Feit Network Infrastructure (2011)
22. DD-WRT, http://www.dd-wrt.com
23. Ecdysis, http://ecdysis.viagenie.ca/
24. Bind, https://www.isc.org/software/bind
25. Unbound, http://www.unbound.net/

Author Index

CPSIA information can be obtained at www.ICGtesting.com
Printed in the USA
LVOW01s1945270713

344972LV00004B/80/P

9 783642 371684